Skeletal Muscle Physiology

Christopher Myers
Editors

Skeletal Muscle Physiology

An Update to Anatomy and Function

Springer

Editor
Christopher Myers
3 Sports Science
3 Sports Science LLC
Akron, OH, United States of America

ISBN 978-3-031-47064-6 ISBN 978-3-031-47065-3 (eBook)
https://doi.org/10.1007/978-3-031-47065-3

This Springer imprint is published by the registered company Springer Nature Switzerland AG
The registered company address is: Gewerbestrasse 11, 6330 Cham, Switzerland

Skeletal Muscle Physiology: An Update to Anatomy and FunctionAPO AE, Germany

Auto-summaries can be generated by either an abstractive or extractive auto-summarization:

- An extraction-based summarizer identifies the most important sentences of a text and uses the original sentences to create the summary.
- An abstraction-based summarizer creates new text based on deep learning. New phrases are created to summarize the content.

The auto-summaries you will find in this book have been generated via an extractive summarization approach.

Each chapter was carefully edited by Dr. Christopher Myers. The editors selected the papers, which were then auto-summarized. The editors have not edited the auto-summaries due to the extraction-based approach, and have not changed the original sentences. You will find the editors' reviews and guidance on the auto-summaries in their chapter introductions.

In machine-generated books, editors are defined as those who curate the content for the book by selecting the papers to be auto-summarized and by organizing the output into a meaningful order. Next to the thoughtful curation of the papers, editors should guide the readers through the auto-summaries and make transparent why they selected the papers.

The ultimate goal is to provide a current literature review of Springer Nature publications on a given topic in order to support readers in overcoming information overload and to help them dive into a topic faster; to identify interdisciplinary overlaps; and to present papers which might not have been on the readers' radar.

Please note that the selected papers are not used to train an LLM while the auto-summaries are created.

Preface

Welcome to "Skeletal Muscle Physiology: An Update to Anatomy and Function"—an engaging, cutting-edge, and comprehensive exploration into the fascinating world of skeletal muscle physiology. This remarkable volume, generated by AI, represents a convergence of diverse and profound insights from biology, sports science, medicine, and computational data science.

Skeletal muscle, comprising approximately 40% of the human body weight, plays an essential role in locomotion and posture and in various critical physiological processes such as respiration, thermogenesis, and metabolism. Hence, an in-depth understanding of its physiology can profoundly impact health and disease, exercise performance, rehabilitation, and overall human function.

This book breaks new ground by being one of the first AI-authored texts in the field. It draws upon a vast reservoir of knowledge in the Springer Publishing library and uses advanced AI modeling to provide a meticulously detailed yet lucid perspective of skeletal muscle function. The AI-based authoring ensures high accuracy and consistency, providing an unbiased account of existing scientific knowledge.

This book builds upon the basic and advanced skeletal muscle topics found in exercise physiology books. The contents within these pages delve into various topics, including the detailed anatomy of skeletal muscle, the mechanics of muscle contraction, the biochemistry of energy provision, the impact of various factors on skeletal muscle function such as age, sex, and disease, and the role of exercise and nutrition in muscle health.

This book does not claim to replace the need for human intelligence and experience, but instead, it hopes to supplement it, providing a resource that can aid both learning and teaching. While AI generation has limitations, this text will be an invaluable stepping stone toward a more profound comprehension of skeletal muscle physiology.

So, whether you are a novice to the field, seeking a refresher course, or a seasoned expert looking for a comprehensive resource, we invite you to embark on this captivating journey of exploration into the world of skeletal muscle physiology. Let's delve into the marvel that is the human muscle and unravel the science that underpins its extraordinary capabilities.

Akron, OH, United States of America Christopher Myers

Introduction by the Editor

In the dynamic and ever-evolving landscape of physiological research, the exploration of skeletal muscle physiology stands as a cornerstone, bridging fundamental biological insights with transformative applications in health, disease, and human performance. This book, a pioneering endeavor co-authored with the assistance of advanced artificial intelligence (AI), embarks on a journey to distill and articulate the complex mechanisms, functions, and adaptabilities of skeletal muscles in unprecedented detail and clarity.

The fusion of human expertise and AI in the creation of this book has enabled a unique and innovative approach to synthesizing and presenting the wealth of knowledge that defines skeletal muscle physiology. By harnessing the analytical power of AI, I have sifted through vast arrays of data, research findings, and theoretical frameworks, selecting and integrating the most critical insights into a coherent and comprehensive narrative. This collaboration has not only enhanced the accuracy and depth of the content but also imbued it with novel perspectives and interpretations that reflect the cutting-edge of scientific discovery.

Our journey through skeletal muscle physiology begins with the advanced principles that govern muscle biology and explores new advances in understanding muscular binding proteins. Also, we explore the remarkable plasticity of skeletal muscles, highlighting how they adapt to various stimuli, including exercise, injury, and disease, and how these adaptations can be leveraged for therapeutic and performance-enhancing interventions.

This book is designed to serve as an invaluable resource for students, researchers, and practitioners across multiple disciplines, including physiology, medicine, sports science, and rehabilitation. It aims to not only impart knowledge but also to inspire curiosity, critical thinking, and further exploration into the vast and vibrant field of skeletal muscle physiology.

The integration of AI in the creation of this book represents a novel paradigm in scientific publishing. It signifies a step toward a future where human creativity and artificial intelligence converge to enhance our understanding of the world. As you turn these pages, I invite you to join me at the forefront of this exciting frontier, equipped with the insights and understandings that will empower you to contribute to the next wave of discoveries and innovations in skeletal muscle physiology.

Welcome to a journey of exploration, discovery, and inspiration. Welcome to the cutting edge of skeletal muscle physiology!

Contents

Basic Structure of Skeletal Muscle 1

Christopher Myers

1.1 Introduction by the Editor

Building on the foundational understanding established in the first chapter, Chap. 2 explores the burgeoning world of contemporary research in basic skeletal muscle physiology. While skeletal muscle physiology is a well-established field, it remains an active area of investigation, with novel findings consistently reshaping our understanding of skeletal muscle's complex biology.

The continuous evolution of technologies, including advanced imaging techniques and high-throughput genomics, has allowed researchers to gain unparalleled insight into the inner workings of skeletal muscle. The granular details unveiled by these novel tools are revising our perceptions of muscle function at the cellular and molecular levels.

This chapter's primary area of focus is research updates to skeletal muscle physiology. The content in this chapter explores the role of the thermoregulatory activity of ATPase, myosin light chain activity and myosin binding to actin, and force production and enhancement between the myofilaments.

We also delve into recent discoveries in the realm of the protein titin. This protein is shown to become more and more critical to muscular contraction. Unofficially, titin is called the "third myofilament" due to the research showing its importance in force production. The chapter discusses some of this research that has come to change our fundamental understanding of the structure and function of titin.

In summary, this chapter provides a deep dive into innovative research in basic skeletal muscle physiology. By embracing the newest findings and integrating them with existing knowledge, this chapter illuminates our relentless progress in deciphering the complex physiology of skeletal muscle.

C. Myers (✉)
3 Sports Science, 3 Sports Science LLC, Akron, OH, United States of America
e-mail: chris@3sportsscience.com

C. Myers (ed.), *Skeletal Muscle Physiology*,
https://doi.org/10.1007/978-3-031-47065-3_1

The study of skeletal muscle physiology has long stood as a cornerstone of biological and medical sciences, bridging our understanding from fundamental biological processes to complex systems-level functions. Skeletal muscles are pivotal in locomotion, posture, metabolic health, disease states, and aging. The field of skeletal muscle physiology is experiencing a renaissance powered by technological advancements and an expanding interest in the nuanced mechanisms that govern muscle function. This chapter explores the forefront of skeletal muscle research, shedding light on the complex interplay of molecular components that orchestrate muscle contraction, force production, and energy efficiency.

One of the most captivating areas of current research lies in investigating the myosin super relaxed state (SRX) and its contribution to muscle thermogenesis. The SRX state, characterized by a tightly packed and inactive configuration of myosin heads, represents a crucial mechanism for adenosine triphosphate preservation in resting muscle fibers. This state's regulatory impact on ATPase activity underscores the efficiency of muscle energy use and its role in generating heat, a vital aspect of thermoregulation. This chapter dives into the intricacies of myosin dynamics and uncovers the layers of complexity in muscle function that extend beyond the conventional understanding of contraction and relaxation via the excitation-coupling mechanism and sliding filament theory.

Furthermore, recent discoveries highlight the significance of the actin-titin interaction in force enhancement within muscle fibers. Traditionally overshadowed by the actin-myosin interaction, the role of titin—often heralded as the "third myofilament"—in muscle elasticity and force production has come to the fore. New research offers groundbreaking insights into how muscles maintain tension and elasticity, especially under stretching. The binding of titin to actin illuminates a novel mechanism of force enhancement and redefines our understanding of muscle flexibility and strength.

The field of muscle research has undergone significant transformation, leading scientists to question the long-standing nonlinear perspective on the active force-length relationship. Recent investigations have introduced a linear model that appears to better reflect muscle behavior under specific scenarios, including submaximal contractions and stretches. This shift towards a linear model marks a significant paradigm change, altering how we conceptualize muscle stretching and contraction. The new linear approach offers a more apparent and predictable framework for studying muscle dynamics. This evolution in understanding has profound implications, potentially reshaping theoretical research and practical methodologies. In the realms of sports science and rehabilitation, the implications are particularly noteworthy. Practitioners and researchers can now employ this model to achieve more accurate predictions and effective interventions. Ultimately, this reevaluation facilitates a more profound comprehension of muscle function, opening new avenues for advancement in our understanding and application of muscle physiology.

In addition to these mechanistic insights, this chapter will explore the contributions of titin filaments and myosin's regulatory and essential light chains to muscle function. The passive forces attributed to titin filaments highlight the protein's role as a molecular spring, critical for muscle elasticity. Meanwhile, the light chains of myosin emerge as crucial modulators of muscle contraction, adding another layer of complexity to the regulation of muscle activity. Together, these components play a defining role in the nuanced and finely tuned-process of muscle movement and force generation.

Leveraging advanced simulations and model extensions, we can now visualize and quantify force enhancement across various muscle levels, seamlessly integrating the roles of myofilaments within a unified framework. These advancements enhance our understanding of muscle function under a broad spectrum of conditions, including submaximal contractions and the intricate forces exerted by titin. As we embrace these innovative methodologies, we venture into a new era of physiology, aiming to demystify the complex mechanisms of skeletal muscle physiology. This chapter embarks on a journey to blend cutting-edge discoveries with established knowledge, guided by the latest developments and insights in muscle research. Our goal is to shed light on the continuous progress in the field, paving the way for breakthroughs in health, disease management, and human performance, thereby opening new frontiers in our quest for a comprehensive comprehension of muscle physiology.

1.2 Machine Generated Summaries

Disclaimer: The summaries in this chapter were generated from Springer Nature publications using extractive AI auto-summarization: An extraction-based summarizer aims to identify the most important sentences of a text using an algorithm and uses those original sentences to create the auto-summary (unlike generative AI). As the constituted sentences are machine selected, they may not fully reflect the body of the work, so we strongly advise that the original content is read and cited. The auto generated summaries were curated by the editor to meet Springer Nature publication standards. To cite this content, please refer to the original papers.

Machine generated keywords: titin, filament, myosin, actin, enhancement, rabbit, contain, interaction, thin, light chain, actin filament, interaction actin, length, tension, inhibit.

The Role of the Myosin ATPase Activity in Adaptive Thermogenesis by Skeletal Muscle [1]

This is a machine-generated summary of:

Cooke, Roger: The role of the myosin ATPase activity in adaptive thermogenesis by skeletal muscle [1]

Published in: Biophysical Reviews (2011)

Link to original: https://doi.org/10.1007/s12551-011-0044-9

Copyright of the summarized publication:

The Author(s) 2011

If you want to cite the papers, please refer to the original.

For technical reasons we could not place the page where the original quote is coming from.

Abstract-Summary "Resting skeletal muscle is a major contributor to adaptive thermogenesis, i.e., the thermogenesis that changes in response to exposure to cold or to overfeeding."

"A new state of myosin, the super relaxed state (SRX), with a very slow ATP turnover rate has recently been observed in skeletal muscle [2]."

"To be compatible with the basal metabolic rate observed in vivo for resting muscle, most myosin heads would have to be in the SRX."

"Modulation of the population of this state, relative to the normal relaxed state, was proposed to be a major contributor to adaptive thermogenesis in resting muscle."

"Transfer of only 20% of myosin heads from the SRX into the normal relaxed state would cause muscle thermogenesis to double."

"Phosphorylation of the myosin regulatory light chain was shown to transfer myosin heads from the SRX into the relaxed state, which would increase thermogenesis."

Introduction

"The SRX has been proposed to play a role in adaptive thermogenesis by resting muscle."

"Thermogenesis by resting muscle and its role in whole body metabolism is considered."

"I then describe how the new state of myosin could provide a mechanism for thermogenesis and integrate this proposed mechanism into our current knowledge of muscle thermogenesis."

"Two fields of research into muscle function, one concerned with the contractile proteins as force generators and one concerned with the role of muscle in adaptive thermogenesis, should now find common ground."

The ATPase Activity of Purified Myosin Is Not Compatible with the Metabolic Rate of Resting Muscle

"The presence of an inhibited state of skeletal myosin in resting muscle was first suggested more than 30 years ago when the ATPase activity of purified myosin was compared for the first time with the resting metabolic rate of the muscle from which it came [3]."

"The degree of inhibition must be considerably greater than the factor of five, as the resting metabolism of skeletal muscle is low and, in addition to the metabolism of myosin, ATPase must also carry out other obligatory cell functions, such as protein synthesis, ion pumping, etc A similar argument was made for rabbit myosin in comparison to living resting rabbit muscle [4, 5]."

"If the population of this state is regulated, with some myosins in the inhibited state while others are in a state with an ATPase activity similar to that of purified myosin, redistribution of myosin between the two states would have a large effect on resting muscle metabolism."

Measuring Single Nucleotide Turnovers in Permeable Muscle Fibers

"The rapid phase, which has a time constant of approximately 20 s, comprises multiple factors, including the release of non-specifically bound nucleotides, the release of nucleotides by a fraction of the normally relaxed myosin heads, and the diffusion of released nucleotides out of the fibers, requiring approximately 10 s [6]."

"The time constant for the second phase is much slower, 230 s, and as discussed below, it arises from the release of nucleotides by a second fraction of myosin heads with a very slow ATP turnover."

"The slow release was only observed if the fiber was relaxed, and myosin is the only nucleotide binding protein that responds to the activation level of the fiber."

"The SRX explains the discrepancy described above between the more rapid ATPase activity of purified myosin and the much slower ATPase activity required to be compatible with the metabolic rate of living fibers [3]."

Structural and Biochemical Data Show Myosin ATP Turnover Is Inhibited by the Binding of Myosin Heads to the Core of the Thick Filament

"A helical array of heads bound to the thick filament was also observed at lower resolution in relaxed filaments from scallop muscle, a muscle controlled by binding of calcium to myosin [7, 8]."

"A similar ordered helical array of myosin heads in the J motif was resolved in myosin filaments from vertebrate cardiac muscle [9]."

"The observation of the folded J structural motif in a variety of myosins, including those from both thick- and thin-filament-regulated muscles, led Craig and coworkers to propose that the structural motif is conserved across different species and muscle types [10–12]."

"Filaments in relaxed insect flight muscle show myosin heads in a different structural motif [13]."

"These correlations led to the conclusion that myosin heads in the SRX are bound to the core of the thick filament and have the J motif."

Active Fibers

"If myosin heads in the SRX are bound to the core of the thick filament, where they are unable to interact with actin, and their lifetime is long, how is skeletal muscle able to achieve maximal force in a few tens of milliseconds in a tetanic activation?"

"This observation suggests that the interaction of some myosin heads with actin destabilizes the binding of adjacent myosin heads to the core of the thick filament."

"There must always be some fraction of the myosin heads in the normal relaxed state, which would sense the activation of the thin filament and initiate the cooperative process leading to full activation."

"During activation, all of the myosin heads leave their relaxed positions on the thick filament and become disordered."

The Dynamic Nature of the Resting Metabolic Rate of Skeletal Muscle

"Early investigations of muscle metabolism revealed that the resting metabolic rate of amphibian muscles could vary widely."

"Stretching a muscle to longer lengths could increase metabolism up to fourfold [14]."

"The resting metabolic rate of frog muscle was also increased by higher external concentrations of potassium or by the addition of hyper-osmotic agents in the medium bathing the muscle [15, 16]."

"Hypoxia caused a decrease in the basal metabolic rate of resting frog muscle to 20% of the control rate and also greatly inhibited the increased thermogenesis seen after stretch and in the presence of higher potassium and hyper-osmotic agents [17]."

"The resting metabolic rates of both amphibian and mammalian skeletal muscles are highly variable, undergoing large increases or decreases depending on the conditions employed, all of which occur in the absence of active force generation."

Adaptive Thermogenesis and the Role of Muscle

"Resting muscle thermogenesis increases following a meal and during exposure to cold (for review, see [18–20]."

"This adaptive thermogenesis is produced by norepinephrine released from sympathetic nerves and by adrenal secretion of epinephrine, and a large portion of this metabolic process, approximately 40%, occurs in resting muscle [21, 22]."

"Leptin has also been shown to directly increase resting skeletal muscle thermogenesis [23]."

"The response to cold exposure is complex and involves both non-shivering and shivering thermogenesis in skeletal muscle."

"Epinephrine has been shown to activate thermogenesis, showing that skeletal muscle is involved in cold-induced thermogenesis."

"Resting muscle thermogenesis responds to cold and to food consumption, and epinephrine and leptin have been identified as being involved in some of the pathways controlling this process."

What Is the "Furnace" Responsible for Increased Thermogenesis in Resting Skeletal Muscle?

"The discovery of an UCP that is highly and selectively expressed in skeletal muscle, UCP3, led to the suggestion that this protein may play a dominant role in muscle thermogenesis (for review, see [24])."

"The over-expression or knockout of UCP3 in mouse muscle had little effect on thermogenesis by resting excised muscles, showing UCP3 did not play a major role in this process [25]."

"Knockout of UCP1, known to play a role in thermogenesis by brown fat, also does not lead to obesity, although the knockout mouse muscle is cold-intolerant [26]."

"The mechanisms discussed above undoubtedly play some role in skeletal muscle thermogenesis."

"The important observation, however, is that they have not been shown to account quantitatively for all observed muscle thermogenesis, leaving room for additional mechanisms to play a role."

The Role of the SRX in Thermogenesis by Resting Muscle

"Changes in the relative populations of the relaxed state and the SRX can lead to dramatic changes in thermogenesis in resting muscle."

"Most myosin heads must be in the SRX in living resting muscle under basal conditions."

"Shifting only 20% of myosin heads from the SRX into the relaxed state would increase muscle thermogenesis by approximately a factor of two, increasing whole body metabolic rate by about 16%."

"There is some evidence implicating myosin in the increased thermogenesis seen during the stretch and osmotic stress discussed above."

"Stretch of resting frog muscle has been shown to diminish the intensity of the myosin layer lines, indicating that the myosin heads become disordered [27]."

"Both heat output and the change in myosin layer lines did not occur for modest stretch, but did occur at larger stretches [27, 28]."

Does Phosphorylation Regulate the SRX In Vivo?

"Phosphorylation of the RLC disorders the array of myosin heads bound to the skeletal thick filament and decreases the population of the SRX [2, 29]."

"There is another kinase that produces low levels of RLC phosphorylation, approximately 10%, in the knockout mouse [30], and it is possible that this kinase is a mediator of thermogenesis."

"A second pathway that may influence the population of the SRX is the phosphorylation of myosin binding protein-C (MBP-C)."

"One report has shown that protein kinase A can phosphorylate MBP-C from slow and fast skeletal muscles [31]."

"If this phosphorylation plays a similar role as in the mouse cardiac muscle, namely, producing disordering of the thick filament array, it would also affect the population of the SRX, thereby providing a direct connection to the thermogenic effects of epinephrine, which is known to activate protein kinase A."

Response to Long-Term Overfeeding

"The largest expenditure of excess calories, and the one that has been proposed to explain much of the wide variation in weight gain, has been non-exercise activity thermogenesis, known as NEAT [18, 32–37]."

"Despite the high efficiency of muscle, a number of low-level activities have been shown to expend a reasonable amount of calories [36, 38–40]."

"This mechanism for thermogenesis would be expected to play a greater role during low levels of activity, where muscle spends an appreciable time in the mechanically relaxed states."

"Below, there is also a stronger genetic component to muscle efficiency at low activity levels than at high ones, which again could be easily explained by the proposed model."

"In fast twitch muscle, factors that increase resting thermogenesis, such as myosin phosphorylation, epinephrine and cold exposure, also lead to potentiation of twitch tensions, showing that the muscle is more easily activated [30, 41, 42]."

The SRX as a Target for Possible Therapeutic Applications

"Activation of the metabolic rate of skeletal muscle would also lead to lower blood glucose levels."

"A major problem in type II diabetes is insulin resistance, in which insulin-mediated glucose uptake by tissues, including skeletal muscle, is impaired."

"Activation of the resting metabolic rate via increased myosin ATPase activity could possibly produce effects similar to exercise."

"Resting muscle consumes fatty acids as a major fuel source, and up-regulation of resting metabolism would consume free fatty acids and reduce the deposition of lipid in muscle, both of which are thought to be involved in insulin resistance [43]."

"Ectopic expression of UCP1 in skeletal muscle increases resting energy use and also increases the metabolism of glucose [44]."

"The two characteristics of metabolic syndrome are obesity and type II diabetes and, as suggested above, both could be alleviated by up-regulation of the myosin ATPase activity in resting skeletal muscle."

Summary

"A new state of myosin in relaxed muscle with a highly inhibited ATPase activity, the SRX, has been identified in skinned skeletal muscle fibers [2]."

"An obvious role of this state is to help achieve the low metabolic rate seen in resting in vivo skeletal muscle."

"Alterations in the population of this state have a very large capacity to alter muscle metabolism, and one physiological regulator, RLC phosphorylation, has been identified."

Myosin Regulatory Light Chain Phosphorylation Inhibits Shortening Velocities of Skeletal Muscle Fibers in the Presence of the Myosin Inhibitor Blebbistatin [45]

This is a machine-generated summary of:

Stewart, Melanie; Franks-Skiba, Kathy; Cooke, Roger: Myosin regulatory light chain phosphorylation inhibits shortening velocities of skeletal muscle fibers in the presence of the myosin inhibitor blebbistatin [45]

Published in: Journal of Muscle Research and Cell Motility (2009)

Link to original: https://doi.org/10.1007/s10974-008-9162-9

If you want to cite the papers, please refer to the original.
For technical reasons we could not place the page where the original quote is coming from.

Abstract-Summary "Phosphorylation of skeletal myosin regulatory light chain (RLC) occurs in fatigue and may play a role in the inhibition of shortening velocities observed in vivo."

"Forces and shortening velocities were measured in permeabilized rabbit psoas fibers with either phosphorylated or dephosphorylated RLCs and in the presence or absence of the myosin inhibitor blebbistatin."

"In blebbistatin maximal shortening velocities (V_{max}) at 30 °C, were decreased by 45% (3.2 ± 0.34 vs. 5.8 ± 0.18 lengths/s) in phosphorylated fibers but were not inhibited in dephosphorylated fibers (6.0 ± 0.30 vs. 5.4 ± 0.30)."

"RLC phosphorylation inhibited velocity in blebbistatin at both 30 and 10 °C, unlike previous reports where RLC phosphorylation only affected shortening velocities at higher temperatures."

Introduction

"The effect of myosin RLC phosphorylation on shortening velocity has been studied extensively, is an active site of investigation and still the subject of debate."

"Crow and Kushmerick initially suggested that shortening velocities were inhibited by myosin RLC phosphorylation, from a correlation observed in mouse skeletal muscle during a sustained tetanus [46, 47]."

"Recent work from our laboratory, has found that under conditions where fiber tensions are partially inhibited, RLC phosphorylation inhibits the inhibition of shortening velocities [48, 49]."

"In the presence of vanadate, for example, RLC phosphorylation inhibits shortening velocity compared with fibers with dephosphorylated RLC; although isometric tensions were not affected [48]."

"At the lower temperature most agents that inhibited tension, e.g. phosphate analogs or lower pH, also inhibited shortening velocity and phosphorylation of the fibers had little or no effect."

Materials and Methods

"Bundles of fibers were tied to supports, placed in glycerol solution (120 mM K-acetate, 5 mM $MgCl_2$, 5 mM EGTA, 50 mM MOPS set at pH 7.0 and 50% glycerol (v/v)) and gently mixed overnight at 4 °C."

"Bundles were then placed in fresh solution (120 mM K-acetate, 5 mM $MgCl_2$, 5 mM EGTA, 50 mM MOPS set at pH 7.0 and 50% glycerol (v/v)) pH 7.0 incubated at 4 °C for 4 h before transferring to −20 °C for storage."

"Following a 2 min equilibration period fibers were periodically and rapidly transferred to and from a third solution containing activating buffer set at a temperature of 30 or 10 °C according to the experiment and load clamps recorded."

Results

"Fiber shortening velocities were measured in the presence and absence of 20 μM blebbistatin."

"Our previous work had found that the inhibition of fiber shortening velocity by phosphorylation occurred only at higher temperatures and not at 10 °C, a temperature where many studies are carried out due to greater fiber stability."

"Addition of 20 μM blebbistatin inhibited tension equally in phosphorylated and dephosphorylated fibers and to the same extent as it did at 30 °C."

"In the absence of blebbistatin, phosphorylation had no effect on either V_{max} or α/Po, as observed previously under other conditions [48–50]."

"In the presence of blebbistatin the fibers have a greater apparent affinity for ATP, shown by the lower value of K_m in phosphorylated fibers, $K_m = 50 \pm 20$ μM, than in dephosphorylated fibers, $K_m = 330 \pm 84$ μM. This effect is similar to that seen in vanadate [48]."

Discussion

"Our previous work has found that, in conditions where sizable populations of myosin heads in states not generating force, myosin phosphorylation inhibits velocity."

"The mechanism by which phosphorylation induces inhibition of shortening velocity can be explained by its effect on the array of myosin heads bound to the thick filament."

"The observations that dephosphorylated fibers inhibited at 10 °C by blebbistatin, are fast, but that phosphorylated fibers are slow, strongly support the stability of the ordered thick filament array as playing a role in the inhibition of velocity."

"The hypothesis discussed above suggests that velocity is inhibited when there are large populations of myosin heads in non force states that are capable of interacting with the thin filament."

"Why would the presence of these non-force generating myosin heads interacting with the thin filament inhibit velocity?"

"In conditions where this non-force generating state is populated, myosin phosphorylation significantly inhibits shortening velocities."

Myosin Flares and Actin Leptomeres as Myofibril Assembly/ Disassembly Intermediates in Sonic Muscle Fibers [51]

This is a machine-generated summary of:

Nahirney, Patrick C.; Fischman, Donald A.; Wang, Kuan: Myosin flares and actin leptomeres as myofibril assembly/disassembly intermediates in sonic muscle fibers [51]

Published in: Cell and Tissue Research (2006)

Link to original: https://doi.org/10.1007/s00441-005-0110-3

If you want to cite the papers, please refer to the original.

For technical reasons we could not place the page where the original quote is coming from.

Abstract-Summary "Each sonic muscle fiber contains a tubular contractile apparatus with radially arranged myofibrillar plates encased in a desmin-rich cytoskeleton that is anchored to broad Z bands (~1.2 μm wide)."

"Leptomeres consist of dense arrays of filaments (~4 nm) with a structure that resembles myofibrillar Z band structure."

"We propose that flares and leptomeres are distinct filamentous arrays representing site-specific processing of myofibrillar components during the assembly and disassembly of the sarcomere."

"Recent reports that myosin assembles into filamentous aggregates before incorporating into the A band in the skeletal muscles of vertebrates and Caenorhabditis elegans suggest that sonic fibers utilize a similar pathway."

"Sonic muscle fibers, with their tubular design and abundant sarcoplasmic space, may provide an attractive muscle model to identify myofibrillar intermediates by structural and molecular techniques."

Introduction

"Type I males of this species use their sonic muscle to produce a hum of ~100 Hz for up to 2 h to attract females during the mating season in early summer [52] and exhibit a unique sonic muscle fiber morphology that differs from those found in the type II (sneaker) male and the female of this species [53–55]."

"The sonic muscle consists of a pure population of specialized muscle fibers under direct somatic control by central nervous system stimuli [56, 57]."

"The most unique feature of type 1 male sonic muscle fibers is the extremely broad Z band of the myofibrils, measuring ~1.2 μm in width [53, 58], a thickness that is ~20 times the width of comparable Z bands in type II males or in females of the same species and of typical vertebrate skeletal muscle [53]."

"Its growth, especially in type 1 males, appears to be hormonally regulated, and hypertrophy of muscle fibers can be induced by seasonal variations and androgen implants [54, 55]."

"We propose that these novel clusters of contractile protein aggregates represent assembly or remodeling intermediates in sonic muscle fibers."

Materials and Methods

"Strips of sonic muscle were then removed from the swimbladder, cut into small pieces (~1 mm^3), immersion-fixed for an additional 2 h on ice, washed, and then postfixed for 1 h with 1% osmium tetroxide in 0.1 M sodium cacodylate, pH 7.2."

"To obtain single fibers from fresh or stored tissue, the strips were further dissected in relaxing solution under a stereo dissecting microscope to produce fiber bundles of approximately 0.3–0.5 mm in diameter and 5–10 mm in length."

"Sections were first fixed on the slide with 2% v/v paraformaldehyde in PBS for 30 min at RT, washed three times with PBS, and then blocked with 1% bovine serum albumin (BSA; Sigma) in PBS (PBS BSA) for 1 h at RT."

Results

"Myosin flares were composed of branching thread-like filaments and appeared to emanate from the contractile tube toward the sarcolemma in three-dimensional arrays."

"Muscle fibers containing the myosin flares ranged in area from 813 μm^2 to 1458 μm^2 (n = 52) with an average area of 1123 μm^2 (SD:±198 μm^2)."

"In these fibers, the area occupied by myosin flares ranged between 9% (84 μm^2) and 42% (463 μm^2) with an average area of 268 μm^2 (SD:±144 μm^2), or 23% of the total muscle fiber cross-sectional area."

"Muscle fibers not containing flares exhibited a diffuse punctate or spotted distribution of myosin staining around the tube in the peripheral sarcoplasm, whereas muscle fibers containing distinct flares exhibited less diffuse staining in the sarcoplasm."

"Thick filaments in sarcomeres of the contractile tube were typical of vertebrate muscle and measured ~1.5 μm in length and contained a prominent M line and bare zone."

Discussion

"Their studies have shown that, when myosin tagged with green fluorescent protein is expressed in cultured C2C12 muscle cells, it accumulates in aggregates containing short filamentous structures that are later replaced by mature myofibrils."

"We propose that myosin flares in midshipman sonic muscle are analogous intermediate structures in thick filament and A band assembly to those found in vertebrate striated muscle cells."

"Several potential functions for leptomeres have been proposed. (1) They are protective structures for Z bands during contraction; however, this is unlikely to be a general feature because of their low frequency in striated muscle. (2) They are an early intermediate of myofibril assembly; nevertheless, they are not observed in de novo formation muscle fibers and subsequent myofibrillogenesis in typical skeletal muscle. (3) They have a tendon-like function between successive Z bands, since they have been observed spanning between Z bands in intrafusal fibers and at branching points of Purkinje fibers where shear stresses are most abundant [59, 60]. (4) They are contractile structures. (5) They are products of an alteration in myofibrillogenesis, especially in specialized or transformed striated muscle such as Purkinje fibers in which the major function of the fiber is not to contract, but rather to conduct membrane potential [61]."

Earning Stripes: Myosin Binding Protein-C Interactions with Actin [62]

This is a machine-generated summary of:

van Dijk, Sabine J.; Bezold, Kristina L.; Harris, Samantha P.: Earning stripes: myosin binding protein-C interactions with actin [62]

Published in: Pflügers Archiv—European Journal of Physiology (2014)

Link to original: https://doi.org/10.1007/s00424-013-1432-8

If you want to cite the papers, please refer to the original.

For technical reasons we could not place the page where the original quote is coming from.

Abstract-Summary "Soon after its discovery, MyBP-C was also shown to bind actin."

"This was in part because interactions of MyBP-C with the thick filament could adequately explain most (but not all) effects of MyBP-C on actomyosin interactions and in part because the specificity of actin binding was uncertain."

"We review current evidence supporting MyBP-C interactions with actin and discuss these findings in terms of their ability to account for the functional effects of MyBP-C. We conclude that the influence of MyBP-C on muscle contraction can be explained equally well by interactions with actin as by interactions with myosin."

"Because data showing that MyBP-C binds to either myosin or actin has come almost exclusively from in vitro biochemical studies, the challenge for future studies is to define which binding partner(s) MyBP-C interacts with in vivo."

Introduction

"The importance of MyBP-C to muscle contraction is further underscored by the discovery that mutations in genes encoding MyBP-C cause myopathies in both skeletal [63, 64] and cardiac [65, 66] muscles and that MyBP-C is involved in cardiac stress pathways during both normal physiologic signaling and in pathological states such as heart failure [67]."

"What is far less certain and what remains a critical unresolved question is the precise mechanism(s) by which MyBP-C affects muscle contraction."

"Because of its ability to bind to myosin through two discrete binding sites, early hypotheses focused on the idea that interactions of MyBP-C with the thick filament alone were sufficient to account for the regulatory effects of MyBP-C [68]."

"The distinction is important not only for better understanding effects of MyBP-C on contraction but also for better insight into how sarcomeres function."

Structure and Position of MyBP-C in Sarcomeres

"MyBP-C is expressed in vertebrate striated muscles where it occurs as distinct isoforms originating from three separate genes: two skeletal genes that correspond to expression predominantly in slow and fast skeletal muscles [69] and a third cardiac gene expressed in the heart [70]."

"Within a sarcomere, MyBP-C is localized to a characteristic set of seven to nine discrete stripes (with the exact number of stripes depending on the type of muscle [71]) that are evenly spaced in the C-zone of the A-band."

"MyBP-C is present at a limited stoichiometry but in discrete positions relative to myosin."

Myosin, the First Binding Partner of MyBP-C

"Given its co-purification with and strong binding to myosin, it is not surprising that early studies investigating the influence of MyBP-C on actomyosin interactions initially focused on its connections to myosin."

"MyBP-C was shown to bind to both the light meromyosin (LMM) and the S2 subfragment of myosin [68, 72]."

"Ababou and others [73, 74] demonstrated that C1 and C2 also bound to the S2Δ segment of myosin, albeit through much weaker interactions."

Actin, the Second Binding Partner of MyBP-C

"The finding that MyBP-C could bind thin filaments was striking in that it marked the discovery of the only myofilament protein (aside from myosin) that could simultaneously link thick and thin filaments together within the region of active cross-bridge cycling."

"Functional roles for a protein that could span the two filament systems were immediately suggested, such as stabilization of the sarcomere lattice through weak coupling of filaments in relaxed muscle or that MyBP-C could make transient contacts with actin such that myosin cross-bridge kinetics could be affected in contracting muscle [75]."

"Renewed interest in the physiological significance of the ability of MyBP-C to bind actin came with the discovery that the same recombinant truncated MyBP-C proteins that were found to bind myosin S2 could also bind to actin and thin filaments [76]."

"The possibility that multiple interaction sites dispersed throughout multiple N-terminal domains contribute to the actin binding properties of MyBP-C is supported by the ability of C1C2 to cross-link F-actin filaments [76, 77] and by 3-D reconstructions of actin decorated with N-terminal MyBP-C domains showing that the recombinant proteins can span multiple actin monomers [77, 78]."

Can Interactions with Myosin or Actin Adequately Explain Effects of MyBP-C on Contraction?

"According to these authors, interactions of MyBP-C with myosin S2 could restrict the outward movement of cross-bridges toward the thin filament."

"Because phosphorylation of cardiac MyBP-C abolishes interactions with myosin S2 [79] and increases the proximity of myosin heads to actin [80], their model provides a straightforward mechanism to account for the ability of MyBP-C to limit cross-bridge interactions with actin."

"By binding to the critical S1/S2 junction, MyBP-C is in an ideal location to influence interactions with other thick filament proteins such as RLC [81, 82] or to otherwise affect myosin S1 head position or head-head interactions."

"The ability of MyBP-C to compete with myosin S1 heads and inhibit actomyosin interactions [75, 83] could provide an explanation for the long-standing puzzle that MyBP-C is present at a limited stoichiometry with respect to myosin: by being present at a low concentration relative to myosin heads, the activating effects of MyBP-C on the thin filament may be optimized while inhibitory competitive effects with myosin cross-bridges are minimized."

MyBP-C Binding Interactions In Vivo

"While there is considerable evidence to conclude that MyBP-C can bind to both myosin S2 and the thin filament, at present there is no direct evidence that MyBP-C interacts with either myosin S2 or actin filaments in sarcomeres."

"Resolution of the central question of whether MyBP-C binds to actin or myosin S2 to affect contraction will thus ultimately require innovations in structural methods that allow more precise visualization of the position(s) of MyBP-C in the sarcomere."

"It is possible that MyBP-C binds to both myosin S2 and the actin filament to affect contraction."

"Time-resolved methods such as FRET and X-ray diffraction that can differentiate dynamic changes in binding to the thick and thin filaments will thus be essential in further defining the mechanism(s) by which MyBP-C affects contraction."

Insights into Myosin Regulatory and Essential Light Chains: A Focus on Their Roles in Cardiac and Skeletal Muscle Function, Development and Disease [84]

This is a machine-generated summary of:

Sitbon, Yoel H.; Yadav, Sunil; Kazmierczak, Katarzyna; Szczesna-Cordary, Danuta: Insights into myosin regulatory and essential light chains: a focus on their roles in cardiac and skeletal muscle function, development and disease [84].

Published in: Journal of Muscle Research and Cell Motility (2019)

Link to original: https://doi.org/10.1007/s10974-019-09517-x

If you want to cite the papers, please refer to the original.

For technical reasons we could not place the page where the original quote is coming from.

Abstract-Summary "We focus on the involvement of myosin regulatory (RLC) and essential (ELC) light chains in striated muscle development, isoform appearance and their function in normal and diseased muscle."

"We review the consequences of isoform switching and knockout of specific MLC isoforms on cardiac and skeletal muscle function in various animal models."

"We discuss how dysregulation of specific RLC/ELC isoforms can lead to cardiac and skeletal muscle diseases and summarize the effects of most studied mutations leading to cardiac or skeletal myopathies."

Introduction

"The striated muscle sarcomere contains a system of interdigitating myosin II-containing thick filaments and actin/tropomyosin (Tm)/troponin(Tn)-containing thin filaments [85]."

"Binding of Ca^{2+} to TnC triggers a series of conformational changes in actin–Tm–Tn thin filaments, allowing for ATP-dependent cyclic interactions of myosin cross-bridges with actin and muscle contraction [86, 87]."

"Important domains of the myosin head (cross-bridge) include the motor domain containing the ATP and actin binding sites, and the lever arm, which undergoes large rotational motions that drive the power stroke [88, 89]."

"To fully understand the function of myosin and how it powers cardiac or skeletal muscle, it is essential to have insight into the role of myosin light chain (MLC) components."

"This review focuses on the involvement of myosin RLC and ELC in striated muscle development and their function in cardiac and skeletal muscle in health and disease."

Diverse Appearance and Functions of MLCs

"Three-dimensional maps of vertebrate muscle thin filaments obtained by cryo-electron microscopy revealed that the N-terminal portion of the myosin ELC is in a position to make molecular contacts with the C-terminus of the actin monomer [90]."

"Despite these multilevel investigations, the questions as to whether these protein–protein contacts between the N-ELC and actin promote or inhibit the affinity of myosin for actin, force production and muscle contraction are still to be determined."

"Studies from the Szczesna-Cordary group revealed that the lack of the N-terminal ELC extension in Tg-Δ43 mice, expressing a 43 aa truncated ELC, led to changes in myosin head orientation, positioning it closer to the actin filaments [91]."

"The important function of the N-terminal extension of the ELC protein was also demonstrated in another study using Tg-Δ43 animals that showed that the length-dependent activation in Tg-Δ43 muscle strips was blunted [92]."

Myogenesis of Striated Muscle Components

"Skeletal myogenesis begins with the expression of Myf-5 at embryonic day 8 (E8) (high levels at E9.25) in the dermomyotome of somites before formation of myotome."

"MyoD is the last transcription factor regulating myogenesis in skeletal muscle and is not detectable in the myotome until E10.5 where its expression rapidly increases until E17.5 and continues to be expressed at high levels in embryonic and fetal development [93, 94]."

"MYL4 (atrial ELC) is the most predominant gene expressed at E9.5, but its expression in skeletal muscle decreases at E15.5."

"According to Lyons and others, the expression of MYL3 (ventricular ELC) in the slow-twitch skeletal muscle starts at E15 and continues throughout adulthood [95]."

Effects of Genetic MLC Ablation/Isoform Switch in RLC and ELC Animal Models

"Lack of the ventricular RLC isoform in knock-out mice resulted in embryonic lethality at E12.5 even though the level of atrial RLC was increased reaching levels comparable to the ventricular RLC."

"The specific functions of the ventricular versus atrial isoforms of the myosin ELC were tested in transgenic mice by the Robbins laboratory [96]."

"Exchanging atrial RLC with ventricular isoform led to enhancement of mechanical properties of transgenic atrial myocytes to the level seen in ventricular myocytes of control animals."

"Using the same animal model, Buck and others showed an increase in shortening velocity in transgenic atrial myocytes expressing ventricular RLC compared to wild-type mice."

"The authors found that wild-type mice did not express the ventricular/slow skeletal RLC isoform until after birth, while it was expressed normally in the embryonic heart."

Involvement of Striated MLC Isoforms in Disease

"Ever since the very first discovery of A13T, E22K and P95A during RLC screening in HCM patients with pronounced mid cavity obstruction, a plethora of mutations in ventricular MYL2 have since been associated with one or more types of familial cardiomyopathies."

"Analyses of skeletal tissues in individuals with mutations in MYL3 have been performed, and a study by Poetter and others [97] found that biopsy obtained from soleus and deltoid muscles of patients with the M149V-ELC mutation showed myopathic changes associated with increased accumulation of mitochondria and ragged red fiber (RRF) pattern with no abnormality of cytochrome oxidase."

"Mutations in the ventricular ELC isoform (MYL3 gene) have been identified by population studies to cause HCM phenotypes."

"By Orr and others, the authors identified a novel E11K mutation in atrial ELC in a family with a previously unreported syndrome characterized by early-onset atrial fibrillation AF (age <35 years), conduction disease and signs of a primary atrial myopathy [98]."

Summary, Conclusions and Future Directions

"The goal of this review was to summarize the role of myosin regulatory and essential light chains in cardiac and skeletal muscle development and function in normal and disease states."

"MLCs are expressed in several isoforms depending on whether they are expressed in atria, ventricles or skeletal muscles where they contribute to tissue-specific function."

"Cardiac and skeletal myogenesis refer to the process of converting progenitor cells to myocytes via expression of specific transcription factors necessary for determination of unique RLC and ELC isoforms."

"Understanding this process is crucial in dissecting specific functions of MLCs in regulating cardiac or skeletal muscle contraction."

"We reviewed the consequences of isoform switching and knockout of specific MLC isoforms in the heart and skeletal muscles on cardiac and skeletal muscle function in various animal models."

A Multi-scale Continuum Model of Skeletal Muscle Mechanics Predicting Force Enhancement Based on Actin–Titin Interaction [99]

This is a machine-generated summary of:

Heidlauf, Thomas; Klotz, Thomas; Rode, Christian; Altan, Ekin; Bleiler, Christian; Siebert, Tobias; Röhrle, Oliver: A multi-scale continuum model of skeletal muscle mechanics predicting force enhancement based on actin–titin interaction [99]

Published in: Biomechanics and Modeling in Mechanobiology (2016)

Link to original: https://doi.org/10.1007/s10237-016-0772-7

If you want to cite the papers, please refer to the original.

For technical reasons we could not place the page where the original quote is coming from.

Abstract-Summary "Although recent research emphasises the possible role of titin in skeletal muscle force enhancement, this property is commonly ignored in current computational models."

"This work presents the first biophysically based continuum-mechanical model of skeletal muscle that considers, in addition to actin–myosin interactions, force enhancement based on actin–titin interactions."

"The mechanical behaviour of titin is included on the microscopic half-sarcomere level of a multi-scale chemo-electro-mechanical muscle model, which is based on the classic sliding-filament and cross-bridge theories."

"To titin stress contributions in the muscle fibre direction, the continuum-mechanical constitutive relation accounts for geometrically motivated, titin-induced stresses acting in the muscle's cross-fibre directions."

"Predicted titin-induced stresses in the muscle's cross-fibre directions are rather insignificant."

Introduction

"Rode et al. [100] proposed a biophysical, molecular model of titin's semi-active behaviour."

"Rode et al. [100] implemented their biophysical model of force enhancement and force depression within a simple Hill-type muscle model."

"To the best knowledge of the authors, the model of Lemos et al. [101] is the only continuum-mechanical muscle model that aims to model force enhancement."

"In Lemos et al. [101], the nonlinear active force–length relationship of the muscle is replaced by a linear model when stretch of an activated muscle starts."

"The stress contributions induced by titin filaments are derived from the biophysical 'sticky-spring' model of Rode et al. [100]."

"Further, to be able to simulate submaximal contractions, in addition to passive and fully activated half-sarcomeres, the model of Rode et al. [100] needs to be extended."

"Although these terms result from purely geometrical considerations (trigonometry), none of the previous models of force enhancement [100–102] have considered this."

Material and Methods

"Rode et al. [100] consider in their model one half-sarcomere and attribute all passive forces in the half-sarcomere to the titin filaments."

"Having introduced the titin forces in the passive and active states, the model of Rode et al. [100] is now extended to account for submaximal activation levels."

"The 1D muscle fibre meshes are only used to solve the monodomain equation (including the biophysical half-sarcomere model of Shorten et al. [103] and the titin model at each discretisation point)."

"The data required for the titin model, i. e., the force–elongation relations of the distinct titin regions and the PEVK stiffness depending on the initial half-sarcomere length preceding stretch are taken from Figures 2 and B1A of Rode et al. [100], respectively."

Results

"The aim of this section is to present simulation results of force enhancement, first, on the microscopic half-sarcomere level, then on the muscle fibre level (mesoscale), and, finally, on the macroscopic whole muscle level."

"To demonstrate the effect of force enhancement on the muscle fibre level, an 8-cm-long fibre model consisting of 301 discrete half-sarcomere models, i. e., models of the excitation–contraction coupling [103], is considered."

"Stretching the muscle while maintaining the level of activation resulted in enhanced muscle forces due to actin–titin binding compared to the isometric force–stretch relation."

"The muscle specimen is passively stretched in both the muscle fibre direction and the constrained XF direction."

"After stimulating the muscle under fixed-length conditions with 50 Hz for 500 ms, the muscle is actively stretched in fibre direction."

Discussion

"The titin model by Nishikawa et al. [102] predicts decreased stiffness of titin with decreasing active force."

"Continuum-mechanical muscle models can potentially predict the 3D geometrical deformation of the muscle and the muscle's influence on other muscle forces [104–106]."

"When stretching the activated muscle on the descending limb of the active force–length relationship, the models by Schappacher-Tilp et al. [107] and by Nishikawa et al. [102] transmit the accumulating titin tension via one actin–titin binding site."

"Force predictions of our generic multi-scale model depend, for example, on the chosen force–length relationship and the model and data describing the titin properties, which change from muscle to muscle (even in the same species, Prado et al. [108])."

"To one-dimensional models [100, 102, 107] and the only existing continuum-mechanical model of force enhancement [101], the presented chemo-electro-mechanical muscle model also predicts titin-induced stresses in the muscle's XF directions."

Interaction of Formin FH2 with Skeletal Muscle Actin. EPR and DSC Studies [109]

This is a machine-generated summary of:

Kupi, Tünde; Gróf, Pál; Nyitrai, Miklós; Belágyi, József: Interaction of formin FH2 with skeletal muscle actin. EPR and DSC studies [109]

Published in: European Biophysics Journal (2013)

If you want to cite the papers, please refer to the original.

For technical reasons we could not place the page where the original quote is coming from.

Abstract-Summary "The formin homology 2 (FH2) domain is responsible for actin binding and acts as an important nucleating factor in eukaryotic cells."

"EPR and DSC were used to investigate the properties of the mDia1-FH2 formin fragment and its interaction with actin."

"In DSC and temperature-dependent EPR experiments we observed that mDia1-FH2 has a flexible structure and observed a major temperature-induced conformational change at 41 °C."

"The results also confirmed the previous observation obtained by fluorescence methods that formin binding can destabilize the structure of actin filaments."

"In the EPR experiments the intermolecular connection between the monomers of formin dimers proved to be flexible."

Introduction

"Experiments using fluorescence spectroscopic and paramagnetic resonance techniques have shown that formin binds to the barbed end of actin filaments and induces a change of their flexibility [110–112]."

"The binding of formins to the sides of the actin filaments is less tight and stabilizes the structure of the filaments, probably by connecting neighboring protomers [110]."

"The DSC transients also revealed the destabilization of actin filaments by formin [110]."

"It is expected that the interaction between actin and formin is mutual in a sense that their binding affects the conformation of both proteins."

"Little is known about the conformational changes in formin accompanying the actin binding."

"This domain is responsible for the interaction between actin and formin."

"We used electron paramagnetic resonance (EPR) to reveal properties of the spin-labeled formin and the consequences of its binding to actin."

Materials and Methods

"F-actin was prepared by the addition of 2 mM $MgCl_2$ and 100 mM KCl to buffer A. The mDia1-FH2 sample was dialyzed in DTT-free buffer (buffer T: 50 mM NaCl, 50 mM Tris–HCl, pH 7.6)."

"Unreacted labels were removed by dialysis in DTT-free buffer T. The amount of bound labels was determined comparing the double integrals of the EPR spectra of the labeled samples and an MSL solution of known concentration."

"The concentration of pyrene-labeled actin was determined photometrically at 344 nm by use of the absorption coefficient 2.2×10^4 M^{-1} cm^{-1} [113]."

"Unlabeled or labeled formin concentrations were 1 μM. Before the polymerization measurements the bound calcium of actin was replaced with magnesium by adding 200 μM EGTA and 50 μM $MgCl_2$ and incubating the samples for 5 min."

"The corresponding buffer solutions of formin or F-actin were used as reference samples."

Results

"MSL-labeled formin did not significantly accelerate actin polymerization."

"When dialysis of NEM-labeled formin was performed against DTT-containing buffer the effect of formin on actin polymerization was preserved."

"MSL labeled formin was mixed with monomeric actin in a low-salt magnesium-free buffer and incubated for 6 h at 4 °C."

"These results also proved that the labeled formin could initiate actin polymerization."

"This conclusion is in agreement with observations from our sedimentation experiments and also with previous results showing that formins could bind to the sides of the actin filaments [110, 111]."

"These results are in agreement with previous reports that addition of formin to actin (at a formin-to-actin molar ratio of 1:20) had a destabilizing effect on actin filament structure and resulted in a shift of approximately 1.5 °C in the transition temperature [110]."

Discussion

"For formin dimers rotating as spherical single entities one would expect values at least two times greater than for the monomers."

"On the basis of these considerations the 25.0 ns rotational correlation time was not attributed to the formin dimers, but to the individual monomers in dimers."

"The EPR signal reflects both the rotation of the entire monomer and that of the smaller part together, and the two components are characteristic of the same formin conformation."

"The rotational correlation time (τ_2) characteristic of the slower rotating component was calculated from the temperature-dependent EPR spectra obtained in the presence of actin."

"The longer rotational correlation time (25 ns) was attributed to the rotation of one formin monomer, indicating that the two formin monomers could wobble almost independently from each other in the dimers."

Conclusions

"We showed that formins can be labeled on a single cysteine residue with spin probes."

"Spin-labeled formin can furnish information about important aspects of the intra-molecular transitions occurring in formins and in formin–actin complexes."

"The data from EPR experiments indicated that formin monomers are connected by flexible links in the dimers, which seems to be important for fulfilling their biological functions."

The Complexity of Titin Splicing Pattern in Human Adult Skeletal Muscles [114]

This is a machine-generated summary of:

Savarese, Marco; Jonson, Per Harald; Huovinen, Sanna; Paulin, Lars; Auvinen, Petri; Udd, Bjarne; Hackman, Peter: The complexity of titin splicing pattern in human adult skeletal muscles [114]

Published in: Skeletal Muscle (2018)

Link to original: https://doi.org/10.1186/s13395-018-0156-z

If you want to cite the papers, please refer to the original.

For technical reasons we could not place the page where the original quote is coming from.

Abstract-Summary "TTN includes 363 coding exons, a repeated region with a high degree of complexity, isoform-specific elements, and metatranscript-only exons thought to be expressed only during fetal development."

"Although three main classes of isoforms have been described so far, alternative splicing events (ASEs) in different tissues or in different developmental and physiological states have been reported."

"We confirmed that the skeletal muscle N2A isoforms are highly prevalent, but we found an elevated number of alternative splicing events, some at a very high level."

"This study provides an extensive picture of the complex TTN splicing pattern in human adult skeletal muscle, which is crucial for a proper clinical interpretation of TTN variants."

Background

"Titin isoforms have traditionally been classified in three main categories based on the presence of the N2A and N2B elements within the I-band region [115–117]."

"N2A isoforms (mainly expressed in the skeletal muscles) contain the N2A element, but not the cardiac-specific N2B element."

"N2B isoforms only include the cardiac-specific N2B element."

"N2BA isoforms, expressed in the heart, include both the N2B and N2A elements."

"N2A and N2BA isoforms also include additional exons, resulting in a higher number of Ig and PEVK domains in the I-band region."

"Two further isoforms, named Novex-1 and Novex-2, are very similar to N2B but each of them includes an isoform-specific exon (exon 45 and exon 46, respectively)."

"Although early work focused on gene-expression analyses, RNA-Seq is a powerful tool for the identification and the study of alternative exon and splice site usage and of novel isoforms."

Methods

"Total RNA was extracted from the selected samples by the TRIzol reagent method, according to the manufacturer's instructions (Invitrogen, Life Technologies, Canada)."

"RNA quality was checked with BioAnalyzer equipment using the RNA 6000 Nano Assay kit (Agilent Technologies, CA, USA)."

"Indexed sequencing libraries were generated from 1 μg of total RNA, using the TruSeq Stranded Total RNA kit according to the manufacturer's instructions (Illumina, CA, USA)."

"Single-end sequencing (86 bp reads) of multiplex libraries was performed on NextSeq500 instrument."

"For experimental validation of RNA-Seq results, cDNA synthesis was performed using SuperScript III First-Strand Synthesis System (Thermo Scientific, USA)."

"Amplified products were separated on 2% agarose gels and specific electrophoresis bands, corresponding to differently spliced products, were extracted using NucleoSpin Gel and PCR Clean-up (Macherey-Nagel, Germany) and analyzed by Sanger sequencing."

Results

"Before focusing on alternative splice events, we analyzed the canonical junctions, which are present in the previously reported isoforms, to evaluate their relative expression in human adult skeletal muscles."

"We proceeded with a multistep analysis, based on two different categories of splicing events: (1) ASEs involving canonical splice sites (out of the repeated region and within this area) and (2) ASEs involving alternative splicing sites."

"Most of the canonical N2A exons, reported to be expressed in adult skeletal muscle, have a high inclusion value."

"Exon 148 has a similar expression in fetal and adult muscles, and it is mostly skipped in fetal and adult hearts."

"Meta-only exons 213–217 are almost constitutively expressed in fetal muscles and their expression is halved in adult muscles."

Discussion

"Exon 11, included in the canonical adult skeletal muscle isoform N2A, is mostly skipped in adult skeletal muscles."

"Most of the so-called metatranscript-only exons are expressed in adult skeletal muscle at a variable level."

"Most of the ASEs occur in the I-band region of titin, where a large number of exons are alternatively spliced [115, 116]."

"Our experimental data as well as publicly available data suggests a significant expression of the meta-only exons 213, 214, 215, 216, and 217 in adult skeletal muscle, although their inclusion is higher in fetal muscles."

"In line with previous data, exon 363 is skipped in about 10% of TTN transcripts in human adult skeletal muscle."

"Our data clearly shows a variable expression for most of the meta-only exons (148, 150, 159, 167–171, 213–217), confirming, however, that some of them (160–166) are not expressed at all in human adult skeletal muscles."

Conclusions

"We have identified and partly characterized a large number of alternative splicing events in titin, providing the first RNA-Seq-based, accurate and comprehensive picture of TTN splicing pattern in adult human skeletal muscle."

"This same approach will probably unveil similar complex splicing patterns for other muscle transcripts."

The Effects of a Skeletal Muscle Titin Mutation on Walking in Mice [118]

This is a machine-generated summary of:

Pace, Cinnamon M.; Mortimer, Sarah; Monroy, Jenna A.; Nishikawa, Kiisa C.: The effects of a skeletal muscle titin mutation on walking in mice [118]

Published in: Journal of Comparative Physiology A (2016)

Link to original: https://doi.org/10.1007/s00359-016-1137-5

If you want to cite the papers, please refer to the original.

For technical reasons we could not place the page where the original quote is coming from.

Abstract-Summary "The mdm mouse exhibits a small deletion in the titin gene resulting in dystrophic mutants and phenotypically normal heterozygotes."

"We examined the effects of this mutation on locomotion to assess how, and if, changes to muscle phenotype explain observed locomotor differences."

"Mutant mice are much smaller in size than their siblings and gait abnormalities may be driven by differences in limb proportions and/or by changes to muscle phenotype caused by the titin mutation."

"We quantified differences in walking gait among mdm genotypes and also determined whether genotypes vary in limb morphometrics."

"Morphological differences among genotypes in hindlimb proportions were small and do not explain the locomotor differences."

"We suggest that differences in locomotion among mdm genotypes are due to changes in muscle phenotype caused by the titin mutation."

Introduction

"In cardiac muscle, having different titin isoforms affects cardiac function and biomechanics [119] and the same relationship is likely true in skeletal muscle."

"Numerous studies have demonstrated that the size of the titin molecule affects the passive tension of single muscle fibers, muscle bundles, and whole muscle (i.e., [108, 120–124]."

"The rabbit soleus, with its longer titin protein, has lower passive tension than muscles with shorter titin proteins [108]."

"It varies among muscles whether titin or collagen is the primary determiner of passive tension within the physiological range of muscle length [108]."

"It is possible that their gait abnormalities are driven by morphological differences (such as limb proportions) rather than by changes to the muscle phenotype induced by the titin mutation."

"We examined how titin-induced muscle phenotypic differences among genotypes contribute to locomotor biomechanics in the mdm mouse."

Methods

"To mass and total length (measured from nose to anus and abbreviated TL), the hindlimbs of each mouse were measured in lateral view as a series of four segments: toes (from toe tip to the joint between the toes and metatarsals); metatarsals (from the metatarsal joint to the center of the ankle); shank (from the center of the ankle to the center of the knee); thigh (from the center of the knee to the center of the hip)."

"Walking data were collected from 9 wildtype mice, 10 heterozygous mice, and 8 mutant mice that were 28–44 days old."

"We selected a walking gait for analysis because at faster speeds the mutant mouse gait becomes idiosyncratic and difficult to compare to the other genotypes."

"We controlled for speed by filming many locomotor trials per mouse, calculated the speed of each trial, and then analyzed only those trials that were speed matched."

Results

"Wildtype and heterozygous mice increased in mass with age; whereas, mutant mice showed little change in mass with age."

"The genotype*sex interaction was also significant for body mass such that mice formed three statistically different groups from largest to smallest: wildtype and heterozygous males, wildtype and heterozygous females, and mutants (there was no difference between mutant males and females)."

"Wildtype and heterozygous mice (but not mutant) showed a difference between sexes in proximal limb segment lengths."

"Average speed (%TL) had a significant effect on stance duration, total stride duration, duty factor, and stride frequency."

"Corresponding to the increase in stance duration, the duty factors for wildtype and heterozygous mice (~0.70) were significantly larger than for mutants (~0.61)."

"For all genotypes, during the stance phase the ankle was first flexed and then extended, reaching maximum extension at the end of stance."

Discussion

"A complete kinematic study of walking among the mdm genotypes would determine the full impact of the titin mutation on walking and how mutants may compensate for changes to muscle phenotype."

"Unlike the mutant mice, who exhibit different locomotor patterns at the ankle joint in addition to overall stride timing differences, heterozygote differences are more subtle and their ankle kinematics are the same as the wildtype mice."

"At both the transcriptome and protein levels there have been differences found in potential titin-related signaling proteins between mutant and wildtype mice [125–128] that likely affect aspects of whole muscle phenotype either directly or indirectly."

"Heterozygote mice may prove particularly interesting for examining muscle compensatory mechanisms as in most instances they have been found to not differ from wildtype mice; yet, they express mutant titin and subtle locomotor differences have been observed both by Huebsch and others [129] and in this study."

Conclusions

"Changes in titin expression may be one way that muscle is modified."

"Eccentrically trained rats demonstrated an increase in passive stiffness (likely attributable to titin) in their triceps brachii muscle [130]."

"Mice have shown increased expression of titin in their gastrocnemius muscle in response to endurance training [131]."

"This study shows that the mdm mutation in titin changes walking kinematics in both mutant and heterozygote mice, possibly by affecting both the signaling and mechanical properties of titin resulting in a changed muscle phenotype."

Bibliography

1. Cooke R (2011) The role of the myosin ATPase activity in adaptive thermogenesis by skeletal muscle. Biophys Rev 3:33–45. https://doi.org/10.1007/s12551-011-0044-9
2. Stewart MA, Franks-Skiba K, Chen S, Cooke R (2010) Myosin ATP turnover rate is a mechanism involved in thermogenesis in resting skeletal muscle fibers. Proc Natl Acad Sci USA 107:430–435
3. Ferenczi MA, Homsher E, Simmons RM, Trentham DR (1978) Reaction mechanism of the magnesium ion-dependent adenosine triphosphatase of frog muscle myosin and subfragment 1. Biochem J 171(1):165–175
4. Gutierrez G, Pohil RJ, Narayana P (1989) Skeletal muscle O2 consumption and energy metabolism during hypoxemia. J Appl Physiol 66(5):2117–2123
5. Myburgh KH, Franks-Skiba K, Cooke R (1995) Nucleotide turnover rate measured in fully relaxed rabbit skeletal muscle myofibrils. J Gen Physiol 106(5):957–973
6. Cooke R, Pate E (1985) The effects of ADP and phosphate on the contraction of muscle fibers. Biophys J 48(5):789–798
7. Al-Khayat HA, Morris EP, Squire JM (2009) The 7-stranded structure of relaxed scallop muscle myosin filaments: support for a common head configuration in myosin-regulated muscles. J Struct Biol 166(2):183–194
8. Vibert P, Craig R (1985) Structural changes that occur in scallop myosin filaments upon activation. J Cell Biol 101(3):830–837
9. Zoghbi ME, Woodhead JL, Moss RL, Craig R (2008) Three-dimensional structure of vertebrate cardiac muscle myosin filaments. Proc Natl Acad Sci USA 105(7):2386–2390
10. Craig R, Woodhead JL (2006) Structure and function of myosin filaments. Curr Opin Struct Biol 16(2):204–212
11. Jung HS, Komatsu S, Ikebe M, Craig R (2008) Head-head and head-tail interaction: a general mechanism for switching off myosin II activity in cells. Mol Biol Cell 19(8):3234–3242
12. Zhao FQ, Craig R, Woodhead JL (2009) Head-head interaction characterizes the relaxed state of limulus muscle myosin filaments. J Mol Biol 385(2):423–431
13. Al-Khayat HA, Hudson L, Reedy MK, Irving TC, Squire JM (2003) Myosin head configuration in relaxed insect flight muscle: x-ray modeled resting cross-bridges in a pre-powerstroke state are poised for actin binding. Biophys J 85(2):1063–1079
14. Feng TP (1932) The effect of length on the resting metabolism of muscle. J Physiol 74(4):441–454
15. Solandt DY (1936) The effect of potassium on the excitability and resting metabolism of frog's muscle. J Physiol 86(2):162–170
16. Yamada K (1970) The increase in the rate of heat production of frog's skeletal muscle caused by hypertonic solutions. J Physiol 208(1):49–64
17. West TG, Boutilier RG (1998) Metabolic suppression in anoxic frog muscle. J Comp Physiol B 168(4):273–280
18. Johannsen DL, Ravussin E (2008) Spontaneous physical activity: relationship between fidgeting and body weight control. Curr Opinion Endocrinol Diabet Obes 15(5):409–415
19. Lowell BB, Spiegelman BM (2000) Towards a molecular understanding of adaptive thermogenesis. Nature 404(6778):652–660
20. Wijers SL, Saris WH, van Marken Lichtenbelt WD (2009) Recent advances in adaptive thermogenesis: potential implications for the treatment of obesity. Obes Rev 10(2):218–226
21. Astrup A, Bulow J, Christensen NJ, Madsen J, Quaade F (1986) Facultative thermogenesis induced by carbohydrate: a skeletal muscle component mediated by epinephrine. Am J Phys 250(2 Pt 1):E226–E229
22. Astrup A, Simonsen L, Bulow J, Madsen J, Christensen NJ (1989) Epinephrine mediates facultative carbohydrate-induced thermogenesis in human skeletal muscle. Am J Phys 257(3 Pt 1):E340–E345

23. Dulloo AG, Stock MJ, Solinas G, Boss O, Montani JP, Seydoux J (2002) Leptin directly stimulates thermogenesis in skeletal muscle. FEBS Lett 515(1–3):109–113, 113.
24. Costford SR, Seifert EL, Bezaire V, Gerritz M, Bevilacqua L, Gowing A, Harper ME (2007) The energetic implications of uncoupling protein-3 in skeletal muscle. Appl Physiol Nutr Metab 32(5):884–894
25. Barclay CJ, Woledge RC, Curtin NA (2009) Effects of UCP3 genotype, temperature and muscle type on energy turnover of resting mouse skeletal muscle. Pflugers Arch 457(4):857–864
26. Enerback S, Jacobsson A, Simpson EM, Guerra C, Yamashita H, Harper ME, Kozak LP (1997) Mice lacking mitochondrial uncoupling protein are cold-sensitive but not obese. Nature 387(6628):90–94
27. Haselgrove JC (1975) X-ray evidence for conformational changes in the myosin filaments of vertebrate striated muscle. J Mol Biol 92(1):113–143
28. Clinch NF (1968) On the increase in rate of heat production caused by stretch in frog's skeletal muscle. J Physiol 196(2):397–414
29. Levine RJ, Kensler RW, Yang Z, Stull JT, Sweeney HL (1996) Myosin light chain phosphorylation affects the structure of rabbit skeletal muscle thick filaments. Biophys J 71(2):898–907
30. Zhi G, Ryder JW, Huang J, Ding P, Chen Y, Zhao Y, Kamm KE, Stull JT (2005) Myosin light chain kinase and myosin phosphorylation effect frequency-dependent potentiation of skeletal muscle contraction. Proc Natl Acad Sci USA 102(48):17519–17524
31. Lim MS, Walsh MP (1986) Phosphorylation of skeletal and cardiac muscle C-proteins by the catalytic subunit of cAMP-dependent protein kinase. Biochem Cell Biol 64(7):622–630
32. Dauncey MJ (1990) Activity and energy expenditure. Can J Physiol Pharmacol 68(1):17–27
33. Esparza J, Fox C, Harper IT, Bennett PH, Schulz LO, Valencia ME, Ravussin E (2000) Daily energy expenditure in Mexican and USA Pima indians: low physical activity as a possible cause of obesity. Int J Obes Relat Metab Disord 24(1):55–59
34. Levine JA (2004) Nonexercise activity thermogenesis (NEAT): environment and biology. Am J Physiol Endocrinol Metab 286(5):E675–E685
35. Levine JA, Eberhardt NL, Jensen MD (1999) Role of nonexercise activity thermogenesis in resistance to fat gain in humans. Science 283(5399):212–214
36. Levine J, Melanson EL, Westerterp KR, Hill JO (2001a) Measurement of the components of nonexercise activity thermogenesis. Am J Physiol Endocrinol Metab 281(4):E670–E675
37. Schoeller DA (2001) The importance of clinical research: the role of thermogenesis in human obesity. Am J Clin Nutr 73(3):511–516
38. Levine JA, Schleusner SJ, Jensen MD (2000) Energy expenditure of nonexercise activity. Am J Clin Nutr 72(6):1451–1454
39. Levine JA, Lanningham-Foster LM, McCrady SK, Krizan AC, Olson LR, Kane PH, Jensen MD, Clark MM (2005) Interindividual variation in posture allocation: possible role in human obesity. Science 307(5709):584–586
40. Ravussin E, Lillioja S, Anderson TE, Christin L, Bogardus C (1986) Determinants of 24-hour energy expenditure in man. Methods and results using a respiratory chamber. J Clin Invest 78(6):1568–1578
41. Lannergren J, Westerblad H (1987) The temperature dependence of isometric contractions of single, intact fibres dissected from a mouse foot muscle. J Physiol 390:285–293
42. Williams JH, Barnes WS (1989) The positive inotropic effect of epinephrine on skeletal muscle: a brief review. Muscle Nerve 12(12):968–975
43. Kovacs P, Stumvoll M (2005) Fatty acids and insulin resistance in muscle and liver. Best Pract Res Clin Endocrinol Metab 19(4):625–635
44. Klaus S, Rudolph B, Dohrmann C, Wehr R (2005) Expression of uncoupling protein 1 in skeletal muscle decreases muscle energy efficiency and affects thermoregulation and substrate oxidation. Physiol Genomics 21(2):193–200
45. Stewart M, Franks-Skiba K, Cooke R (2009) Myosin regulatory light chain phosphorylation inhibits shortening velocities of skeletal muscle fibers in the presence of the myosin inhibitor blebbistatin. J Muscle Res Cell Motil 30:17–27. https://doi.org/10.1007/s10974-008-9162-9

46. Crow MT, Kushmerick MJ (1982a) Correlated reduction of velocity of shortening and the rate of energy utilization in mouse fast-twitch muscle during a continuous tetanus. J Gen Physiol 82:703–722
47. Crow MT, Kushmerick MJ (1982b) Myosin light chain phosphorylation is associated with a decrease in the energy cost for contraction in fast twitch mouse muscle. J Biol Chem 257:2121–2124
48. Franks-Skiba K, Lardelli R, Goh G, Cooke R (2007) Myosin light chain phosphorylation inhibits muscle fiber shortening velocity in the presence of vanadate. Am J Physiol Regul Integr Comp Physiol 292:R1603–R1612
49. Karatzaferi C, Franks-Skiba K, Cooke R (2007) Inhibition of shortening velocity of skinned skeletal muscle fibers in conditions that mimic fatigue. Am J Physiol Regul Integr Comp Physiol 294:R948–R955
50. Sweeney HL, Kushmerick MJ (1985) Myosin phosphorylation in permeabilized rabbit psoas fibers. Am J Phys 249:C362–C365
51. Nahirney PC, Fischman DA, Wang K (2006) Myosin flares and actin leptomeres as myofibril assembly/disassembly intermediates in sonic muscle fibers. Cell Tissue Res 324:127–138. https://doi.org/10.1007/s00441-005-0110-3
52. Ibara RM, Penny LT, Ebeling AW, Dykhuizen G van, Cailliet G (1983) The mating call of the plainfin midshipman fish, Porichthys notatus. In: Noakes DGL, Lindquist DG, Helfman GS, Ward JA (eds) Predators and prey in fishes. Junk, The Hague, pp. 205–212.
53. Bass AH, Marchaterre MA (1989) Sound–generating (sonic) motor system in a teleost fish (Porichthys notatus)—Sexual polymorphism in the ultrastructure of myofibrils. J Comp Neurol 286:141–153
54. Brantley RK, Marchaterre MA, Bass AH (1993a) Androgen effects on vocal muscle structure in a teleost fish with inter– and intra–sexual dimorphism. J Morphol 216:305–318
55. Brantley RK, Wingfield JC, Bass AH (1993c) Sex steroid levels in Porichthys notatus, a fish with alternative reproductive tactics, and a review of the hormonal bases for male dimorphism among teleost fishes. Horm Behav 27:332–347
56. Bass AH (1989) Evolution of vertebrate motor systems for acoustic and electric communication: peripheral and central elements. Brain Behav Evol 33:237–247
57. Bass AH, Baker R (1990) Sexual dimorphisms in the vocal control system of a teleost fish: morphology of physiologically identified neurons. J Neurobiol 21:1155–1168
58. Lewis MK, Nahirney PC, Chen V, Adhikari BB, Wright J, Reedy MK, Bass AH, Wang K (2003) Concentric intermediate filament lattice links to specialized Z–band junctional complexes in sonic muscle fibers of the type I male midshipman fish. J Struct Biol 143:56–71
59. Katz B (1961) The terminations of the afferent nerve fiber in the muscle spindle of the frog. Philos Trans R Soc Lond B 243:221–240
60. Thornell LE, Sjostrom M, Andersson KE (1976) The relationship between mechanical stress and myofibrillar organization in heart Purkinje fibres. J Mol Cell Cardiol 8:689–695
61. Oliphant LW, Loewen RD (1976) Filament systems in Purkinje cells of the sheep heart: possible alterations of myofibrillogenesis. J Mol Cell Cardiol 8:679–688
62. van Dijk SJ, Bezold KL, Harris SP (2014) Earning stripes: myosin binding protein-C interactions with actin. Pflugers Arch Eur J Physiol 466:445–450. https://doi.org/10.1007/s00424-013-1432-8
63. Gurnett CA, Desruisseau DM, McCall K et al (2010) Myosin binding protein C1: a novel gene for autosomal dominant distal arthrogryposis type 1. Hum Mol Genet 19:1165–1173
64. Markus B, Narkis G, Landau D et al (2012) Autosomal recessive lethal congenital contractural syndrome type 4 (LCCS4) caused by a mutation in MYBPC1. Hum Mutat 33:1435–1438
65. Bonne G, Carrier L, Richard P et al (1998) Familial hypertrophic cardiomyopathy: from mutations to functional defects. Circ Res 83:580–593
66. Harris SP, Lyons RG, Bezold KL (2011) In the thick of it: HCM-causing mutations in myosin binding proteins of the thick filament. Circ Res 108:751–764
67. Sadayappan S, de Tombe PP (2012) Cardiac myosin binding protein-C: redefining its structure and function. Biophys Rev 4:93–106

68. Starr R, Offer G (1978) The interaction of C-protein with heavy meromyosin and subfragment-2. Biochem J 171:813–816
69. Weber FE, Vaughan KT, Reinach FC, Fischman DA (1993) Complete sequence of human fast-type and slow-type muscle myosin-binding-protein C (MyBP-C). Differential expression, conserved domain structure and chromosome assignment. Eur J Biochem 216:661–669
70. Carrier L, Bonne G, Bahrend E et al (1997) Organization and sequence of human cardiac myosin binding protein C gene (MYBPC3) and identification of mutations predicted to produce truncated proteins in familial hypertrophic cardiomyopathy. Circ Res 80:427–434
71. Luther PK, Bennett PM, Knupp C et al (2008) Understanding the organisation and role of myosin binding protein C in normal striated muscle by comparison with MyBP-C knockout cardiac muscle. J Mol Biol 384:60–72
72. Moos C, Offer G, Starr R, Bennett P (1975) Interaction of C-protein with myosin, myosin rod and light meromyosin. J Mol Biol 97:1–9
73. Ababou A, Gautel M, Pfuhl M (2007) Dissecting the N-terminal myosin binding site of human cardiac myosin-binding protein C. Structure and myosin binding of domain C2. J Biol Chem 282:9204–9215
74. Ababou A, Rostkova E, Mistry S, Le Masurier C, Gautel M, Pfuhl M (2008) Myosin binding protein C positioned to play a key role in regulation of muscle contraction: structure and interactions of domain C1. J Mol Biol 384(3):615–630
75. Moos C, Mason CM, Besterman JM et al (1978) The binding of skeletal muscle C-protein to F-actin, and its relation to the interaction of actin with myosin subfragment-1. J Mol Biol 124:571–586
76. Shaffer JF, Kensler RW, Harris SP (2009) The myosin-binding protein C motif binds to F-actin in a phosphorylation-sensitive manner. J Biol Chem 284:12318–12327
77. Mun JY, Gulick J, Robbins J et al (2011) Electron microscopy and 3D reconstruction of F-actin decorated with cardiac myosin-binding protein C (cMyBP-C). J Mol Biol 410:214–225
78. Orlova A, Galkin VE, Jeffries CM et al (2011) The N-terminal domains of myosin binding protein C can bind polymorphically to F-actin. J Mol Biol 412:379–386
79. Gruen M, Prinz H, Gautel M (1999) cAPK-phosphorylation controls the interaction of the regulatory domain of cardiac myosin binding protein C with myosin-S2 in an on–off fashion. FEBS Lett 453:254–259
80. Colson BA, Bekyarova T, Locher MR, Fitzsimons DP, Irving TC, Moss RL (2008) Protein kinase A-mediated phosphorylation of cMyBP-C increases proximity of myosin heads to actin in resting myocardium. Circ Res 103(3):244–251
81. Margossian SS (1985) Reversible dissociation of dog cardiac myosin regulatory light chain 2 and its influence on ATP hydrolysis. J Biol Chem 260:13747–13754
82. Ratti J, Rostkova E, Gautel M, Pfuhl M (2011) Structure and interactions of myosin-binding protein C domain C0: cardiac-specific regulation of myosin at its neck? J Biol Chem 286:12650–12658
83. Shaffer JF, Razumova MV, Tu AY et al (2007) Myosin S2 is not required for effects of myosin binding protein-C on motility. FEBS Lett 581:1501–1504
84. Sitbon YH, Yadav S, Kazmierczak K, Szczesna-Cordary D (2019) Insights into myosin regulatory and essential light chains: a focus on their roles in cardiac and skeletal muscle function, development and disease. J Muscle Res Cell Motil 41:313–327. https://doi.org/10.1007/s10974-019-09517-x
85. Schiaffino S, Reggiani C (1996) Molecular diversity of myofibrillar proteins: gene regulation and functional significance. Physiol Rev 76(2):371–423
86. Huxley HE (1985) The crossbridge mechanism of muscular contraction and its implications. J Exp Biol 115(1):17–30
87. Geeves MA, Holmes KC (2005) The molecular mechanism of muscle contraction. Adv Protein Chem 71:161–193
88. Rayment I, Rypniewski WR, Schmidt-Base K, Smith R, Tomchick DR, Benning MM, Winkelmann DA, Wesenberg G, Holden HM (1993) Three-dimensional structure of myosin subfragment-1: a molecular motor. Science 261(5117):50–58

89. Geeves MA (2002) Molecular motors: stretching the lever-arm theory. Nature 415(6868):129–131
90. Milligan RA, Whittaker M, Safer D (1990) Molecular structure of F-actin and location of surface binding sites. Nature 348(6298):217–221
91. Muthu P, Wang L, Yuan CC, Kazmierczak K, Huang W, Hernandez OM, Kawai M, Irving TC, Szczesna-Cordary D (2011) Structural and functional aspects of the myosin essential light chain in cardiac muscle contraction. FASEB J 25(12):4394–4405
92. Michael JJ, Gollapudi SK, Ford SJ, Kazmierczak K, Szczesna-Cordary D, Chandra M (2013) Deletion of 1-43 amino acids in cardiac myosin essential light chain blunts length dependency of ca(2+) sensitivity and cross-bridge detachment kinetics. Am J Physiol Heart Circ Physiol 304(2):H253–H259
93. Sassoon D, Lyons G, Wright WE, Lin V, Lassar A, Weintraub H, Buckingham M (1989) Expression of two myogenic regulatory factors myogenin and MyoD1 during mouse embryogenesis. Nature 341(6240):303–307
94. Ott MO, Bober E, Lyons G, Arnold H, Buckingham M (1991) Early expression of the myogenic regulatory gene, myf-5, in precursor cells of skeletal muscle in the mouse embryo. Development 111(4):1097–1107
95. Lyons GE, Schiaffino S, Sassoon D, Barton P, Buckingham M (1990b) Developmental regulation of myosin gene expression in mouse cardiac muscle. J Cell Biol 111(6 Pt 1):2427–2436
96. Fewell JG, Hewett TE, Sanbe A, Klevitsky R, Hayes E, Warshaw D, Maughan D, Robbins J (1998) Functional significance of cardiac myosin essential light chain isoform switching in transgenic mice. J Clin Invest 101(12):2630–2639
97. Poetter K, Jiang H, Hassanzadeh S, Master SR, Chang A, Dalakas MC, Rayment I, Sellers JR, Fananapazir L, Epstein ND (1996) Mutations in either the essential or regulatory light chains of myosin are associated with a rare myopathy in human heart and skeletal muscle. Nat Genet 13(1):63–69
98. Orr N, Arnaout R, Gula LJ, Spears DA, Leong-Sit P, Li Q, Tarhuni W, Reischauer S, Chauhan VS, Borkovich M, Uppal S, Adler A, Coughlin SR, Stainier DY, Gollob MH (2016) A mutation in the atrial-specific myosin light chain gene (MYL4) causes familial atrial fibrillation. Nat Commun 7:11303
99. Heidlauf T, Klotz T, Rode C, Altan E, Bleiler C, Siebert T, Röhrle O (2016) A multi-scale continuum model of skeletal muscle mechanics predicting force enhancement based on actin–titin interaction. Biomech Model Mechanobiol 15:1423–1437. https://doi.org/10.1007/s10237-016-0772-7
100. Rode C, Siebert T, Blickhan R (2009) Titin-induced force enhancement and force depression: a 'sticky-spring' mechanism in muscle contractions? J Theor Biol 259(2):350–360. https://doi.org/10.1016/j.jtbi.2009.03.015
101. Lemos RR, Epstein M, Herzog W, Wyvill B (2001) Realistic skeletal muscle deformation using finite element analysis. XIV Brazilian symposium on computer graphics and image processing, pp 192–199.
102. Nishikawa KC, Monroy JA, Uyeno TE, Yeo SH, Pai DK, Lindstedt SL (2012) Is titin a 'winding filament'? A new twist on muscle contraction. Proc R Soc Lond Ser B Biol Sci 279(1730):981–990
103. Shorten PR, O'Callaghan P, Davidson JB, Soboleva TK (2007) A mathematical model of fatigue in skeletal muscle force contraction. J Muscle Res Cell Motil 28(6):293–313. https://doi.org/10.1007/2Fs10974-007-9125-6
104. Maas H, Baan GC, Huijing PA (2001) Intermuscular interaction via myofascial force transmission: effects of tibialis anterior and extensor hallucis longus length on force transmission from rat extensor digitorum longus muscle. J Biomech 34(7):927–940
105. Yucesoy CA, Koopman BH, Grootenboer HJ, Huijing PA (2007) Finite element modeling of aponeurotomy: altered intramuscular myofascial force transmission yields complex sarcomere length distributions determining acute effects. Biomech Model Mechanobiol 6(4):227–243

106. Blemker SS, Pinsky PM, Delp SL (2005) A 3D model of muscle reveals the causes of nonuniform strains in the biceps brachii. J Biomech 38(4):657–665. https://doi.org/10.1016/j.jbiomech.2004.04.009
107. Schappacher-Tilp G, Leonard T, Desch G, Herzog W (2015) A novel three-filament model of force generation in eccentric contraction of skeletal muscles. PLoS One 10(3):e0117634
108. Prado LG, Makarenko I, Andresen C, Krüger M, Opitz CA, Linke WA (2005) Isoform diversity of giant proteins in relation to passive and active contractile properties of rabbit skeletal muscles. J Gen Physiol 126(5):461–480
109. Kupi T, Gróf P, Nyitrai M, Belágyi J (2013) Interaction of formin FH2 with skeletal muscle actin. EPR and DSC studies. Eur Biophys J 42:757–765. https://doi.org/10.1007/s00249-013-0922-0
110. Bugyi B, Papp G, Hild G, Lorinczy D, Nevalainen EM, Lappalainen P, Somogyi B, Nyitrai M (2006) Formins regulate actin filament flexibility through long range allosteric interactions. J Biol Chem 281:10727–10736
111. Kupi T, Grof P, Nyitrai M, Belagyi J (2009) The uncoupling of the effects of formins on the local and global dynamics of actin filaments. Biophys J 96:2901–2911
112. Papp G, Bugyi B, Ujfalusi Z, Barko S, Hild G, Somogyi B, Nyitrai M (2006) Conformational changes in actin filaments induced by formin binding to the barbed end. Biophys J 91:2564–2572
113. Kouyama T, Mihashi K (1981) Fluorimetry study of N-(1-pyrenyl) iodoacetamide-labelled F-actin. Local structural change of actin protomer both on polymerization and on binding of heavy meromyosin. Eur J Biochem 114:33–38
114. Savarese M, Jonson PH, Huovinen S, Paulin L, Auvinen P, Udd B, Hackman P (2018) The complexity of titin splicing pattern in human adult skeletal muscles. Skelet Muscle 8:11. https://doi.org/10.1186/s13395-018-0156-z
115. Linke WA, Rudy DE, Centner T, Gautel M, Witt C, Labeit S, Gregorio CC (1999) I-band titin in cardiac muscle is a three-element molecular spring and is critical for maintaining thin filament structure. J Cell Biol 146(3):631–644
116. Freiburg A, Trombitas K, Hell W, Cazorla O, Fougerousse F, Centner T, Kolmerer B, Witt C, Beckmann JS, Gregorio CC et al (2000) Series of exon-skipping events in the elastic spring region of titin as the structural basis for myofibrillar elastic diversity. Circ Res 86(11):1114–1121
117. Labeit S, Lahmers S, Burkart C, Fong C, McNabb M, Witt S, Witt C, Labeit D, Granzier H (2006) Expression of distinct classes of titin isoforms in striated and smooth muscles by alternative splicing, and their conserved interaction with filamins. J Mol Biol 362(4):664–681
118. Pace CM, Mortimer S, Monroy JA, Nishikawa KC (2016) The effects of a skeletal muscle titin mutation on walking in mice. J Comp Physiol A 203:67–76. https://doi.org/10.1007/s00359-016-1137-5
119. Anderson BR, Granzier HL (2012) Titin-based tension in the cardiac sarcomere: molecular origin and physiological adaptations. Prog Biophys Mol Biol 110:204–217
120. Wang K, McCarter R, Wright J, Beverly J, Ramirez-Mitchell R (1991) Regulation of skeletal muscle stiffness and elasticity by titin isoforms: a test of the segmental extension model of resting tension. Proc Natl Acad Sci 88:7101–7105
121. Fukuda N, Granzier HL, Ishiwata S, Kurihara S (2008) Physiological functions of the giant elastic protein titin in mammalian striated muscle. J Physiol Sci 58:151–159
122. Lopez MA, Pardo PS, Cox GA, Boriek AM (2008) Early mechanical dysfunction of the diaphragm in the muscular dystrophy with myositis (Ttnmdm) model. Am J Physiol Cell Physiol 295:C1092–C1102
123. Ottenheijm CAC, Knottnerus AM, Buck D, Luo X, Greer K, Hoying A, Labeit S, Granzier H (2009) Turning passive mechanics through differential splicing of titin during skeletal muscle development. Biophys J 97:2277–2286
124. Cornachione AS, Leite F, Bagni MA, Rassier DE (2016) The increase in non-cross-bridge forces after stretch of activated striated muscle is related to titin isoforms. Am J Physiol Cell Physiol 310:C19–C26

125. Haravuori H, Vihola A, Straub V, Auranen M, Richard I, Marchand S, Voit T, Labeit S, Somer H, Peltonen L, Beckmann JS, Udd B (2001) Secondary calpain3 deficiency in 2q-linked muscular dystrophy: titin is the candidate gene. Neurology 56:869–877
126. Garvey SM, Rajan C, Lerner AP, Frankel WN, Cox GA (2002) The muscular dystrophy with myositis (mdm) mouse mutation disrupts a skeletal muscle-specific domain of titin. Science 79:146–148
127. Witt CC, Ono Y, Puschmann E, McNabb M, Wu Y, Gotthardt M, Witt SH, Haak M, Labeit D, Gregorio CC, Sorimachi H, Granzier H, Labeit S (2004) Induction and myofibrillar targeting of CARP, and suppression of the Nkx2.4 pathway in the mdm mouse with impaired titin-based signaling. J Mol Biol 336:145–154
128. Hayashi C, Ono Y, Doi N, Kitamura F, Tagami M, Mineki R, Arai T, Taguchi H, Yanagida M, Hirner S, Labeit D, Labeit S (2008) Multiple molecular interactions implicate the connectin/titin N2A region as a modulating scaffold for p94/calpain 3 activity in skeletal muscle. J Biol Chem 283:14801–14814
129. Huebsch KA, Kudryashova E, Wooley CM, Sher RB, Seburn KL, Spencer MJ, Cox GA (2005) Mdm muscular dystrophy: interactions with calpain 3 and a novel functional role for titin's N2A domain. Hum Mol Genet 14:2801–2811
130. Reich TE, Lindstedt SL, LaStayo PC, Pierotti DJ (2000) Is the spring quality of muscle plastic? Am J Physiol Regul Integr Comp Physiol 278:R1661–R1666
131. Bellafiore M, Cappello F, Palumbo D, Macaluso F, Bianco A, Palma A, Farina F (2007) Increased expression of titin in mouse gastrocnemius muscle in response to an endurance-training program. Eur J Histochem 51:119–124
132. Biewener A, Baudinette R (1995) In vivo muscle force and elastic energy storage during steady-speed hopping of tammar wallabies (macropus eugenii). J Exp Biol 198(9):1829–1841

Binding Proteins

2

Christopher Myers

2.1 Introduction by the Editor

This chapter pivots to the microscopic world of skeletal muscle-binding proteins—these molecular powerhouses that perform many tasks essential to the structure and function of skeletal muscles. Recent research has brought these proteins into sharp focus, and this chapter explores the cutting-edge discoveries and their implications for our understanding of skeletal muscle physiology.

In skeletal muscles, binding proteins are central to the contractile mechanism, allowing the complex interplay of actin and myosin that fuels muscle contraction. While the role of these proteins has been acknowledged for years, recent studies continue to shed light on their precise functions and modes of action.

One of the areas explored is how age and disease affect myosin-binding protein C. This protein is an essential structural protein that assists in anchoring actin and myosin filaments. Deterioration or alterations to this structural protein has deleterious effects on how the muscle contracts and produces force. This chapter will explore these effects.

Another area we explore is the role of insulin and insulin resistance to the binding protein C. Immerging evidence suggests that the inactivation of the C protein may have a role in insulin resistance. The information gathered from rat models may be emerging evidence on how to further combat insulin resistance in humans.

In conclusion, this chapter explores the pioneering research on skeletal muscle-binding proteins, reinforcing these proteins' pivotal role in muscle function. By harnessing the power of advanced technologies and innovative methodologies, researchers are gaining a deeper understanding of these complex biological systems, leading to transformative impacts on muscle physiology.

C. Myers (✉)
3 Sports Science, 3 Sports Science LLC, Akron, OH, United States of America
e-mail: chris@3sportsscience.com

C. Myers (ed.), *Skeletal Muscle Physiology*,
https://doi.org/10.1007/978-3-031-47065-3_2

Skeletal muscle, a tissue characterized by its ability to contract and produce force, is a marvel of biological engineering. As described in the previous chapter, an intricate network of proteins, each playing a pivotal role in the biomechanics of muscle fibers, plays a crucial role in skeletal muscle function. Among these critical proteins, Myosin Binding Protein-C (MyBP-C) stands out for its significant contributions to muscle performance and regulation. This chapter delves into the complex world of skeletal muscle-binding proteins, focusing on the various paralogs of MyBP-C found in Type 2x muscle fibers and their regulation through phosphorylation and carbonylation processes.

Recent research reveals that MyBP-C proteins in Type 2x muscle fibers are highly phosphorylated. This finding suggests a regulatory mechanism that enhances muscle contraction speed and efficiency. This phosphorylation state is crucial for the optimal performance of fast-twitch muscles primarily engaged in rapid and high-intensity movements. Intriguingly, the modifications of MyBP-C through carbonylation and phosphorylation appear unaffected by the aging process, indicating the robustness of muscle function regulation across the lifespan. However, the complexity of MyBP-C regulation and its implications for muscle physiology are far from fully understood. More research is needed to understand the underlying mechanisms driving age-related modifications in MyBP-C.

The functional significance of MyBP-C extends beyond its phosphorylation state. Models suggest that MyBP-C contributes to the internal load influencing muscle contraction's maximum velocity (Vmax) at low activation levels, following a certain degree of active shortening. This internal loading, possibly mediated by MyBP-C's interaction with actin and myosin or its influence on the mechanical properties of myosin cross-bridges, highlights the multifaceted role of MyBP-C in muscle function. Increasingly, evidence suggests that MyBP-C may play a critical role in modulating the internal mechanical environment of muscle fibers, thereby affecting overall muscle performance.

Further underscoring the complexity of MyBP-C's role in skeletal muscle, research indicates that the N terminus of each skeletal MyBP-C isoform is functionally distinct, with modulatory capacities that vary depending on the muscle type in which they are expressed. This discovery opens up fascinating avenues for the molecular tuning of skeletal muscle performance, where differential expression of MyBP-C isoforms could be leveraged to optimize muscle function for specific needs or conditions. However, our understanding of this functionality is still in its infancy. The scientific community needs more research to understand better the underlying mechanisms that guide this functionality.

Embarking on this exploration of skeletal muscle binding proteins reveals that understanding muscle physiology is still in its infancy. The nuanced interactions between MyBP-C and other muscle components, its phosphorylation and carbonylation implications, and the potential for age-related modifications present a complex puzzle. This chapter unravels these intricacies, illuminating the sophisticated mechanisms governing skeletal muscle performance. Integrating current research findings with foundational knowledge aims to pave the way for future breakthroughs in muscle biology, with the ultimate goal of enhancing human health and athletic performance through a deeper understanding of skeletal muscle binding proteins.

Skeletal muscle, an integral component of the human body, is characterized by its remarkable ability to contract and produce force, making it a focal point of interest in studying biological engineering and physiology. This tissue's unique capabilities allow it to perform various functions, from facilitating movement to playing a pivotal role in metabolic regulation. As discussed in the preceding chapter, skeletal muscle functionality is made possible through an intricate network of proteins, each of which plays a critical role in the biomechanics of muscle fibers. Among the plethora of proteins involved in muscle function, Myosin Binding Protein-C (MyBP-C) is particularly significant due to its substantial contributions to muscle performance and regulatory mechanisms.

This chapter aims to delve deeper into the complex universe of skeletal muscle-binding proteins, focusing on the role and regulation of MyBP-C within Type 2x muscle fibers. Type 2x fibers, known for their fast-twitch properties, are crucial for performing rapid and high-intensity movements, such as sprinting or lifting heavy objects. The regulation of MyBP-C in these fibers, primarily through phosphorylation and carbonylation processes, is paramount for understanding muscle physiology and optimizing muscle function.

Recent advancements in research have highlighted that MyBP-C proteins in Type 2x muscle fibers exhibit a high degree of phosphorylation. This discovery points towards a sophisticated regulatory mechanism that significantly enhances the speed and efficiency of muscle contractions. The phosphorylated state of MyBP-C is vital for the peak performance of fast-twitch muscles, underscoring the importance of this post-translational modification in regulating muscle dynamics. Moreover, the fact that modifications of MyBP-C through both carbonylation and phosphorylation processes appear unaffected by aging suggests remarkable robustness in regulating muscle function across an individual's lifespan. However, the intricacies of MyBP-C's regulation and its broader implications for muscle health and disease remain primarily elusive, signaling a pressing need for further research to decipher the mechanisms underlying age-related changes in MyBP-C functionality.

Beyond its phosphorylation state, the functional significance of MyBP-C extends to its contribution to the internal load that influences muscle contraction's maximum velocity (Vmax) under conditions of low activation. This internal loading mechanism, potentially mediated by MyBP-C's interaction with actin and myosin or its influence on the mechanical properties of myosin cross-bridges, highlights the complex and multifaceted role that MyBP-C plays in muscle function. Increasing evidence suggests that MyBP-C is instrumental in modulating the internal mechanical environment of muscle fibers, thereby profoundly impacting overall muscle performance. This modulation is crucial for fine-tuning muscle responses to a wide range of physical activities and demands.

Furthermore, MyBP-C isoforms play a pivotal role in regulating thin filament activity in skeletal muscle fibers, exerting their influence in a calcium (Ca2+)-dependent manner. By binding to both myosin and actin, these isoforms can modulate cross-bridge cycling and thus fine-tune muscle contraction in response to varying Ca2+ levels. Through this interaction, MyBP-C isoforms influence the sensitivity of the muscle to Ca2+, altering the force production in a manner that is directly linked to intracellular Ca2+ concentration. This regulatory mechanism is crucial for the efficient functioning of skeletal muscles, enabling precise control over muscle contraction and relaxation processes. The Ca2 + -dependent modulation by MyBP-C isoforms highlights the sophisticated biochemical and mechanical interplay necessary for the nuanced control of muscle dynamics. Such regulation is vital for maintaining muscle performance and adaptability, especially under conditions that demand rapid adjustments in contraction strength. This chapter explores these mechanisms and provides insights into muscle physiology and potential targets for addressing muscle-related diseases.

Another aspect of MyBP-C characteristics is the occurrence of myopathies and age. As individuals age, changes in the expression, phosphorylation, and overall function of MyBP-C can significantly impact muscle performance, potentially contributing to age-related declines in muscle strength and cardiac function. The same changes are seen within myopathies such as Duchenne Dystrophy and Arthrogryposis Myopathy Arthrogryposis Myopathy. Research has shown that alterations in MyBP-C, whether through genetic mutations or age-related modifications, can lead to cardiomyopathies and diminished muscle function, underscoring its importance in maintaining muscle health over the lifespan. The aging process may affect the post-translational modifications of MyBP-C, such as phosphorylation, thereby influencing its regulatory functions and potentially leading to reduced muscle efficiency and increased susceptibility to damage. Furthermore, age-related oxidative stress and inflammation can exacerbate alterations in MyBP-C, further impairing muscle function and contributing to sarcopenia, the progressive loss of muscle mass and strength seen in older adults. Efforts to enhance MyBP-C function or mitigate its age-related changes could offer promising strategies for preserving muscle health and combating age-related muscle deterioration. As discussed in this chapter, understanding the complex interactions between MyBP-C, aging, and muscle function highlights the potential for targeted interventions that address the molecular underpinnings of muscle aging, offering hope for improved quality of life in the aging population.

As we embark on this detailed exploration of skeletal muscle-binding proteins, it becomes evident that our understanding of muscle physiology is still developing. The nuanced interactions between MyBP-C and other muscle components, its phosphorylation and carbonylation implications, and the potential for adaptive or maladaptive changes with age present a rich tapestry of scientific inquiry. This chapter unravels these complexities, shedding light on the sophisticated mechanisms governing skeletal muscle performance. By integrating the latest research findings with foundational knowledge, this discussion aims to pave the way for future breakthroughs in the field of muscle biology. The ultimate goal is to enhance human health and athletic performance by understanding the roles and regulation of skeletal muscle-binding proteins, offering new insights into preventing and treating muscle-related diseases and optimizing athletic prowess.

2.2 Machine Generated Summaries

Disclaimer: The summaries in this chapter were generated from Springer Nature publications using extractive AI auto-summarization: An extraction-based summarizer aims to identify the most important sentences of a text using an algorithm and uses those original sentences to create the auto-summary (unlike generative AI). As the constituted sentences are machine selected, they may not fully reflect the body of the work, so we strongly advise that the original content is read and cited. The auto generated summaries were curated by the editor to meet Springer Nature publication standards. To cite this content, please refer to the original papers.

Machine generated keywords: mybpc, bind proteinc, proteinc, myosin bind, cardiac, myosin, insulin, rna, insulin resistance, filament, insulin signal, sarcomeric, isoform, arthrogryposis, phosphorylate

Fast and Slow Skeletal Myosin Binding Protein-C and Aging [1]

This is a machine-generated summary of:

Perazza, L. R.; Wei, G.; Thompson, L. V.: Fast and slow skeletal myosin binding protein-C and aging [1]

Published in: GeroScience (2022)

Link to original: https://doi.org/10.1007/s11357-022-00689-y

If you want to cite the papers, please refer to the original.

For technical reasons we could not place the page where the original quote is coming from.

Abstract-Summary "The role of fast (fMyBP-C) and slow (sMyBP-C) skeletal muscle MyBP-C remains to be elucidated."

"We aimed to characterize MyBP-C and its paralogs in the fast tibialis anterior (TA) muscle from adult and old mice."

"We further found that the expression level of cardiac myosin binding protein-C (cMyBP-C), an important modulator of cardiac output, was lowered by age."

"Standard SDS-PAGE along with Pro-Q Diamond phosphoprotein staining did not identify age-related changes in phosphorylated MyBP-C proteins from TA and cardiac muscles; however, it revealed that MyBP-C paralogs in fast skeletal and cardiac muscle were highly phosphorylated."

"MyBP-C protein-bound carbonyls were determined using anti-DNP immunostaining and found the carbonyl level of fMyBP-C, sMyBP-C, and cMyBP-C to be similar between old and adult animals."

"Our data showed some differences regarding the MyBP-C paralog expression and identified an age-related reduction of cMyBP-C expression."

Introduction

"It is unknown whether the expression of the skeletal muscle MyBP-C paralogs (slow and fast skeletal MyBP-C) changes in the fast tibialis anterior (TA) muscle with aging."

"Although the MyBP-C paralogs have yet to be fully described in aging skeletal muscles, there is differential phosphorylation of specific residues in the sMyBP-C paralog within the slow soleus and the fast flexor digitorum brevis muscles early in the lifespan (between the ages of 2 and 14 months) from WT mice and from mice with dystrophy (mdx mouse model) providing sufficient rationale for further investigation of this sarcomeric protein and its paralogs as potential factors contributing to the complex nature of sarcopenia [2, 3]."

"Understanding the post-translational oxidative state of skeletal muscle MyBP-C paralogs with age may offer a mechanism contributing to sarcopenia."

"According to age-related changes in protein isoform compositions within skeletal muscles, we hypothesized that the expression levels of the fMyBP-C paralog would decrease with age."

Materials and Methods

"Before immunoblotting, muscle homogenates were separated by SDS-PAGE using a 4–15% Mini-PROTEAN® TGX Stain-Free™ Protein Gel (12 well, 20 µl) at RT, 100 V for ~ 1 h. Gels were transferred to nitrocellulose membranes (0.2 µm, Bio-Rad) on a Mini Trans-Blot cell at 30 V overnight in a buffer containing 192-mM glycine, 25-mM Tris, 0.025% SDS, and 20% methanol."

"To determine total proteins within the TA muscle, the blots were stained with REVERT™ Total Protein Stain and Wash Solution Kit (LI-COR Biosciences) and imaged at 700 nm (red) with Odyssey imaging system (LI-COR Biosciences)."

"To determine total proteins within the heart muscle, the blots were stained with Ponceau S (red stain), rinsed, and imaged (700 nm, red)."

"Phosphorylation of TA and heart muscle myofibrillar proteins was estimated by Pro-Q® Diamond phosphoprotein staining method according to the manufacturer's instructions."

Results

"To gain insight into the overall expression profile of the specific MyBP-C paralogs present in the fast TA muscles from adult and old mice, we analyzed the relative content of both the sMyBP-C (slow paralog) and fMyBP-C (fast paralog)."

"We found both paralogs, sMyBP-C and fMyBP-C, constitutively expressed in the fast TA muscles from adult and old mice."

"Because sarcomeric protein stoichiometry is tightly maintained, MyBP-C is reported to impact contractility [4–6], and we found both skeletal MyBP-C paralogs (fMyBP-C and sMyBP-C) expressed in the fast TA muscles, we further analyzed the relative content of each specific MyBP-C paralog to that of the fast myosin heavy chain isoforms (MHC) and to α-actin expressed in the TA muscles."

"To determine MyBP-C protein-bound carbonyls in fast TA and cardiac muscles from adult and old mice, we used immunostaining (anti-DNP, anti-fMyBP-C, anti-sMyBP-C, and anti-cMyBP-C)."

Discussion

"Age-associated changes in the overall phosphorylation and carbonylation levels in MyBP-C from the TA and cardiac muscles are not observed, whereas an age-related reduction in the expression of the cMyBP-C paralog in cardiac muscle is observed."

"Our results, investigating age-associated alterations in phosphorylation levels in the sMyBP-C paralog and in the fMyBP-C, reveal no change in phosphorylation levels in the fast TA muscles from adult and old mice."

"Because of the emerging studies that provide insight into the multiple roles of MyBP-C paralogs in sarcomere structure/function and the evidence of oxidative stress in the skeletal muscle with aging, we sought to determine the carbonylation status of MyBP-C. Our investigation finds that under basal conditions, the MyBP-C paralogs exhibit protein carbonylation, which do not show an increase with age in the fast TA and cardiac muscles."

Conclusion

"Not much is known about the skeletal MyBP-C paralogs in the fast TA muscles and whether their expression levels and post-translational modifications are affected by aging."

"Besides being highly phosphorylated, MyBP-C is carbonylated and these two post-translational modifications are not influenced by aging."

"We cannot rule out other age-driven modifications nor the role of site-specific post-translational modifications in the MyBP-C paralogs; thus, future studies are needed."

Myosin Binding Protein-C Slow Phosphorylation Is Altered in Duchenne Dystrophy and Arthrogryposis Myopathy in Fast-Twitch Skeletal Muscles [7]

This is a machine-generated summary of:

Ackermann, Maegen A.; Ward, Christopher W.; Gurnett, Christina; Kontrogianni-Konstantopoulos, Aikaterini: Myosin Binding Protein-C Slow Phosphorylation is Altered in Duchenne Dystrophy and Arthrogryposis Myopathy in Fast-Twitch Skeletal Muscles [7].

Published in: Scientific Reports (2015).

Link to original: https://doi.org/10.1038/srep13235

If you want to cite the papers, please refer to the original.

For technical reasons we could not place the page where the original quote is coming from.

Abstract-Summary "Myosin Binding Protein-C slow (sMyBP-C), encoded by MYBPC1, comprises a family of regulatory proteins of skeletal muscles that are phosphorylated by PKA and PKC."

"Our results indicate both constitutive and differential phosphorylation of sMyBP-C in aged and diseased muscles."

"We report a 7–35% reduction in the phosphorylation levels of select sites in old wild type and young or old mdx FDB mouse muscles, compared to young wild type tissue."

"We observe a 30–70% decrease in the phosphorylation levels of all PKA and PKC phospho-sites in the DA-1 AH, but not gastrocnemius, muscle."

Introduction

"The human transcriptome contains fourteen sMyBP-C transcripts, which encode fourteen distinct variants, differing by small segments of amino acids within the Pro/Ala rich motif, the M-motif, the Ig domain C7 and the extreme COOH-terminus [8]."

"The effects of either of the DA-1 mutations on the phosphorylation profile of sMyBP-C are still unknown."

"Using a panel of phospho-specific antibodies in conjunction with standard and phosphate affinity SDS-PAGE, we examined the phosphorylation profile of sMyBP-C in different mouse and human skeletal muscles."

"We compared the phosphorylation profile of sMyBP-C in young (~2 months old) and old (~14 months old) fast-twitch Flexor Digitorum Brevis (FDB) muscles obtained from wild type and mdx mice and human Abductor Hallucis (AH) and gastrocnemius fast-twitch muscles carrying the Y856H and W236R mutations, respectively."

"The phosphorylation profile of sMyBP-C is significantly altered in human AH muscle, which is severely affected in DA-1, but not in gastrocnemius muscle, which shows no phenotypic manifestation of the disease."

Results

"Using standard SDS-PAGE along with our panel of phospho-specific antibodies (Ackermann and others 2014; Companion paper), we examined the phosphorylation levels of each known phospho-site within the NH_2-terminus of sMyBP-C in young and aged wild type and mdx FDB muscles."

"To study the effects of each of these mutations on the phosphorylation profile of sMyBP-C, we used human biopsies of the abductor hallucis (AH) and gastrocnemius muscles carrying the Y856H and W236R mutations, respectively; AH and gastrocnemius muscles that did not contain the DA-1 mutations served as controls."

"Similar examination of the phosphorylation profile of sMyBP-C in control gastrocnemius muscle showed eight different species, three of low (red, yellow and blue dots), four of intermediate (light yellow, olive green, light green and dark blue dots) and one of high (light purple dot) mobility."

"Using standard SDS-PAGE along with our panel of phospho-specific antibodies, we examined the phosphorylation levels of each phospho-site within the NH_2-terminus of sMyBP-C in AH and gastrocnemius muscles carrying the Y856H and W236R mutations, respectively."

Discussion

"Of the seven unique sMyBP-C forms expressed in young wild type FDB muscle, a low mobility species (yellow dot) is the most abundant one (~35%) and is phosphorylated at all four known residues."

"SMyBP-C species exhibiting similar electrophoretic mobility among different FDB muscles may contain different numbers of phosphorylation events (e.g. bands denoted with purple and pink dots)."

"These findings suggest that these species either correspond to sMyBP-C variants of different molecular weights and phosphorylation profiles that exhibit similar electrophoretic mobility, or to the same variant that is differentially phosphorylated between muscles, at known and/or unknown residues."

"Similar analysis of control and DA-1 gastrocnemius muscle carrying the W236R mutation revealed no significant differences in the relative abundance or phosphorylation profile of individual sMyBP-C species."

"Our findings show that the distal AH muscle carrying the Y856H mutation exhibits altered expression and phosphorylation of sMyBP-C, while the proximal gastrocnemius muscle carrying the W236R mutation does not."

Methods

"Using standard SDS-PAGE blots and ImageJ software (NIH, Bethesda, MD), we quantified the relative content of total sMyBP-C and the relative expression of each phosphorylated residue within a sample."

"Percent (%) expression was normalized to the loading control, Hsp90 and subsequently to the levels of sMyBP-C present in the mouse young wild type FDB or the appropriate human control clubfoot muscle."

"The quantified relative content of total sMyBP-C for each sample calculated above, served as a representation of total sMyBP-C protein within that sample."

"The relative abundance of each immuno-reactive band identified in the phosphate affinity SDS-PAGE blots by the pan-sMyBP-C antibody was calculated as a percentage of the total protein within that sample."

"We used the phosphate affinity SDS-PAGE western blots probed with each of the phospho-specific antibodies to ascertain the phosphorylation events within a specific immuno-reactive band detected with the pan-sMyBP-C antibody."

Axial Distribution of Myosin Binding Protein-C Is Unaffected by Mutations in Human Cardiac and Skeletal Muscle [9]

This is a machine-generated summary of:

Vydyanath, Anupama; Gurnett, Christina A.; Marston, Steve; Luther, Pradeep K.: Axial distribution of myosin binding protein-C is unaffected by mutations in human cardiac and skeletal muscle [9].

Published in: Journal of Muscle Research and Cell Motility (2012).

Link to original: https://doi.org/10.1007/s10974-012-9286-9

If you want to cite the papers, please refer to the original.

For technical reasons we could not place the page where the original quote is coming from.

Abstract-Summary "We report the first detailed electron microscopy studies on human cardiac and skeletal tissues carrying MyBP-C gene mutations, using samples obtained from HCM and DA-1 patients."

"We have used established image averaging methods to identify and study the axial distribution of MyBP-C on the thick filament by averaging profile plots of the A-band of the sarcomere from electron micrographs of human cardiac and skeletal myopathy specimens."

"Due to the difficulty of obtaining normal human tissue, we compared the distribution to the A-band structure in normal frog skeletal, rat cardiac muscle and in cardiac muscle of MyBP-C-deficient mice."

"Very similar overall profile averages were obtained from the C-zones in cardiac HCM samples and skeletal DA-1 samples with MyBP-C gene mutations, suggesting that mutations in MyBP-C do not alter its mean axial distribution along the thick filament."

Introduction

"Myosin binding protein-C (MyBP-C) is a 140 kDa sarcomeric protein that binds to the thick filament in vertebrate striated muscle and can be visualised as 7–9 stripes of separation 43 nm in the C-zone of the sarcomere."

"The interaction between MyBP-C, myosin, and titin is believed to be responsible for maintaining the ordered arrangement of the sarcomere (Freiburg and Gautel 15)."

"The importance of MyBP-C for the normal functioning of the sarcomere is further emphasised by the occurrence of cardiac and skeletal myopathies due to missing or aberrant MyBP-C proteins."

"Mutations in MYBPC3 encoding cardiac MyBP-C (cMyBP-C) are a major cause of hypertrophic cardiomyopathy (HCM), a disease that is estimated to affect 1 in 500."

"Mutations in MYBPC1 encoding slow-skeletal MyBP-C (ssMyBP-C) have recently been implicated in a disease of skeletal muscle—distal arthrogryposis type 1 (DA-1)—characterised by congenital contracture of distal limbs."

Experimental Procedures

"Myectomy samples were obtained either as freshly excised tissue or snap frozen in liquid nitrogen."

"Freshly excised cardiac tissue obtained directly after the myectomy procedure were incubated straightaway for 30 min in oxygenated Krebs solution (94.5 mM sodium chloride, 5 mM potassium chloride, 25 mM sodium hydrogen carbonate, 1 mM disodium hydrogen phosphate, 1 mM magnesium sulphate·7H_2O, 10 mM sodium acetate, 10 mM glucose and 1 mM calcium chloride, pH 7.4) with 30 mM 2,3-butanedione monoxime (BDM) and then fixed in 3 % gultaraldehyde in Krebs solution for 1 h, followed by secondary fixation in 1 % osmium tetroxide for 30 min."

"These samples were thawed and homogenised in cold relaxing solution (100 mM potassium chloride, 20 mM imidazole, 2 mM magnesium chloride, 2 mM K_2EGTA and 4 mM ATP) with 30 mM BDM (10 ml) using a Polytron homogeniser and the resultant suspension spun at 1200 rpm for 1 min."

Results

"Axial distribution of MyBP-C in these samples was analysed by averaging profile plots of the A-band over several electron micrographs of half-sarcomere regions by cross-correlation to produce an average axial density profile."

"Like the rat cardiac sample, this sample was prepared by cryosectioning as described in Luther and others [16]; this technique gives particularly high resolution structural detail e.g. it shows the presence of the two layers of myosin crowns between each pair of 43 nm stripe (the 43 nm stripes are a summation of MyBP-C and one layer of myosin crowns)."

"This is in contrast to the distinct, regular peaks observed in the C-zone of the cardiac mutant samples that coincide with the densities corresponding to MyBP-C stripe positions in the rat cardiac and human myectomy control sample."

"This suggests that although mutant cMyBP-C is not detected in human cardiac samples, the heterozygous allele contributes sufficient cMyBP-C protein to the C-zone banding pattern for this type of structural analysis."

Discussion

"The following observations emerge for both the cardiac and skeletal muscle specific mutation: 1) The ultrastructure of the sarcomere in longitudinal sections appears to be well preserved despite the phenotype of muscle disorder. 2) Image analysis confirms MyBP-C incorporation into the A-band and sharpness of peaks further indicates that it is confined to a narrow disc axially. 3) Comparison with the cardiac MyBP-C ko model provides no indication of altered axial distribution in the HCM samples."

"The state of N-terminus cannot be commented upon from these results and electron tomography of transverse sections of the muscle is required to investigate the disposition of N-terminal half of MyBP-C. The 1-D Fourier transforms of the cardiac profile plots and comparison with cardiac MyBP-C ko model further confirms the notion that cMyBP-C is incorporated and axially distributed correctly."

"We describe here the first EM based analysis of human cardiac and skeletal muscle samples with MyBP-C gene mutations."

Skeletal Myosin Binding Protein-C Isoforms Regulate Thin Filament Activity in a Ca^{2+}-Dependent Manner [10]

This is a machine-generated summary of:

Lin, Brian Leei; Li, Amy; Mun, Ji Young; Previs, Michael J.; Previs, Samantha Beck; Campbell, Stuart G.; dos Remedios, Cristobal G.; Tombe, Pieter de P.; Craig, Roger; Warshaw, David M.; Sadayappan, Sakthivel: Skeletal myosin binding protein-C isoforms regulate thin filament activity in a Ca^{2+}-dependent manner [10].

Published in: Scientific Reports (2018).

Link to original: https://doi.org/10.1038/s41598-018-21053-1

If you want to cite the papers, please refer to the original.

For technical reasons we could not place the page where the original quote is coming from.

Abstract-Summary "The interactions between myosin and actin are finely tuned by three isoforms of myosin binding protein-C (MyBP-C): slow-skeletal, fast-skeletal, and cardiac (ssMyBP-C, fsMyBP-C and cMyBP-C, respectively), each with distinct N-terminal regulatory regions."

"The skeletal MyBP-C isoforms are conditionally coexpressed in cardiac muscle, but little is known about their function."

"Addition of the fragments to in vitro motility assays demonstrated that ssMyBP-C and cMyBP-C activate thin filament sliding at low Ca^{2+}."

"At higher Ca^{2+}, addition of fsMyBP-C and cMyBP-C fragments reduced sliding velocities in the in vitro motility assays and increased force production in cardiac muscle fibers."

"Skeletal MyBP-C isoforms may be tuned to meet the needs of specific skeletal muscles."

Introduction

"Myosin binding protein-C (MyBP-C) is a striated muscle protein that regulates contraction and consists of three isoforms known as slow-skeletal, fast-skeletal, and cardiac (ssMyBP-C, fsMyBP-C, and cMyBP-C), encoded by MYBPC1, MYBPC2 and MYBPC3, respectively [11–13]."

"As their names would suggest, ssMyBP-C and fsMyBP-C are predominantly expressed in adult skeletal muscle tissue, while cMyBP-C is the predominant isoform in the heart [14]."

"cMyBP-C is the best-characterized isoform, and its N-terminal region regulates contractility through its interaction with both myosin subfragment 2 (S2) and actin [15–17]."

"The cMyBP-C N-terminal regulatory region activates the thin filament by shifting the position of tropomyosin (Tm) from a "blocked" position to a "closed" position [16], exposing myosin-binding sites on F-actin."

"Our studies demonstrate that ssMyBP-C and cMyBP-C share similar regulatory function at low Ca^{2+}."

"Steady-state functional experiments demonstrate that all MyBP-C isoforms sensitize the thin filaments to Ca^{2+} and that the extent of sensitization is isoform-specific."

Results

"To investigate how skeletal MyBP-C isoforms modulate contraction at low Ca^{2+} levels compared to maximal Ca^{2+} levels, we applied our N-terminal fragments to in vitro motility assays."

"The greater ability of C0C2 and ssC1C2 to displace Tm suggests that they would be more potent activators of thin filaments at low Ca^{2+} levels than fsC1C2, underscoring the Ca^{2+}-dependent manner in which MyBP-C isoforms differentially regulate contraction."

"The different effects of each isoform on Tm shift strongly suggest that thin filament activation is one mechanism by which MyBP-C isoforms regulate contraction at low Ca^{2+}."

"The differential thin filament activation capacity of each MyBP-C isoform regulates contraction, but may also impact relaxation."

"To determine whether the greater capacity of cMyBP-C to activate the thin filament also explains its regulation of relaxation kinetics during contraction, we utilized an online muscle simulator (OMS) that simulates unloaded shortening of a single cardiomyocyte [18]."

Discussion

"The capacity of each MyBP-C isoform to regulate within a particular range of [Ca^{2+}] is dependent on the ability of each to bind to, and activate the thin filament."

"The present studies focus on MyBP-C N-terminal interactions with the thin filament and do not preclude regulation by MyBP-C through its interactions with myosin."

"Both the binding affinity of MyBP-C to actin and its effects on the thin filament underlie the ability of MyBP-C to regulate contraction at various ranges of Ca^{2+}, suggesting, in turn, that MyBP-C may both inhibit and facilitate myosin binding."

"Various slow-twitch and fast-twitch skeletal muscles may be fine-tuned for specific functions by expressing different amounts of slow- and fast-skeletal MyBP-C. In contrast, the function of cMyBP-C is highly tunable via post-translational modification which modulates its interactions with actin and myosin [19]."

Experimental Methods

"V8 Stereo, PlanAPO S 0.63 × FWD 81 mm) and permeabilized overnight in 1% Triton X-100 (Sigma) in mounting relaxing solution (6.3 mM ATP, 6.48 mM Mg Cl_2, 10 mM EGTA, 10 mM Na_2CrP, 49.76 mM Kprop, 100 mM BES, pH 7) at 4 °C, removing cell membrane and membrane-bound proteins."

"The permeabilized fibers were incubated with 10 μM of the ssC1C2, fsC1C2, and C0C2 recombinant proteins or unincubated for 3 minutes in relaxing solution, followed by 3 min in preactivating solution, before measuring force in activating solution."

"Cardiac myosin from mice (100 µg/mL) were incubated in a nitrocellulose-coated flow cell (2 minutes), followed by blocking with two aliquots of BSA (1 mg/mL) in actin buffer (AB: 25 mM KCl, 1 mM EGTA, 10 mM DTT, 25 mM imidazole, 4 mM $MgCl_2$, pH 7.4)."

Control of mRNA Stability Contributes to Low Levels of Nuclear Poly(A) Binding Protein 1 (PABPN1) in Skeletal Muscle [20]

This is a machine-generated summary of:

Apponi, Luciano H; Corbett, Anita H; Pavlath, Grace K: Control of mRNA stability contributes to low levels of nuclear poly(A) binding protein 1 (PABPN1) in skeletal muscle [20].

Published in: Skeletal Muscle (2013).

Link to original: https://doi.org/10.1186/2044-5040-3-23

If you want to cite the papers, please refer to the original.

For technical reasons we could not place the page where the original quote is coming from.

Abstract-Summary "Given the general function of PABPN1 in RNA metabolism,intrinsic characteristics of skeletal muscle may make this tissue susceptible tothe effects of mutant PABPN1."

"To begin to understand the muscle specificity of OPMD, we investigated thesteady-state levels of PABPN1 in different tissues of humans and mice."

"We analyzed the levels of PABPN1 during muscle regeneration afterinjury in mice."

"We show that the steady-state levels of both PABPN1 mRNA and protein aredrastically lower in mouse and human skeletal muscle, particularly those impactedin OPMD, compared to other tissues."

"PABPN1 levels are increasedduring muscle regeneration, suggesting a greater requirement for PABPN1 functionduring tissue repair."

"Our results demonstrate that PABPN1 steady-state levels and likely control ofexpression differ significantly in skeletal muscle as compared to other tissues,which could have important implications for understanding the muscle-specificnature of OPMD."

Background

"PABPN1 has critical roles at multiple steps in post-transcriptionalregulation of gene expression."

"To playing a key role in RNA metabolism, PABPN1 is of significant clinicalinterest as mutations in the PABPN1 gene lead to oculopharyngeal musculardystrophy (OPMD) [21]."

"Given the ubiquitous expression and general function of PABPN1 in RNA metabolism, how mutations of this post-transcriptional regulatory factor cause amuscle-specific disease is unclear [22]."

"To begin to understand the muscle specificity of OPMD, we investigated the steady-statelevels of the PABPN1 protein in different tissues."

"We find that the steady-state levelsof PABPN1 are drastically lower in skeletal muscle compared to other tissues."

"Studies of mRNA stability indicate that regulation of PABPN1 expression is likely due todistinct post-transcriptional mechanisms in different tissues."

"Ourresults demonstrate that PABPN1 steady-state levels and likely control of expressiondiffer significantly in skeletal muscle as compared to other tissues, which could haveimportant implications for understanding the muscle-specific nature of OPMD."

Methods

"Total RNA from tissues, primary cell cultures and fluorescence-activated cell sorting(FACS)-sorted cells was isolated using TRIzol (Invitrogen, Carlsbad, CA) according tothe manufacturer's protocol."

"DNA probes were generated by polymerase chain reaction (PCR) using customprimers (PABPN1-F 5′-CCCAGGCAATGCTGGCCCAGTGATCATGTCTC-3′ and PABPN1-R5′-CTAGCCCGGCCCCTGTAGATTCGACCCCGGGGC-3′, c-Myc-F5′-GAACTTCACCAACAGGAACTATGACCTCG-3′ and c-Myc-R5′-GGTGTCTCCTCATGCAGCACTAGG-3′), primers from SA Biosciences, Valencia,CA (peroxisome proliferator-activated receptor gamma coactivator 1α(PGC1α); PPM03360E, glyceraldehyde 3-phosphate dehydrogenase (GAPDH); PPM02946E)or by Ambion, Austin, TX (QuantumRNA Classic II 18S)."

"PCR products were labeled with[α-^{32}P]dCTP using a random primer DNA labeling system (Invitrogen,Carlsbad, CA)."

"Mononucleated cells were enzymatically isolated from gastrocnemius muscles 3 daysafter $BaCl_2$ injury and fluorescently labeled with antibodies to CD31 andCD45 (PE), Sca-1 (PE-Cy7), and alpha-7-integrin (AlexaFluor 649)."

"cDNA synthesis from 100 ng RNA was performed using M-MLV reverse transcriptase(Invitrogen, Carlsbad, CA)."

"Total RNA was extracted fromtissues or cells and analyzed by northern blot and half-lives were determined bydensitometry."

"In order to determine the 5′ and 3′UTRs of PABPN1 transcripts, we usedthe 5′ and 3′ rapid amplification of cDNA ends (RACE) system (Invitrogen,Carlsbad, CA), respectively."

Results

"Our results from immunoblotting and northern blottingreveal low steady-state levels of PABPN1 mRNA and protein in skeletal muscle, whichis indicative of either a decrease in PABPN1 transcription or altered mRNA stabilityin this tissue."

"As we were unable to perform immunoblots forPABPN1 on the small amount of cells isolated by flow cytometry, we used quantitativeRT-PCR to examine PABPN1 transcript levels in sorted cells compared to muscle tissue."

"Theseresults demonstrate a strong correlation between the high steady-state levels ofPABPN1 protein and the stable transcript in kidney, whereas in skeletal muscle, thelow steady-state levels of PABPN1 correlate with the unstable PABPN1 transcript."

"As observed in skeletal muscle tissue,the 2.1 kb PABPN1 transcript was the predominant transcript in myoblasts (data notshown)."

"These resultsindicate that the high levels of PABPN1 in cultured myoblasts, and likely inmyoblasts present during muscle regeneration, is due at least in part to the increasein PABPN1 mRNA stability in those cells compared to muscle tissue."

Discussion

"Wereport that steady-state levels of PABPN1 mRNA and protein are low in skeletal muscleand that expression of PABPN1 in this tissue is controlled, at least in part, bypost-transcriptional regulation of RNA levels."

"The low complexity of the skeletal muscle transcriptomeassociated with low turnover of a significant fraction of its transcripts may explainwhy skeletal muscle has low requirements for a protein involved in mRNA metabolism suchas PABPN1."

"As the increased levels ofPABPN1 in regenerating muscle correlate with an increased transcript stability inmyoblasts and subsequent increase in the steady-state levels of PABPN1 transcript, wesuggest that skeletal muscle employs a post-transcriptional mechanism to control PABPN1levels according to the tissue requirements."

"The specific PABPN1 expression pattern observed in skeletal muscle may be an importantfeature that makes this tissue more susceptible than others to the mutations in PABPN1that cause the muscle-specific disease OPMD."

Conclusions

"Our results demonstrate that PABPN1 steady-state levels and likely control of expressiondiffer significantly in skeletal muscle as compared to other tissues, which could haveimportant implications for understanding the muscle-specific nature of OPMD."

"We suggestthe low levels of PABPN1 observed in skeletal muscle may be an important aspect of thistissue that underlies OPMD pathology."

"Further studies are necessary to better comprehendthe mechanisms and implications of the regulation of PABPN1 expression in skeletalmuscle, which could open avenues for potential therapeutic approaches for OPMD."

Co-activator Binding Protein PIMT Mediates TNF-α Induced Insulin Resistance in Skeletal Muscle via the Transcriptional Down-Regulation of MEF2A and GLUT4 [23]

This is a machine-generated summary of:

Kain, Vasundhara; Kapadia, Bandish; Viswakarma, Navin; Seshadri, Sriram; Prajapati, Bhumika; Jena, Prasant K; Teja Meda, Chandana Lakshmi; Subramanian, Maitreyi; Kaimal Suraj, Sashidhara; Kumar, Sireesh T; Prakash Babu, Phanithi; Thimmapaya, Bayar; Reddy, Janardan K; Parsa, Kishore V. L.; Misra, Parimal: Co-activator binding protein PIMT mediates TNF-α induced insulin resistance in skeletal muscle via the transcriptional down-regulation of MEF2A and GLUT4 [23].

Published in: Scientific Reports (2015).

Link to original: https://doi.org/10.1038/srep15197

If you want to cite the papers, please refer to the original.

For technical reasons we could not place the page where the original quote is coming from.

Abstract-Summary "We report that the expression of PIMT, a transcriptional co-activator binding protein, was up-regulated in the soleus muscle of high sucrose diet (HSD) induced insulin resistant rats and TNF-α exposed cultured myoblasts."

"Whereas, PIMT knock down relieved TNF-α inhibited insulin signaling."

"Mechanistic analysis revealed that PIMT differentially regulated the expression of GLUT4, MEF2A, PGC-1α and HDAC5 in cultured cells and skeletal muscle of Wistar rats."

"We conclude that PIMT mediates TNF-α induced insulin resistance at the skeletal muscle via the transcriptional modulation of GLUT4, MEF2A, PGC-1α and HDAC5 genes."

Introduction

"It is well documented that the expression of pro-inflammatory cytokines like TNF-α is locally enhanced in skeletal muscle and adipose tissues of humans and animals with insulin resistance and/or diabetes [24–31]."

"We observed that the expression of PIMT was up-regulated in the soleus muscle of high sucrose diet (HSD) induced insulin resistant rats and TNF-α treated cultured skeletal muscle cells."

"Knockdown of PIMT in L6 myoblasts enhanced insulin sensitivity and relieved TNF-α induced inhibition of glucose uptake."

"Data obtained from cultured skeletal muscle cells and adenoviral infected rat skeletal muscle tissue demonstrate that PIMT inhibited insulin stimulated glucose uptake via the transcriptional modulation of several genes associated with skeletal muscle glucose uptake, particularly GLUT4, MEF2A, PGC-1α and HDAC5."

"We show that PIMT/phospho PIMT facilitates TNF-α induced insulin resistance in skeletal muscle."

Results

"We used high sucrose diet (HSD) induced insulin resistant rats and TNF-α induced L6 cells (skeletal muscle cell line) as the model systems to study the involvement of PIMT in skeletal muscle insulin resistance."

"The above findings demonstrate that PIMT mediates TNF-α-induced insulin resistance in skeletal muscle cells."

"To do this, we overexpressed wild type (WT) PIMT or phospho-deficient (Ser298Ala) or phospho-mimic (Ser298Asp) mutants of PIMT in L6 myoblasts and examined their effect on insulin-stimulated glucose uptake."

"Data indicate that TNF-α induced phosphorylation of PIMT at Ser^{298} is required for inhibition of insulin stimulated glucose uptake."

"PIMT overexpression caused a robust inhibition of insulin-stimulated glucose uptake by L6 cells suggesting that a key player such as GLUT4 may be the likely downstream target of PIMT."

"Our data establish that the transcriptional co-activator binding protein, PIMT, mediates TNF-α induced insulin resistance in the skeletal muscle via the transcriptional modulation of several genes associated with glucose uptake."

Discussion

"We identified that the transcriptional co-activator binding protein, PIMT, is an important mediator of TNF-α induced skeletal muscle insulin resistance."

"We observed that 1) the expression of PIMT was enhanced in TNF-α exposed cultured skeletal muscle cells and soleus muscle of HSD fed rats 2) PIMT expression regulates insulin receptor signaling cascade 3) TNF-α stimulated phosphorylation of PIMT at Ser^{298} 4) PIMT suppressed insulin stimulated glucose uptake via the modulation of the expression of GLUT4, MEF2A and HDAC5 in cultured cells and skeletal muscle of rats in phosphorylation dependent manner."

"Our findings that PIMT disrupts insulin signaling potentially through the induction of TNF-α mediated phosphorylation of JNK and p38 suggests that PIMT may have a broader role in glucose control by skeletal muscle."

"Treatment of L6 cells with the MEK/ERK inhibitor, U0126, blunted the inhibitory effect of PIMT on the glucose uptake suggesting a negative regulatory role for ERK-mediated phosphorylation of PIMT in skeletal muscle insulin sensitivity."

Methods

"The plasma triglyceride was measured using a diagnostic kit from Accucare India Ltd. Levels of TNF-α in sera collected from various animal groups were estimated using Multiplex CBA Rat Inflammation kit (BD Biosciences, CA, USA)."

"Immunoblotting was performed as described earlier [32] using anti-PIMT (1:1000; Santa Cruz Biotechnology, CA, USA) anti-GLUT4 (1:1000; Novus Biologicals, MO, USA), anti-MEF2A (1:1000; Cell Signaling Technology, MA, USA), anti-PGC-1 (1:1000; Santa Cruz Biotechnology, CA, USA),anti-pSer307-IRS1 (1:1000; Cell Signaling Technology, MA, USA),anti-pTyr608/612-IRS (1:1000; Abcam, MA, USA), anti-IRS1 (1:1000; Cell Signaling Technology, MA, USA), anti-pJNK1/2 (1:1000; Cell Signaling Technology, MA, USA), anti-pAkt (Ser473) (1:1000; Cell Signaling Technology, MA, USA), anti-Akt (1:1000; Cell Signaling Technology, MA, USA),anti-ERK1/2 (1:1000; Cell Signaling Technology, MA, USA) and anti-pERK1/2 (1:1000; Cell Signaling Technology, MA, USA) using specific antibodies and appropriate HRP-conjugated secondary antibody."

Role for Sterol Regulatory Element Binding Protein-1c Activation in Mediating Skeletal Muscle Insulin Resistance via Repression of Rat Insulin Receptor Substrate-1 Transcription [33]

This is a machine-generated summary of:

Bi, Yan; Wu, Wenjun; Shi, Junfeng; Liang, Hua; Yin, Wenwen; Chen, Yingying; Tang, Sunyinyan; Cao, Shu; Cai, Mengyin; Shen, Shanmei; Gao, Qian; Weng, Jianping; Zhu, Dalong: Role for sterol regulatory element binding protein-1c activation in mediating skeletal muscle insulin resistance via repression of rat insulin receptor substrate-1 transcription [33].

Published in: Diabetologia (2013).

Link to original: https://doi.org/10.1007/s00125-013-3136-1

If you want to cite the papers, please refer to the original.

For technical reasons we could not place the page where the original quote is coming from.

Abstract-Summary "We investigated the role of SREBP-1c in the regulation of IRS-1 in skeletal muscle cells."

"Adenovirus vectors expressing Srebp-1c (also known as Srebf1) and small interfering RNA (siRNA) against Srebp-1c were transfected into the L6 cells."

"We found that both gene and protein expression of SREBP-1c was increased in contrast to IRS-1 expression in PA-treated L6 cells."

"SREBP-1c overproduction decreased Irs-1 mRNA and IRS-1 protein expression in a dose-dependent manner, and suppressed the resultant insulin signalling, whereas SERBP-1c knockdown by Serbp-1c siRNA blocked the downregulation of IRS-1 induced by PA."

"Protein–DNA interaction studies demonstrated that SREBP-1c was able to bind to the rat Irs-1 promoter region, thereby repressing its gene transcription."

"Of particular importance, we found that metformin treatment downregulated Srebp-1c promoter activity, decreased the specific binding of SREBP-1c to Irs-1 promoter and upregulated Irs-1 promoter activity in PA-cultured L6 cells."

"Our data indicate for the first time that SREBP-1c activation participates in skeletal muscle insulin resistance through a direct effect of suppressing Irs-1 transcription."

Introduction

"To the regulation of lipid synthesis, studies reveal expanding roles for SREBP-1c in controlling pathways for insulin resistance, in which the pathological process involves a series of cascades, including defective activation of insulin receptor substrates (IRS), phosphatidlyinositol-3-kinase (PI3K) and Akt [34, 35]."

"In the liver, nuclear active SREBP-1c can directly bind to the promoter region of Irs-2, suppressing Irs-2 expression and the subsequent insulin signalling pathway, thereby leading to hepatic insulin resistance [36, 37]."

"The link between SREBP-1c activation and IRS-1-associated insulin signalling in skeletal muscle is unclear."

"This finding led us to speculate that Irs-1 expression might be negatively regulated by SREBP-1c in skeletal muscle, but this remains to be determined."

"The aim of the present study, therefore, was to assess the role of SREBP-1c in the regulation of IRS-1-associated insulin signalling in skeletal muscle."

Methods

"To study the effect of PA or liver X receptor (LXR) agonist (TO901317 from Sigma-Aldrich, Saint Louis, MO, USA), cells were incubated for 8 h in serum-free medium and treated with 0.5 mmol/l PA, 0.5 mmol/l oleate or 5 mmol/l TO901317 for 24 h. After PA incubation, cells were incubated for a further 12 h with 100 nmol/l of insulin or 10 mmol/l of metformin if indicated."

"To determine the time course effects of PA, L6 myobutes were incubated with 0.5 mmol/l of PA for 0.5, 1, 3, 6, 12, 24 or 48 h. For transient transfection assays, L6 myotubes were added with adenovirus vectors expressing SREBP-1c or green fluorescent protein (GFP) and then transfected for 24 h. The media were renewed and cells were treated with or without 100 nmol/l insulin for 12 h. The cells were then harvested for protein and mRNA immediately."

"L6 cells were transfected with Srebp-1c adenovirus vectors for 48 h or treated with PA for 24 h and cross-linked 1% formaldehyde (vol /vol.) for 10 min at room temperature."

Results

"To investigate whether LXR-induced SREBP-1c activation correlates with IRS-1-associated insulin signalling, SREBP-1c and IRS-1 expression was compared in L6 myotubes following LXR agonist intervention."

"Protein expression of IRS-1, insulin-stimulated p-IRS-1(Tyr608/612) and p-Akt (Ser473)/Akt was decreased by LXR agonist treatment, compared with controls, further indicating a close relationship between SREBP-1c and IRS-1."

"The data suggest that SREBP-1c has a direct effect on the expression of Irs-1."

"Metformin has been widely used in the management of insulin resistance and type 2 diabetes, but it is unknown whether SREBP-1c inhibition is required for the beneficial effect of metformin on IRS-1-associated insulin signalling in muscle."

"We performed luciferase reporter assays and ChIP assays to further investigate whether increased IRS-1 expression following metformin intervention is associated with the inhibition of SREBP-1c activity."

"These results suggest that inhibition of SREBP-1c expression contributes to the beneficial effects of metformin on IRS-1-associated insulin signalling in muscle."

Discussion

"The accumulation of SREBP-1c, a pivotal transcription factor regulating de novo lipogenesis, can result in muscular insulin resistance by a direct effect of suppressing Irs-1 promoter transcription and gene expression."

"Our results are consistent with the hypothesis that SREBP-1c is as important for fatty acid-induced insulin resistance in skeletal muscle as it is in hepatocytes."

"Our data suggest, for the first time to our knowledge, that inhibition of SREBP-1c contributes to the beneficial effect of metformin on IRS-1-associated insulin signalling in insulin resistant muscle cells induced by fatty acid."

"The current study indicates that a high level of SREBP-1c results in impaired insulin signalling in skeletal muscle cells via a direct effect of suppressing Irs-1 promoter transcription and gene expression."

"The data reveal a key role for SREBP-1c in the development of skeletal muscle insulin resistance correlated to abnormal fatty acid metabolism."

Bibliography

1. Perazza LR, Wei G, Thompson LV (2022) Fast and slow skeletal myosin binding protein-C and aging. GeroScience. https://doi.org/10.1007/s11357-022-00689-y
2. Ackermann MA, Ward CW, Gurnett C, Kontrogianni-Konstantopoulos A (2015) Myosin binding protein-C slow phosphorylation is altered in Duchenne dystrophy and arthrogryposis myopathy in fast-twitch skeletal muscles. Sci Rep 5:13235
3. Ackermann MA, Kerr JP, King B, Ward CW, Kontrogianni-Konstantopoulos A (2018) Corrigendum: the phosphorylation profile of myosin binding protein-C slow is dynamically regulated in slow-twitch muscles in health and disease. Sci Rep 8:46969
4. Li A, Nelson SR, Rahmanseresht S, Braet F, Cornachione AS, Previs SB, O'Leary TS, McNamara JW, Rassier DE, Sadayappan S, Previs MJ, Warshaw DM (2019) Skeletal MyBP-C isoforms tune the molecular contractility of divergent skeletal muscle systems. Proc Natl Acad Sci U S A 116:21882–21892

5. Hofmann PA, Greaser ML, Moss RL (1991) C-protein limits shortening velocity of rabbit skeletal muscle fibres at low levels of Ca2+ activation. J Physiol 439:701–715
6. Lin BL, Li A, Mun JY, Previs MJ, Previs SB, Campbell SG, Dos Remedios CG, Tombe PP, Craig R, Warshaw DM, Sadayappan S (2018) Skeletal myosin binding protein-C isoforms regulate thin filament activity in a Ca. Sci Rep 8:2604
7. Ackermann MA, Ward CW, Gurnett C, Kontrogianni-Konstantopoulos A (2015) Myosin binding protein-C slow phosphorylation is altered in Duchenne dystrophy and arthrogryposis myopathy in fast-twitch skeletal muscles. Sci Rep. https://doi.org/10.1038/srep13235
8. Ackermann MA, Kontrogianni-Konstantopoulos A (2013) Myosin binding protein-C slow: a multifaceted family of proteins with a complex expression profile in fast and slow twitch skeletal muscles. Front Physiol 4:391. https://doi.org/10.3389/fphys.2013.00391
9. Vydyanath A, Gurnett CA, Marston S, Luther PK (2012) Axial distribution of myosin binding protein-C is unaffected by mutations in human cardiac and skeletal muscle. J Muscle Res Cell Motil. https://doi.org/10.1007/s10974-012-9286-9
10. Lin BL, Li A, Mun JY, Previs MJ, Previs SB, Campbell SG, dos Remedios CG, de Tombe PP, Craig R, Warshaw DM, Sadayappan S (2018) Skeletal myosin binding protein-C isoforms regulate thin filament activity in a Ca^{2+}-dependent manner. Sci Rep. https://doi.org/10.1038/s41598-018-21053-1
11. Offer G, Moos C, Starr R (1973) A new protein of the thick filaments of vertebrate skeletal myofibrils. J Mol Biol 74:653–376
12. Weber FE, Vaughan KT, Reinach FC, Fischman DA (1993) Complete sequence of human fast-type and slow-type muscle myosin-binding-protein C (MyBP-C). Differential expression, conserved domain structure and chromosome assignment. Eur J Biochem 216:661–669
13. Carrier L et al (1993) Mapping of a novel gene for familial hypertrophic cardiomyopathy to chromosome 11. Nat Genet 4:311–313. https://doi.org/10.1038/ng0793-311
14. Ackermann MA, Kontrogianni-Konstantopoulos A (2011) Myosin binding protein-C: a regulator of actomyosin interaction in striated muscle. J Biomed Biotechnol 2011:636403
15. Previs MJ, Beck Previs S, Gulick J, Robbins J, Warshaw DM (2012) Molecular mechanics of cardiac myosin-binding protein C in native thick filaments. Science 337:1215–1218. https://doi.org/10.1126/science.1223602
16. Mun JY et al (2014) Myosin-binding protein C displaces tropomyosin to activate cardiac thin filaments and governs their speed by an independent mechanism. Proc Natl Acad Sci U S A 111:2170–2175. https://doi.org/10.1073/pnas.1316001111
17. Previs MJ et al (2016) Phosphorylation and calcium antagonistically tune myosin-binding protein C's structure and function. Proc Natl Acad Sci U S A 113:3239–3244. https://doi.org/10.1073/pnas.1522236113
18. Kuo IY et al (2014) Decreased polycystin 2 expression alters calcium-contraction coupling and changes beta-adrenergic signaling pathways. Proc Natl Acad Sci U S A 111:16604–16609. https://doi.org/10.1073/pnas.1415933111
19. Barefield D, Sadayappan S (2010) Phosphorylation and function of cardiac myosin binding protein-C in health and disease. J Mol Cell Cardiol 48(5):866–875
20. Apponi LH, Corbett AH, Pavlath GK (2013) Control of mRNA stability contributes to low levels of nuclear poly(A) binding protein 1 (PABPN1) in skeletal muscle. Skeletal Mus. https://doi.org/10.1186/2044-5040-3-23
21. Brais B, Bouchard JP, Xie YG, Rochefort DL, Chrétien N, Tomé FM, Lafrenière RG, Rommens JM, Uyama E, Nohira O, Blumen S, Korczyn AD, Heutink P, Mathieu J, Duranceau A, Codère F, Fardeau M, Rouleau GA (1998) Short GCG expansions in the PABP2 gene cause oculopharyngeal musculardystrophy. Nat Genet 18:164–167. https://doi.org/10.1038/ng0298-164
22. Banerjee A, Apponi LH, Pavlath GK, Corbett AH (2013) PABPN1: molecular function and muscle disease. FEBS J 280:4230–4250. https://doi.org/10.1111/febs.12294
23. Kain V, Kapadia B, Viswakarma N, Seshadri S, Prajapati B, Jena PK, Teja Meda CL, Subramanian M, Kaimal Suraj S, Kumar ST, Prakash Babu P, Thimmapaya B, Reddy JK, Parsa KVL, Misra P (2015) Co-activator binding protein PIMT mediates TNF-α induced

insulin resistance in skeletal muscle via the transcriptional down-regulation of MEF2A and GLUT4. Sci Rep. https://doi.org/10.1038/srep15197
24. Dandona P, Aljada A, Bandyopadhyay A (2004) Inflammation: the link between insulin resistance, obesity and diabetes. Trends Immunol 25:4–7
25. Uysal KT, Wiesbrock SM, Marino MW, Hotamisligil GS (1997) Protection from obesity-induced insulin resistance in mice lacking TNF-alpha function. Nature 389:610–614. https://doi.org/10.1038/39335
26. Hotamisligil GS, Shargill NS, Spiegelman BM (1993) Adipose expression of tumor necrosis factor-alpha: direct role in obesity-linked insulin resistance. Science 259:87–91
27. Kern PA et al (1995) The expression of tumor necrosis factor in human adipose tissue. Regulation by obesity, weight loss and relationship to lipoprotein lipase. J Clin Investig 95:2111–2119. https://doi.org/10.1172/JCI117899
28. Dandona P et al (1998) Tumor necrosis factor-alpha in sera of obese patients: fall with weight loss. J Clin Endocrinol Metab 83:2907–2910. https://doi.org/10.1210/jcem.83.8.5026
29. Saghizadeh M, Ong JM, Garvey WT, Henry RR, Kern PA (1996) The expression of TNF alpha by human muscle. Relationship to insulin resistance. J Clin Investig 97:1111–1116. https://doi.org/10.1172/JCI118504
30. Mingrone G et al (2002) Skeletal muscle triglycerides lowering is associated with net improvement of insulin sensitivity, TNF-alpha reduction and GLUT4 expression enhancement. Intl J Obes Relat Metab Disord 26:1165–1172. https://doi.org/10.1038/sj.ijo.0802053
31. Hotamisligil GS (1999) The role of TNFalpha and TNF receptors in obesity and insulin resistance. J Int Med 245:621–625
32. Kapadia B et al (2013) ERK2-mediated phosphorylation of transcriptional coactivator binding protein PIMT/NCoA6IP at Ser298 augments hepatic gluconeogenesis. PLoS One 8:e83787. https://doi.org/10.1371/journal.pone.0083787
33. Bi Y, Wu W, Shi J, Liang H, Yin W, Chen Y, Tang S, Cao S, Cai M, Shen S, Gao Q, Weng J, Zhu D (2013) Role for sterol regulatory element binding protein-1c activation in mediating skeletal muscle insulin resistance via repression of rat insulin receptor substrate-1 transcription. Diabetologia. https://doi.org/10.1007/s00125-013-3136-1
34. Sesti G, Federici M, Hribal ML et al (2001) Defects of the insulin receptor substrate (IRS) system in human metabolic disorders. FASEB J 15:2099–2111
35. Taniguchi CM, Ueki K, Kahn R (2005) Complementary roles of IRS-1 and IRS-2 in the hepatic regulation of metabolism. J Clin Investig 115:718–727
36. Shimomura I, Matsuda M, Hammer RE et al (2000) Decreased IRS-2 and increased SREBP-1c lead to mixed insulin resistance and sensitivity in livers of lipodystrophic and ob/ob mice. Mol Cell 6:77–86
37. Ide T, Shimano H, Yahagi N et al (2004) SREBP-1c suppress IRS-2-mediated insulin signalling in the liver. Nat Cell Biol 6:351–357
38. Ackermann MA, Hu L-YR, Bowman AL, Bloch RJ, Kontrogianni-Konstantopoulos A (2009) Obscurin interacts with a novel isoform of MyBP-C slow at the periphery of the sarcomeric M-band and regulates thick filament assembly. Mol Biol Cell 20:2963–2978. https://doi.org/10.1091/mbc.E08-12-1251
39. Ackermann MA et al (2013) Loss of actomyosin regulation in distal arthrogryposis myopathy due to mutant myosin binding protein-C slow. FASEB J 27:3217–3228, fj.13-228882
40. de Tombe PP (2006) Myosin binding protein C in the heart. Circ Res 98:1234–1236
41. James J, Robbins J (2011) Signaling and myosin-binding protein C. J Biol Chem 286:9913–9919., R110.171801
42. Martyn DA (2004) Myosin binding protein-C: structural and functional complexity. J Mol Cell Cardiol 37:813–815. https://doi.org/10.1016/j.yjmcc.2004.07.005S0022-2828(04)00230-5
43. McClellan G, Kulikovskaya I, Flavigny J, Carrier L, Winegrad S (2004) Effect of cardiac myosin-binding protein C on stability of the thick filament. J Mol Cell Cardiol 37:823–835. https://doi.org/10.1016/j.yjmcc.2004.05.023S0022282804001646
44. Oakley CE, Chamoun J, Brown LJ, Hambly BD (2007) Myosin binding protein-C: enigmatic regulator of cardiac contraction. Int J Biochem Cell Biol 39(12):2161–2166

45. Freiburg A, Gautel M (1996) A molecular map of the interactions between titin and myosin-binding protein C. Implications for sarcomeric assembly in familial hypertrophic cardiomyopathy. Eur J Biochem 235(1–2):317–323
46. Luther PK, Bennett PM, Knupp C, Craig R, Padron R, Harris SP, Patel J, Moss RL (2008) Understanding the organisation and role of myosin binding protein C in normal striated muscle by comparison with Mybp-C knockout cardiac muscle. J Mol Biol 384(1):60–72
47. Ackermann MA, Kontrogianni-Konstantopoulos A (2011) Myosin binding protein-C slow is a novel substrate for protein kinase A (PKA) and C (PKC) in skeletal muscle. J Proteome Res 10:4547–4555. https://doi.org/10.1021/pr200355w

The Neuromotor Junction

3

Christopher Myers

3.1 Introduction by the Editor

The neuromuscular junction (NMJ) is a pivotal point of interest in skeletal muscle physiology. This chapter dives into the innovative research in the field, particularly emphasizing the formation and function of this crucial connection between the motor neuron and muscle fiber. The NMJ is the site of electrical impulse transmission from the motor neuron to the muscle, leading to contraction, making its study crucial to understanding motor function.

At the heart of NMJ function is acetylcholine, a neurotransmitter the nerve releases into the junction. Acetylcholine binds to receptors on the postsynaptic muscle membrane, eliciting an electrical response that initiates muscle contraction. Recent research has broadened our understanding of the detailed mechanics of this process, including the role of various proteins in the clustering and stabilization of acetylcholine receptors, a focus that this chapter examines in depth.

One fascinating area that this chapter delves into is the role of the Wnt signaling pathway in the formation and maintenance of the NMJ. Wnt proteins are critical players in the embryonic development of the neuromuscular junction, influencing the clustering of acetylcholine receptors and the overall organization of the postsynaptic apparatus. We'll explore the recent discoveries that underscore the importance of the Wnt pathway and its implications for the functionality and adaptability of the neuromuscular junction.

Motor neurons play a fundamental role in neuromuscular transmission. New research illuminates the intimate interplay between motor neurons and muscle fibers, and how the health and stability of motor neurons can significantly impact NMJ functionality. This chapter dissects the latest findings on the intricate

C. Myers (✉)
3 Sports Science, 3 Sports Science LLC, Akron, OH, United States of America
e-mail: chris@3sportsscience.com

C. Myers (ed.), *Skeletal Muscle Physiology*,
https://doi.org/10.1007/978-3-031-47065-3_3

relationship between motor neurons and their target muscle fibers, further highlighting the complexity of NMJ formation and function.

In summary, this chapter bridges the knowledge gap in the study of the neuromuscular junction, focusing on novel research findings on NMJ formation, the role of Wnt proteins, acetylcholine receptor clustering, and the intimate relationship between motor neurons and muscle fibers. By deciphering these complexities, we understand how nerve impulses translate into muscle action and how disruptions in this process can lead to disease.

As we delve deeper into the molecular choreography that defines the neuromuscular junction (NMJ), the roles of PGC-1α, R-spondin 2 (Rspo2), and the dihydropyridine receptor (DHPR) in describing the critical interface between nerve and muscle fiber become increasingly significant. The NMJ is the epicenter of the excitation-coupling mechanism of neuromuscular communication. The NMJ is where neuronal communication is translated into muscular movement and strength. This extended discussion aims to unravel the complexities of these molecular players further, elucidating their synergistic actions in maintaining the integrity and dynamism of the NMJ.

The peroxisome proliferator-activated receptor gamma coactivator 1-alpha (PGC-1α) extends its influence beyond the realms of mitochondrial function and metabolic control to play a pivotal role in the adaptive responses of skeletal muscle to various stimuli. The PGC-1α's involvement in neuromuscular junction remodeling highlights an aspect of muscle plasticity, where PGC-1α's modulation of gene expression significantly impacts NMJ formation and function. This chapter explores the recent research investigating how PGC-1α contributes to maintaining neuromuscular health and preventing age-associated decline, underscoring its potential as a therapeutic target for enhancing muscle function and combating neuromuscular diseases.

In parallel, the contributions of Rspo2 and DHPR to NMJ architecture and function merit further exploration. Rspo2 acts as a molecular bridge linking motor neurons and muscle fibers in a bidirectional dialogue essential for the precise assembly and maturation of the NMJ. This dialogue encompasses the clustering of acetylcholine receptors (AChRs) and the intricate arrangement of synaptic folds and active zones, which are critical for effective synaptic transmission. This chapter looks at recent research that delves into the signaling pathways activated by Rspo2, detailing how its differential expression in motor neurons versus muscle fibers orchestrates a coordinated response essential for the structural and functional refinement of the NMJ.

New research into the involvement of the DHPR in NMJ pre-patterning signifies a revolutionary understanding of synaptic development. The conclusions from the latest research on the DHPR's functionality within skeletal muscle to establish prepatterning during the NMJ's formation make a strong case for the muscle's active involvement in determining synaptic structure. This new viewpoint challenges the long-held neurocentric models by suggesting that muscle electrical activity, facilitated by DHPR, plays a pivotal role in defining NMJ architecture. Researchers delve into the ramifications of DHPR malfunctioning on the stability of the NMJ, emphasizing the critical pathways through which DHPR contributes to synaptic organization. This new research highlights how DHPR's operational defects can disrupt the normal formation and maintenance of the NMJ. The exploration reveals that DHPR's influence extends to the modulation of AChR clustering and the alignment of synaptic components. The impairment of DHPR function is shown to have profound effects on neuromuscular health, potentially leading to a spectrum of neuromuscular disorders. Through this detailed examination of new research, the indispensable role of DHPR in maintaining NMJ integrity and its broader implications on neuromuscular functionality come to the forefront, suggesting avenues for targeted therapeutic interventions.

In conclusion, this chapter weaves together new research demonstrating the complex molecular tapestry governing the NMJ. The studies highlight a complex network, fueled by intrinsic muscle cues and extrinsic neuronal signals, ensures the adaptability and functionality of the NMJ across diverse physiological and pathological scenarios. By delving into the interactions between vital molecular players such as PGC-1α, Rspo2, and DHPR, researchers illuminate the fundamental processes that underpin neuromuscular communication, thereby paving the way for innovative therapeutic interventions aimed at preserving or enhancing neuromuscular function. This exploration not only showcases the critical roles of these molecules in sustaining neuromuscular integrity but also positions them as promising targets for developing strategies to improve muscle performance and tackle neuromuscular disorders. Through a detailed examination of the NMJ's molecular underpinnings, the chapter sheds light on a research landscape rich in biological innovation and therapeutic potential. The research calls for further investigation into the NMJ, a crucial nerve and muscle interaction nexus. It reveals the depth of complexity and opportunity for advancing our understanding and treatment of neuromuscular health issues. In doing so, the research underscores the importance of exploring this vital junction and unraveling its complexities to harness the full potential of molecular insights for clinical application.

3.2 Machine Generated Summaries

Disclaimer: The summaries in this chapter were generated from Springer Nature publications using extractive AI auto-summarization: An extraction-based summarizer aims to identify the most important sentences of a text using an algorithm and uses those original sentences to create the auto-summary (unlike generative AI). As the constituted sentences are machine selected, they may not fully reflect the body of the work, so we strongly advise that the original content is read and cited. The auto generated summaries were curated by the editor to meet Springer Nature publication standards. To cite this content, please refer to the original papers.

Machine generated keywords: nmj, neuromuscular, motor, postsynaptic, motor neuron, formation, cluster, wnt, junction, neuromuscular junction, electrical, acetylcholine, neuron, nerve, embryonic.

Morphological and Functional Remodelling of the Neuromuscular Junction by Skeletal Muscle PGC-1α [1]

This is a machine-generated summary of:

Arnold, Anne-Sophie; Gill, Jonathan; Christe, Martine; Ruiz, Rocío; McGuirk, Shawn; St-Pierre, Julie; Tabares, Lucía; Handschin, Christoph: Morphological and functional remodelling of the neuromuscular junction by skeletal muscle PGC-1α [1]

Published in: Nature Communications (2014)

Link to original: https://doi.org/10.1038/ncomms4569

If you want to cite the papers, please refer to the original.

For technical reasons we could not place the page where the original quote is coming from.

Abstract-Summary "In the mature muscle, the relative levels of physical activity are the major determinants of NMJ function."

"Motor neuron-mediated activation patterns of skeletal muscle have been thought of as the major drivers of NMJ plasticity and the ensuing fibre-type determination in muscle."

"We find that muscle-specific expression of PGC-1α promotes a remodelling of the NMJ, even in the absence of increased physical activity."

"Our data indicate that skeletal muscle significantly contributes to the adaptation of the NMJ subsequent to physical activity."

Introduction

"Skeletal muscle-specific receptor tyrosine-protein kinase (MuSK) and low-density lipoprotein receptor-related protein 4, two key components involved in post-synaptic agrin action, mediate retrograde signaling from the muscle to the motor neuron that is crucial for presynaptic differentiation of the NMJ during embryonic and postnatal development [2]."

"Skeletal muscle-specific gain- and loss-of-function mouse models for PGC-1α exhibit altered expression of postsynaptic NMJ genes in muscle in vivo [3]."

"Besides controlling postsynaptic NMJ gene expression, PGC-1α is a key regulatory nexus of endurance exercise adaptation in the skeletal muscle."

"To assess the contribution of skeletal muscle to the plasticity of adult NMJ morphology and function, we now studied the muscle-specific PGC-1α transgenic mice as a genetic model for endurance exercise."

"This allows a dissociation of the effects from skeletal muscle and motor neurons on the NMJ in a mouse model of endurance exercise."

Results

"In the transgenic animals, the expression of PGC-1α is driven by the muscle creatine kinase (MCK) promoter."

"To ensure muscle specificity of expression, we measured PGC-1α expression in the MCK mice using the WT samples as reference."

"Important for the present study, neuronal tissue, including spinal cord and brain, exhibited indistinguishable levels of PGC-1α transcript in WT and MCK animals, thus confirming the muscle specificity of the transgenic PGC-1α expression in this mouse line."

"The functional data obtained by the single motor neuron–muscle fibre-based electrophysiology strongly implies that the nerve endplates in the MCK animals with muscle-specific overexpression of PGC-1α not only undergo a post-synaptic but also a pre-synaptic remodelling."

"These findings indicate a change in the pre-to-post-synaptic coupling in the PGC-1α muscle-specific transgenics."

"It was previously shown that the overexpression of PGC-1α in the muscle increased the frequency of motor neuron expressing the synaptic vesicle protein SV2A [4]."

Discussion

"In mice with skeletal muscle-specific overexpression of PGC-1α, the NMJ is altered morphologically, structurally and functionally despite unchanged locomotion and hence motor neuron-mediated activation of muscle fibres."

"Our finding that muscle PGC-1α not only remodels the post-synapse, but also triggers a significant pre-synaptic adaptation of the NMJ obviously suggest the production of a PGC-1α-dependent neurotrophic factor that mediates a local, retrograde signal from muscle to the motor neuron."

"It is conceivable that PGC-1α, a regulatory nexus in metabolic and myofibrillar endurance exercise adaptation in muscle, also controls activity-dependent remodeling of the NMJ and contributes to the attenuation of the age-related deterioration of neuromuscular morphology and function by exercise."

"While the exact mechanism by which skeletal muscle-specific overexpression of PGC-1α improves muscle fibre integrity and function in mdx animals, the mouse model for Duchenne muscular dystrophy, PGC-1α-mediated stabilization of the NMJ could certainly be a part of this therapeutic effect."

Methods

"To be able to measure the expression of pre-synaptic markers from the muscle samples, the synaptic part was stained with Alexa Fluor 488-coupled α-bungarotoxin and micro-dissected under a binocular microscope."

"The whole muscles were then incubated overnight at 4 °C with a mixture of primary antibody and Alexa Fluor 488-coupled α-bungarotoxin (Molecular Probe, Ref B13422) diluted 1/10,000 in blocking solution."

"Synaptic fold lengths and synaptic fold number per NMJ (N=9–16 per mouse) were averaged per mouse and subsequently between mice at the same genotype (WT=4, MCK=4 in SCM muscle)."

"NMJ data for cristae and volume density were averaged per mouse and subsequently between mice at the same genotype (WT=4, MCK=3 in SCM muscle)."

"EPP and mEPPs were recorded from different NMJs within the muscle as described previously."

Myosin VI in Skeletal Muscle: Its Localization in the Sarcoplasmic Reticulum, Neuromuscular Junction and Muscle Nuclei [5]

This is a machine-generated summary of:

Karolczak, Justyna; Sobczak, Magdalena; Majewski, Łukasz; Yeghiazaryan, Marine; Jakubiec-Puka, Anna; Ehler, Elisabeth; Sławińska, Urszula; Wilczyński, Grzegorz M.; Rędowicz, Maria Jolanta: Myosin VI in skeletal muscle: its localization in the sarcoplasmic reticulum, neuromuscular junction and muscle nuclei [5]

Published in: Histochemistry and Cell Biology (2012)

Link to original: https://doi.org/10.1007/s00418-012-1070-9

If you want to cite the papers, please refer to the original.

For technical reasons we could not place the page where the original quote is coming from.

Abstract-Summary "An important role for MVI in striated muscle functioning was suggested in a report showing that a point mutation (H236R) within the MVI gene was associated with cardiomyopathy [6]."

"We have addressed MVI function in striated muscle by examining its expression and distribution in rat hindlimb skeletal muscle."

"We found that MVI was present predominantly at the muscle fiber periphery, and it was also localized within muscle nuclei."

"MVI was detected in the sarcoplasmic reticulum fractions isolated from skeletal and cardiac muscle."

"In denervated muscle, the defined MVI distribution pattern was abolished and accompanied by significant increase in its amount in the muscle fibers."

"We have identified several novel potential MVI-binding partners, which seem to aid our observations that in striated muscle MVI could be involved in postsynaptic trafficking as well as in maintenance of and/or transport within the sarcoplasmic reticulum and non-sarcomeric cytoskeleton."

Introduction

"The first observation on the possible involvement of myosin VI (MVI) in muscle function was made by Mohiddin and others who showed that in humans a point mutation within the MVI gene was associated not only with deafness, but also with mild symptoms of cardiac hypertrophy [6]."

"It has been widely accepted that MVI fulfills its functions through interactions with actin and a set of its interaction partners bound to the MVI C-terminal globular tail domain (also termed the cargo domain) in tissue- and cell type-specific manners."

"A positively charged region of the MVI C-terminal globular tail was found to bind to PIP_2-containing liposomes, which could aid in the partner binding [7]."

"This and other observations described above (including the association of the MVI gene mutation with hypertrophic cardiomyopathy) further indicate that MVI could be engaged in striated muscle functioning."

Materials and Methods

"To obtain the skeletal SR fraction, the white back muscle and the leg muscle (fast twitch muscle) were isolated from Wistar rats weighing approximately 250 g. To prepare the cardiac SR fraction, the ventricular tissue was excised from hearts also isolated from Wistar rats weighing between 250 and 350 g. The fractions after separation on 12 % SDS polyacrylamide gel (about 20 μg of protein of each sample was loaded) were subjected to western blot analysis with anti-MVI and anti-calreticulin antibodies."

"Homogenates of control and denervated soleus and EDL muscles (10–20 μg of protein) or sarcoplasmic reticulum fractions of skeletal and cardiac muscles were separated using 10 or 12% polyacrylamide SDS gels and then transferred to a nitrocellulose membrane."

Results

"Western blot and RT-PCR techniques were used to detect myosin VI (MVI) in rat hindlimb muscles (soleus, EDL, and GM) and to assess its expression levels as well as its isoform pattern."

"The pattern of distribution of MVI in both EDL and GM muscles was almost identical to that observed in the soleus muscle."

"The changes in MVI expression level and its distribution in the denervated and control hindlimb muscles suggest the presence of innervation-dependent factor(s) that could determine its synthesis and localization within the muscle fiber."

"The observation that MVI distribution within the fiber is associated with the state of muscle innervation prompted us to check whether MVI was present at the neuromuscular junction."

"This observation seems to suggest that MVI is rather not involved in neuromuscular junction integrity but could be engaged in transport of particles and vesicles at the muscle side of the junction."

Discussion

"MVI was detected in fibers of rat hindlimb muscles, where it localized predominantly to the sarcoplasmic reticulum (also in cardiac muscle), muscle nuclei and the postsynaptic region of the neuromuscular junction."

"Our study indicates that in adult muscles MVI could be involved in neuromuscular transmission (probably by transporting particles and vesicles at the fiber side of the muscle synapse) and not in junction integrity since its postsynaptic presence was significantly reduced in the ALS rat muscle junction, where its architecture was still preserved even though neuronal transmission was most probably impaired [8]."

"We have not detected MVI in the nuclei of the Schwann cells adjacent to the neuromuscular junction thus indicating that its nuclear presence may be specific for muscle cells."

Conclusions

"We have addressed, for the first time, the role of MVI in striated muscles by characterization of its expression and splice variant formation in rat hindlimb muscle, its distribution in both normal and denervated skeletal muscles as well as in ventricular cardiac muscle, and by identification of proteins binding to the MVI cargo domain (potential MVI partners) in muscle homogenates."

"Our observations indicate that in striated muscles MVI could be involved in intra-fiber trafficking, in the SR and sarcomere maintenance, and possibly in neuromuscular signal transmission and in transcription."

"We believe that this study aids in understanding the role of MVI in striated muscle, and in explanation of the clinical phenotype associated with a point mutation (H246R) within MVI motor domain manifested not only with deafness (as is observed for all vertebrate MVI mutations), but also with cardiac hypertrophy [6]."

Low-Intensity Electrical Stimulation Ameliorates Disruption of Transverse Tubules and Neuromuscular Junctional Architecture in Denervated Rat Skeletal Muscle Fibers [9]

This is a machine-generated summary of:

Tomori, Kounosuke; Ohta, Yukiko; Nishizawa, Tomie; Tamaki, Hiroyuki; Takekura, Hiroaki: Low-intensity electrical stimulation ameliorates disruption of transverse tubules and neuromuscular junctional architecture in denervated rat skeletal muscle fibers [9]

Published in: Journal of Muscle Research and Cell Motility (2010)

Link to original: https://doi.org/10.1007/s10974-010-9223-8

If you want to cite the papers, please refer to the original.

For technical reasons we could not place the page where the original quote is coming from.

Abstract-Summary "We determine the effects of direct electrical stimulation (ES) on the histological profiles in atrophied skeletal muscle fibers after denervation caused by nerve freezing."

"Direct ES was performed on the tibialis anterior (TA) muscle after denervation in 7-week-old rats divided into groups as follows: control (CON), denervation (DN), or denervation with direct ES (subdivided into a 4 mA (ES4), an 8 mA (ES8), or a 16 mA stimulus (ES16)."

"These membrane disruptions seemed to be ameliorated by relatively low intensity ES (4 mA and 8 mA), and the area of longitudinally oriented t-tubules and the number of pentads and heptads decreased significantly ($P < 0.01$) in ES4 and ES8 compared to the DN."

"There were no significant differences in the $_{\alpha 1s}$DHPR and RyR1 mRNA expression among CON, DN, and all ES groups."

"After 3 weeks of denervation all nerve terminals had disappeared from the neuromuscular junctions (NMJs) in the CON and ES16 groups."

"The relatively low-intensity ES ameliorates disruption of membrane system architecture in denervated skeletal muscle fibers, but that it is necessary to select the optimal stimulus intensities to preserve the structural integrity of denervated muscle fibers."

Introduction

"Direct electrical stimulation (ES) using transcutaneous electrodes (i.e., neuromuscular ES and functional ES) is a procedure generally used in human physical rehabilitation in the denervated muscle to maintain and/or recover mass and force in the denervated muscle."

"It has been demonstrated that restoration of normal muscle contractile activity following direct ES inhibits axonal sprouting in partially denervated rat muscles (Lømo and Slater 8; Brown and Holland 9)."

"These results suggest that direct ES does not cause a uniform improvement effect for each structural and/or functional profile in denervated muscle fibers, and direct ES may become harmful in some situations."

"In order to accomplish this purpose, we performed direct ES using three different current intensities on denervated rat tibialis anterior (TA) muscles, and we observed the effects of ES on the recovery process in ultrastructural profiles of membrane systems involved in e-c coupling and neuromuscular junction (NMJ) architectures in degenerated skeletal muscle fibers following denervation."

Materials and Methods

"To quantitatively estimate the mRNA expressions of several target genes, real time PCR amplification was performed on a Light-Cycler instrument system (Roche, Tokyo, Japan)."

"In order to visualize the neuromuscular junctions (NMJs), the fixed TA muscle was incubated in a cholinesterase staining solution which consisted of 30 ml of Tris–HCl buffer (100 mM, pH 7.2) containing 0.3 ml of 5′-bromoindoxyl acetate dissolved in absolute ethanol, 50 mg of $K_3Fe(CN)_6$, and 63 mg of $K_4Fe(CN)_6$ at 37°C for 10–15 min [10]."

"After careful trimming, small muscle bundles containing cholinesterase-positive NMJs were washed several times in the 100 mM sodium cacodylate buffer (pH 7.25), postfixed in 2% OsO_4 (in 100 mM sodium cacodylate buffer) for 2 h, and stained en bloc with saturated aqueous uranyl acetate at 60°C for 4 h. Tissues were dehydrated in graded ethanol and acetone, infiltrated with an Epon-acetone mixture (1:1) overnight and embedded in Epon."

Results

"Relative TA muscle weight decreased significantly ($P < 0.01$) following denervation, and there were significant ($P < 0.05$) differences between the CON and DN groups."

"Another visible change in the t-tubules in denervated muscle fibers is the appearance of double and triple parallel t-tubule profiles accompanied by junctional SR cisternae, resulting in the formation of pentads and heptads [11]."

"Although the disposition of t-tubules and triads is dramatically altered following denervation and/or direct ES, there were no significant differences in the ${}_{\alpha 1s}$DHPR and RyR1 mRNA expression among the CON, DN, and ES groups."

"Synaptic vesicles in the terminal clustered in an active zone lying opposite the mouth of a junctional fold in the muscle fiber surface, while the mitochondrial clusters were located at the center of the nerve terminal."

"The basal lamina ensheathed the muscle fiber and Schwann cell, passed through the synaptic cleft, and extended into secondary junctional folds."

Discussion

"The expression of IGF-1 mRNA in denervated muscle was increased significantly following ES at 16 mA. On the other hand, it is well known that IGF-1 has a role in tissue regeneration after mechanical damage [12]."

"Although IGF-1 mRNA expression is increased depending on stimulation intensity, there is no difference in muscle mass between ES at 8 and 16 mA. Therefore, the stimulation of denervated muscle at 8 mA was enough to retard progressive atrophy of muscle, and expression of IGF-1 mRNA after stimulation at 16 mA increased more than required."

"High-intensity (16 mA) ES was effective in retarding denervated muscle atrophy; however, it may have adverse effects on regeneration of nerve terminals and/or the membrane systems involved in e-c coupling."

"Low-intensity (4–8 mA) ES was effective in regenerating membrane systems involved in e-c coupling and nerve terminals, but ineffective in retarding denervated muscle atrophy."

Effect of Chronic Intermittent Hypoxia (CIH) on Neuromuscular Junctions and Mitochondria in Slow- and Fast-Twitch Skeletal Muscles of Mice—The Role of iNOS [13]

This is a machine-generated summary of:

Bannow, L. I.; Bonaterra, G. A.; Bertoune, M.; Maus, S.; Schulz, R.; Weissmann, N.; Kraut, S.; Kinscherf, R.; Hildebrandt, W.: Effect of chronic intermittent hypoxia (CIH) on neuromuscular junctions and mitochondria in slow- and fast-twitch skeletal muscles of mice—the role of iNOS [13]

Published in: Skeletal Muscle (2022)

Link to original: https://doi.org/10.1186/s13395-022-00288-7

If you want to cite the papers, please refer to the original.

For technical reasons we could not place the page where the original quote is coming from.

Abstract-Summary "We hypothesize that CIH-related morphological alterations in neuromuscular junctions (NMJ) and mitochondrial integrity might be the cause of functional disorders in skeletal muscles."

"In WT soleus muscle, CIH vs. NOX reduced NMJ size (– 37.0%, $p < 0.001$) and length (– 25.0%, $p < 0.05$) together with fiber CSA of type IIa fibers (– 14%, $p < 0.05$) and increased centronucleated fiber fraction ($p < 0.001$)."

"CIH vs. NOX increased the fraction of damaged mitochondria (1.8-fold, $p < 0.001$)."

"Compared to WT, iNOS$^{-/-}$ similarly decreased NMJ area and length with NOX (– 55%, $p < 0.001$ and – 33%, $p < 0.05$, respectively) or with CIH (– 37%, $p < 0.05$ and – 29%, $p < 0.05$), however, prompted no fiber atrophy."

"Increased fractions of damaged (2.1-fold, $p < 0.001$) or swollen (> 6-fold, $p < 0.001$) mitochondria were observed with $iNOS^{-/-}$ vs. WT under NOX and similarly under CIH."

"Both, CIH- and $iNOS^{-/-}$ massively upregulated suppressor-of-cytokine-signaling-3 (SOCS3) > 10-fold without changes in IL6 mRNA expression."

"None of these morphological alterations with CIH- or $iNOS^{-/-}$ were detected in the gastrocnemius muscle."

"INOS expression was undetectable in WT muscle, unlike the liver, where it was massively decreased with CIH."

"CIH leads to NMJ and mitochondrial damage associated with fiber atrophy/centronucleation selectively in slow-twitch muscle of WT."

"In the absence of muscular iNOS expression in WT, this damage may arise from extramuscular, e.g., motoneuronal iNOS deficiency (through CIH or knockout) awaiting functional evaluation."

Introduction

"Studies in OSA patients as well as in CIH animal models detected an overexpression of inducible nitric oxide synthase (iNOS) via inflammatory triggers involving NF-κB activation, especially in neuronal or cardiovascular tissues, e.g., activated macrophages [14–16]."

"The present study used CIH as compared to normoxic control (NOX) as a model of OSA in wildtype (WT) mice to investigate the long-term effect of OSA on skeletal muscle histomorphological measures of NMJ integrity and fiber size and composition as well as on transition electron microscopy (TEM) parameters of ultrastructural mitochondria integrity."

"This is the first report on CIH-induced muscle-specific damage to NMJ, fiber histomorphology, or mitochondrial ultrastructure in WT mice locomotor muscle, which unexpectedly is not inhibited by iNOS deficiency, but strikingly mimicked already at NOX."

Materials and Methods

"Type-specific fiber cross-sectional area (CSA) was determined by manually encircling each cross-section of at least 100 fibers using standard imaging software ImageJ (Scion Image, National Institutes of Health, Bethesda, USA)."

"Single staining was performed by incubation of polyclonal anti- IL1ß (1:100: Abcam; Cambridge, UK) or monoclonal anti-CD68 primary antibody (1:50; AbD Serotec, Kidlington, UK) with goat polyclonal HRP-conjugated anti-rabbit (1:200; LINARIS GmbH, Dossenheim, Germany) or anti-rat (1:100; AbD Serotec, Kidlington, UK) antibody; endogenous peroxidase activity was suppressed with 3 % H_2O_2 in PBS; afterwards, the sections were incubated with DAB solution (Roche Diagnostics)."

"In the soleus muscle, we additionally included a fourth staining with anti-CD68 antibody (1:50; AbD Serotec, Kidlington, UK), detected by Alexa Fluor® 647 labeled donkey-anti rat IgG (1:200, MoBiTec GmbH, Göttingen, Germany), to visualize macrophages next to the NMJ."

Results

"A significant inverse correlation was found between the fraction of centronucleated fibers and the CSA of the fiber population in WT undergoing CIH or NOX ($r = -0.417$; $p = 0.043$)."

"In both conditions, NOX and CIH, $iNOS^{-/-}$ compared to WT was without any significant effect on fiber CSA of the total fiber population or that of fiber type 1, 2a, or 2x."

"Unlike the soleus muscle, the gastrocnemius muscle in $iNOS^{-/-}$ mice revealed no significant alterations in postsynaptic NMJ area, and this was also true for both abovementioned normalization of NMJ area or length for fiber CSA or perimeter, respectively."

"In WT mice gastrocnemius muscle, CIH compared to NOX led to a 50% decrease of IL6 mRNA expression while in $iNOS^{-/-}$ no relevant difference between CIH and NOX was found."

"In WT mice, CIH compared to NOX led to an almost 50% decrease in mtSOD mRNA expression in soleus muscle."

Discussion

"The similarity between CIH (compared to NOX in WT) and $iNOS^{-/-}$ (compared to WT in NOX) was limited to NMJ and mitochondrial damage, while decreases in fiber CSA (including its correlation to NMJ) and centronucleation observed with CIH vs. NOX in WT were absent in $iNOS^{-/-}$ mice, i.e., they revealed no atrophy despite signs of denervation."

"The moderate decrease in mtSOD expression presently observed in soleus muscle with CIH and, to a lesser extent, with iNOS (NOX or CIH) might also play a role in fiber atrophy and mitochondrial deterioration: $Sod1^{-/-}$ mice, used as a murine model of neuromuscular impairment in age-related muscle atrophy (sarcopenia), exhibit reduction in myofiber CSA of type IIa fibers [17, 18]."

"Importantly, $iNOS^{-/-}$ at NOX largely mimicked the CIH effects, i.e., NMJ and mitochondrial damage together with SOCS3 upregulation and mtSOD downregulation and was despite the fact that iNOS mRNA expression in soleus or gastrocnemius muscles in WT was neither detectable with NOX nor with CIH exposure."

Conclusion

"This is the first study to demonstrate that CIH as a model of moderate to severe OSA triggers NMJ and mitochondrial damage accompanied by fiber atrophy in the slow-twitch muscle of WT mice, all of which may contribute to reduced exercise (aerobic) capacity in patients suffering from OSA."

"We furthermore demonstrate that iNOS deficiency, rather than yielding protection of skeletal muscle against CIH stress, leads to similar structural impairments of NMJ and mitochondria under normoxia and might contribute to the CIH effects in WT, putatively through compromised innervation."

Neuromuscular Electrical Stimulation for Preventing Skeletal-Muscle Weakness and Wasting in Critically Ill Patients: A Systematic Review [19]

This is a machine-generated summary of:

Maffiuletti, Nicola A; Roig, Marc; Karatzanos, Eleftherios; Nanas, Serafim: Neuromuscular electrical stimulation for preventing skeletal-muscle weakness and wasting in critically ill patients: a systematic review [19].

Published in: BMC Medicine (2013)

Link to original: https://doi.org/10.1186/1741-7015-11-137

If you want to cite the papers, please refer to the original.

For technical reasons we could not place the page where the original quote is coming from.

Abstract-Summary "The objective of this systematic review was to evaluate the effectiveness of NMES for preventing skeletal-muscle weakness and wasting in critically ill patients, in comparison with usual care."

"We searched PubMed, CENTRAL, CINAHL, Web of Science, and PEDro to identify randomized controlled trials exploring the effect of NMES in critically ill patients, which had a well-defined NMES protocol, provided outcomes related to skeletal-muscle strength and/or mass, and for which full text was available."

"Two independent reviewers extracted data on muscle-related outcomes (strength and mass), and participant and intervention characteristics, and assessed the methodological quality of the studies."

"When means and SDs were provided, the effect sizes of individual outcomes were calculated, and otherwise, a qualitative analysis was performed."

"Five studies reported an increase in strength or better preservation of strength with NMES, with one study having a large effect size."

"Two studies found better preservation of muscle mass with NMES, with small to moderate effect sizes, while no significant benefits were found in two other studies."

"NMES added to usual care proved to be more effective than usual care alone for preventing skeletal-muscle weakness in critically ill patients."

Background

"A large majority of patients admitted to the intensive care unit (ICU) after the very acute phase of a critical illness exhibit major defects in skeletal-muscle strength (weakness) and mass (wasting) [20, 21]."

"ICU-based NMES has recently been introduced for the treatment of ICUAW, as it does not require active patient cooperation, has an acute beneficial systemic effect on muscle microcirculation [22], and seems to provide some structural and functional benefits to critically ill patients [23]."

"Owing to the heterogeneity of the critically ill patient group and also of the NMES procedures implemented in ICUs, the effectiveness of this rehabilitation procedure for ICUAW prevention remains to be clearly proven [24–26]."

"Given the potential use of NMES among patients with a limited capacity to engage in voluntary muscle work, assessment of the evidence for the use of NMES in critically ill patients is urgently needed."

"We therefore undertook a formal systematic review of the literature to determine the rehabilitative effect of NMES on skeletal-muscle strength and mass in critically ill patients, in comparison with standard care."

Methods

"Both reviewers then read the full text and applied the following inclusion criteria: RCTs 1) exploring the effect of NMES in critically ill patients; 2) with a well-defined NMES protocol (that is, the main stimulation parameters were provided) for at least one intervention group; 3) with NMES applied to skeletal muscles with an intensity equal to or greater than motor threshold (that is, evoking a visible muscle contraction); 4) including outcomes related to muscle strength and/or mass; (5) and whose full text was available."

"Two of the authors (EK, MR) independently extracted the data from the studies included in the review."

"Because data were not normally distributed, the majority of the studies included in the review reported continuous outcomes using medians with interquartile range instead of means with standard deviation (SD)."

"When means and SDs at baseline and after the intervention were not provided, individual effect sizes were not calculated and, instead, a qualitative analysis of the data was performed."

Results

"Two studies that investigated the effects of NMES on patients with COPD reported spirometry and blood gas values as measures of disease severity [27, 28]."

"One study found a significantly larger MRC score increase in the NMES group compared with the control group, with a large effect size (d = 1.44) [28]."

"Two studies reported greater MRC scores in the NMES group than in the control group, although baseline measurements were not provided [29, 30]."

"One study, in which quadriceps muscle strength was assessed by dynamometry, reported a significantly larger strength increase in the NMES group compared with the control group [27]."

"In one study, muscle thickness decreased less in the NMES group than in the control group, and effect sizes (d) ranged from 0.11 to 0.39, depending on the muscle group assessed [23]."

Discussion

"These findings suggest that NMES may have the potential to prevent skeletal-muscle weakness in critically ill patients, which could confer many important physical, psychosocial, and economic benefits for these patients after discharge from ICU."

"Considering the limited or absent cooperation of patients at admission into the ICU, and the considerable influence of central factors (including motivation) on maximal voluntary efforts, it would be preferable if evaluation of muscle function in these patients relied on artificially-evoked muscle responses [31]."

"Compared with other rehabilitation strategies, the unique aspects of NMES are that it is relatively cost-effective (one multiple-user NMES unit costs less than €400), does not require patient cooperation (it can be applied to sedated patients) or stable cardiac or respiratory function, can be implemented during the first few days after ICU admission, and provokes considerable central effects, both acute and chronic, which could also contribute to preventing the occurrence of muscle weakness in critically ill patients [32]."

Conclusions

"This systematic review provides evidence that adding NMES therapy to usual care is more effective than usual care alone or sham NMES in preventing ICUAW."

"There is inconclusive evidence about the effectiveness of NMES for the preservation of muscle mass in ICU patients."

"The effects of NMES we found were probably underestimated because of the non-stratification of patients according to main diagnosis and disease severity."

"More studies are needed to explore the long-term effects of NMES therapy during ICU stay on physical function and quality of life in ICU survivors, in order to identify the optimal NMES dosage for ICUAW prevention (both in terms of frequency, intensity and volume), and to describe the feasibility, safety, and cost-effectiveness of NMES in different subpopulations of critically ill patients."

Open-Chamber Co-culture Microdevices for Single-Cell Analysis of Skeletal Muscle Myotubes and Motor Neurons with Neuromuscular Junctions [33]

This is a machine-generated summary of:

Yamaoka, Nao; Shimizu, Kazunori; Imaizumi, Yu; Ito, Takuji; Okada, Yohei; Honda, Hiroyuki: Open-Chamber Co-Culture Microdevices for Single-Cell Analysis of Skeletal Muscle Myotubes and Motor Neurons with Neuromuscular Junctions [33]

Published in: BioChip Journal (2019)

Link to original: https://doi.org/10.1007/s13206-018-3202-3

If you want to cite the papers, please refer to the original.
For technical reasons we could not place the page where the original quote is coming from.

Abstract-Summary "Although various microdevices for the co-culture of skeletal muscle myotubes and motor neurons have been developed to investigate neuromuscular diseases in vitro, it remains difficult to isolate single myotubes and motor neurons from the device for single-cell analyses, such as gene expression analysis."

"Given the small chamber width (0.2 mm), the device significantly prevented the overlap among myotubes within the chamber."

"The device with the 0.2-mm chamber promoted myotube maturation, as indicated by the longer widths and lengths of the myotubes relative to those in the control chamber."

"Single C2C12 myotubes and human induced pluripotent stem cell (hiPSC)-derived motor neurons were successfully collected from the device with the 0.2-mm chamber using a micromanipulator equipped with a glass capillary."

"Myotubes and hiPSC-derived motor neurons were co-cultured in the device with the 0.2- mm chamber, and the formation of NMJs were observed."

Introduction

"Conventional methods for inducing the formation of NMJs in vitro include the co-culture of motor neurons and myotubes in the same dish or plate [34, 35]."

"Recent studies have reported the use of microdevices that can control the positions of motor neurons and myotubes [36, 37]."

"When the motor neurons are cultured in the chamber, axons but not the cell bodies of the motor neurons, pass through the microgrooves or microtunnels and eventually extend to the myotubes in the other chamber to form NMJs."

"Collecting single myotubes and motor neurons from the device for single-cell analyses, such as gene expression analysis, remains difficult because of the following: 1) the cell culture chamber of the microdevices are closed with ceilings; or/and 2) the myotubes are aligned randomly and overlap with each other within the chamber."

"In the present study, we developed the open-chamber co-culture microdevices that allows the isolation of individual myotubes and motor neurons for subsequent single-cell analysis."

Materials and Methods

"To measure the height of the microtunnels, we vertically cut the PDMS device, observed the device with microscope, and measured using ImageJ. The C2C12 cell line, a skeletal muscle myoblast cell line, were cultured as previously described [38]."

"After blocking with PBS containing 10% goat serum and 0.01% Triton X-100, the cells were incubated with the primary antibodies [monoclonal anti-α-actinin antibodies (Sigma-Aldrich, USA)] at 4 °C overnight."

"The cells were then washed twice with PBS and incubated with secondary antibodies [CF™488A goat anti-mouse IgG (H+L)] for 1 h at room temperature."

"The cells were incubated with the primary antibodies [monoclonal anti-β-tubulin III (TuJ1) antibodies (Sigma-Aldrich)] at 4 °C overnight."

"The cells were then washed twice with PBS and incubated with secondary antibodies [CF543™ goat antirabbit IgG (H+L) (BIOTIUM)] for 1 h at room temperature."

Results and Discussion

"The cell bodies of myotubes and motor neurons were cultured separately in each chamber."

"The device was designed to have a chamber with a narrow width, which especially allows the targeted collection of individual myotubes from the chamber."

"We examined whether a micromanipulator equipped with a glass capillary can be used for the collection of individual myotubes and motor neurons cultured in the open chambers."

"We failed to collect individual myotubes using the device with the 7-mm-width chamber because of the high degree of overlap among the myotubes."

"We developed an open-chamber device for co-culturing skeletal muscle myotubes and motor neurons and successfully induced the formation of NMJs on the device."

"Using the device with narrow chambers and chambers without ceilings, we successfully cultured mature myotubes and collected individual myotubes and motor neurons."

Skeletal Muscle Contractile Function and Neuromuscular Performance in Zmpste24$^{-/-}$ Mice, a Murine Model of Human Progeria [39]

This is a machine-generated summary of:

Greising, Sarah M.; Call, Jarrod A.; Lund, Troy C.; Blazar, Bruce R.; Tolar, Jakub; Lowe, Dawn A.: Skeletal muscle contractile function and neuromuscular performance in Zmpste24$^{-/-}$ mice, a murine model of human progeria [39]

Published in: GeroScience (2011)

Link to original: https://doi.org/10.1007/s11357-011-9281-x

If you want to cite the papers, please refer to the original.

For technical reasons we could not place the page where the original quote is coming from.

Abstract-Summary "Skeletal muscle is detrimentally changed by normal aging but whether it is an affected tissue in progeria has not been resolved."

"We hypothesized that mice which mimic Hutchinson–Gilford progeria syndrome would exhibit age-related alterations of skeletal muscle."

"Contractile capacity of the posterior leg muscles were not affected in Zmpste24$^{-/-}$ mice, but muscles of the anterior leg were 30–90% weaker than those of Zmpste24$^{+/+}$ mice ($p < 0.01$)."

"Leg muscles were 32–47% smaller in the Zmpste24$^{-/-}$ mice and contained ~60% greater collagen relative to littermates ($p < 0.01$)."

"The combined factors of muscle atrophy, collagen accumulation, and perturbed joint mechanics likely contributed to poor neuromuscular performance and selective muscle weakness displayed by Zmpste24$^{-/-}$mice."

"These characteristics are similar to those of aged mice indicating accelerated aging of skeletal muscle in progeria."

Introduction

"The age-related disease of skeletal muscle is sarcopenia and it has not been categorized as being present or absent in HGPS."

"Determining whether or not skeletal muscle is an affected tissue in HGPS is critical based not only on clinical importance but also on the idea that stem cells and nuclear abnormalities are fundamentally involved in the segmentation of affected tissues in HGPS."

"Reasons to support the idea that skeletal muscle might not be affected in HGPS include the following."

"These factors are suggestive of skeletal muscle not being an affected tissue in HGPS."

"There are also reasons to believe that skeletal muscle could be affected in HGPS."

"If satellite cells are affected in HGPS, the proposal of stem cell exhaustion is relevant to skeletal muscle in this disease as well."

"These points indicate that skeletal muscle and its major function, contraction and production of force, could be detrimentally affected in HGPS."

Materials and Methods

"At 15 or 20 weeks of age, soleus and extensor digitorum longus (EDL) muscles were dissected and analyzed ex vivo for contractile functions."

"At 20 weeks of age, immediately following testing of joint mechanics and while still anesthetized, mice underwent in vivo contractile testing of the left leg [40, 41]."

"A subset of mice at 15 and 20 weeks of age underwent testing of isolated soleus and EDL muscles [42, 43]."

"Zmpste24$^{+/+}$ mice at 15 and 20 weeks of age were not significantly different as determined by Student's t tests ($p \geq 0.088$) on any measured parameter and therefore these data were collapsed to represent a single control group."

"Data were analyzed by Student's t test to determine differences between genotypes for in vivo joint mechanics, and ex vivo contractility measurements, creatine kinase activity, muscle masses, collagen content, and histological analyses."

Results

"When torque was normalized to the small posterior muscle masses of the Zmpste24$^{-/-}$ mice, strength was not lower than that of Zmpste24$^{+/+}$ mice (122.7 ± 4.5 vs. 104.3 ± 3.3 kN · mm^{-1} kg^{-1} muscle, respectively; $p = 0.091$)."

"This difference could not be explained by smaller muscles because torque normalized to the mass of the anterior muscles was still extremely low in the Zmpste24$^{-/-}$ mice (5.0 ± 1.5 vs. 50.5 ± 2.3 kN · mm^{-1} kg^{-1} muscle for Zmpste24$^{+/+}$ mice; $p < 0.001$)."

"There was no difference in the ratio of prelamin A to histone H3 between soleus and EDL muscles of Zmpste24$^{-/-}$ mice (0.85 ± 0.2 vs. 0.91 ± 0.1, respectively; $p = 0.804$)."

"The mean cross-sectional area of all soleus muscle fibers was not different between Zmpste24$^{-/-}$ and Zmpste24$^{+/+}$ mice (1,431 ± 74 vs. 1,418 ± 156 μm^2, respectively; $p = 0.946$)."

Discussion

"Using a variety of tests for assessing the function of skeletal muscle, we show that Zmpste24$^{-/-}$ mice have impaired neuromuscular performance and selective muscle weakness relative to their Zmpste24$^{+/+}$ littermates."

"All skeletal muscles that were assessed for prelamin A showed that the protein was present in those from Zmpste24$^{-/-}$ mice and not from Zmpste24$^{+/+}$ mice, but accumulation of prelamin A was similar between EDL and soleus muscles and, thus, could not explain why anterior but not posterior leg muscles were weak."

"Bergo and others described a severe effacement of the anterior tibial tuberosity, the origin of the tibialis anterior muscle, in Zmpste24$^{-/-}$ mice and suggested that lower extremity weakness contributed to the bone abnormality (Bergo and others 41)."

"The soleus and EDL muscles in which excessive nuclei were found were from the same Zmpste24$^{-/-}$ mice assessed for contractile function, but because nuclei per se are not involved in the generation of force it is difficult to speculate how excessive nuclei could directly cause muscle weakness."

Wnt Signaling in Skeletal Muscle Dynamics: Myogenesis, Neuromuscular Synapse and Fibrosis [44]

This is a machine-generated summary of:

Cisternas, Pedro; Henriquez, Juan P.; Brandan, Enrique; Inestrosa, Nibaldo C.: Wnt Signaling in Skeletal Muscle Dynamics: Myogenesis, Neuromuscular Synapse and Fibrosis [44]

Published in: Molecular Neurobiology (2013)

Link to original: https://doi.org/10.1007/s12035-013-8540-5

If you want to cite the papers, please refer to the original.
For technical reasons we could not place the page where the original quote is coming from.

Abstract-Summary "We summarize compelling evidence on the role of Wnt signaling on several features of skeletal muscle physiology."

"We briefly review the role of Wnt pathways on the formation of muscle fibers during prenatal and postnatal myogenesis, highlighting its role on the activation of stem cells of the adult muscles."

"Based on previous evidence showing that Wnt pathways are linked to several diseases, such as Alzheimer's and cancer, we address recent studies indicating that Wnt signaling plays a key role in skeletal muscle fibrosis, a disease characterized by an increase in the extracellular matrix components leading to failure in muscle regeneration, tissue disorganization and loss of muscle activity."

"As it has emerged in other pathological conditions, we suggests that muscle fibrosis may be a consequence of alterations of Wnt signaling activity."

Introduction

"Muscle formation, or myogenesis, is mediated by different signaling pathways triggered by extracellular cues, which activate the expression of different myogenic regulator factors (MRFs), such as Myf5, MyoD, and myogenin [45, 46]."

"Signals, studies in prenatal and postnatal muscle development have described that signaling pathways activated by Wnt ligands modulate the expression of several MRFs, thus regulating the formation of functional multinucleated myotubes [47, 48]."

"Recent studies suggest that the activation of Wnt signaling promotes muscle fibrosis [49, 50]."

"The molecular effects of Wnt signaling pathway on the activity of TGF-β and CTGF, as well as the effect of these interactions on the onset of muscle fibrosis are not well understood."

"The aim of this review is to describe the role of the Wnt pathway on skeletal muscle dynamics including myogenesis, NMJ formation, and muscle fibrosis."

Wnt Signaling Pathways

"The canonical Wnt pathway begins with the binding of the Wnt ligand to the widely expressed seven-transmembrane Frizzled (Fzd) receptors, of which ten members have been described in vertebrates [51]."

"Intracellulary, the canonical Wnt pathway requires the intracellular protein β-catenin."

"In the absence of Wnt ligands, β-catenin levels remain relatively low by the action of a so-called "destruction complex" formed by the scaffold protein Axin, adenomatous polyposis coli (APC) and the enzyme glycogen synthase kinase 3β (GSK-3 β), that phosphorylates β-catenin stimulating its destruction by the proteosomal pathway [52]."

"In the planar cell polarity pathway (Wnt/PCP), the Wnt–Fzd dependent recruitment of Dvl leads to activation of small GTPases proteins, such as Rho and Rac, which subsequently activate the c-Jun N-terminal kinase."

Wnt Signaling in Myogenesis

"In chicken embryos, the somite myogenesis in presegmented paraxial mesoderm can be stimulated by activators of Shh pathway together with some Wnt ligands, including Wnt1, Wnt3, and Wnt4."

"At early stages of differentiation, the expression of MyoD is dependent on the simultaneous activation of the Wnt and Shh pathways; in later stages of development, however, Wnt activity alone is able to stimulate the expression of MyoD, suggesting that Wnt signaling is a key factor for late prenatal differentiation [53]."

"The distribution pattern of Fzd and Wnts ligands in different regions of chick embryo support a crucial role for Wnt signaling in early myogenesis, by showing a direct correlation between the expression of some receptors and ligands; however, there is a clear overlapping expression pattern in an embryonic region that could activate both canonical and noncanonical Wnts pathways to regulate myogenesis."

Wnt Signaling in the Function of the Neuromuscular Junction

"The mouse-derived ligands Wnt9a and Wnt11 induce AChR clustering in cultured myotubes through a mechanism dependent on MuSK and Lrp4 [54]."

"These findings reveal a novel key MuSK-dependent mechanism by which Wnt ligands induce the aneural clustering of AChRs on newly formed muscle cells."

"These findings reveal a potential neural-dependent cross-talk of Wnt- and agrin-mediated pathways to induce postsynaptic differentiation at the vertebrate NMJ."

"Even though these findings reveal an inhibitory effect for β-catenin on AChR clustering, the in vivo role of this crucial effector of the Wnt canonical pathway at the NMJ is rather complex."

"Even though our current knowledge regarding the potential role of Wnt signaling on presynaptic differentiation of vertebrate motor neurons is virtually null, the function of diverse Wnt pathways on several steps of presynaptic behavior in central synapses, from the establishment of neuronal cell polarity to synaptogenesis, has been well documented."

Wnt Signaling in Muscle Fibrosis

"Despite the complex mixture of signaling implicated in the onset of skeletal muscle fibrosis until now two key molecules have been described as critical in the development of the disease: the TGF-β and the CTGF [55–59]."

"The injection of TGF-β in vivo in skeletal muscle stimulates the production of ECM in the injected area and the overexpression of TGF-β in myoblasts promotes the formation of myofibroblast cells, suggesting a critical role of the TGF-β in the onset of the fibrotic condition in the skeletal muscle [57, 60]."

"Studies in a fibroblast cell line show that the treatment with the Wnt3a ligand, which signals through the Wnt/β-catenin pathway, increases the expression of CTGF and TGF-β mRNAs, reinforcing the view that these signaling pathways may cross-talk in the context of fibrosis [61]."

"The Wnt/β-catenin signaling pathway could have a key role in triggering the activation of TGF-β and CTGF in activated fibroblasts and maybe in other cells of the fibrotic environment."

General Conclusions

"The Wnt signaling pathway plays a critical role in several processes including the development of muscle tissue and the formation of the motoneuron-muscle synapse."

"In the maturation of the NMJ, the Wnt signaling modulates the localization of several proteins critical for the patterning of the synapse and for the precise control of the muscle contraction."

"In muscle fibrosis, apparently the Wnt signaling plays a critical role, as it is able to functionally interact with other profibrotic molecules, such as CTGF and TGF-β."

Differential Effects of Spinal Motor Neuron-Derived and Skeletal Muscle-Derived Rspo2 on Acetylcholine Receptor Clustering at the Neuromuscular Junction [62]

This is a machine-generated summary of:

Li, Jin; Ito, Mikako; Ohkawara, Bisei; Masuda, Akio; Ohno, Kinji: Differential effects of spinal motor neuron-derived and skeletal muscle-derived Rspo2 on acetylcholine receptor clustering at the neuromuscular junction [62]

Published in: Scientific Reports (2018)

Link to original: https://doi.org/10.1038/s41598-018-31949-7

If you want to cite the papers, please refer to the original.

For technical reasons we could not place the page where the original quote is coming from.

Abstract-Summary "SMN-specific Rspo2 mitigated or over-corrected abnormal features of the NMJs and AChR clusters observed in Rspo2$^{-/-}$ mice including (i) abnormal broadening of enlarged AChR clusters, (ii) three of six abnormal ultrastructural features, and (iii) abnormal expression of nine genes in SMNs and the diaphragm."

"Muscle-specific Rspo2 normalized all six abnormal ultrastructural features, but it had no effect on AChR clustering and NMJ formation at the light microscopy level or on abnormal gene expression in SMNs and the diaphragm."

"These results suggest that SMN-derived Rspo2 plays a major role in AChR clustering and NMJ formation in the postsynaptic region, and muscle-derived Rspo2 also plays a substantial role in juxtaposition of the active zones and synaptic folds."

Introduction

"Genetic depletion of the embryonic AChR γ-subunit in muscle fibers markedly decreases staining for AChR clusters with progressive accumulation of synaptic vesicles in the presynaptic terminal, suggesting that embryonic AChR clustering is required for further pre- and postsynaptic development [63]."

"AChR clustering and formation of the neuromuscular junction (NMJ) are mediated by SMN-derived and muscle-derived molecules."

"Other SMN-derived molecules driving AChR clustering and NMJ formation include neuregulin 1 and ACh [64–70]."

"In cultured C2C12 myotubes, Rspo2 induces MuSK phosphorylation and AChR clustering, which requires Wnt ligands but not agrin."

"We assumed that SMN-derived Rspo2 was anchored at the NMJ and promoted AChR clustering as well as NMJ formation, but Rspo2 is also expressed marginally in the diaphragm at E18.5 and in adults [71]."

"To dissect the differential roles of SMN-derived Rspo2 and muscle-derived Rspo2 on AChR clustering and NMJ formation in mouse embryos, we generated transgenic mice expressing Rspo2 specifically in SMNs or in skeletal muscle on the background of Rspo2$^{-/-}$."

Results

"Transgenic mice carrying MCK-RSPO2 and VAChT-RSPO2 were morphologically normal, viable, active, and fertile without shortened life spans."

"Rspo2 was specifically expressed in the skeletal muscle of MCK-RSPO2 mice and in SMNs of VAChT-RSPO2 mice."

"We next examined AChR clustering at the NMJs of the left hemi-diaphragms at low-magnification in wild-type, Rspo2$^{-/-}$, MCK-RSPO2/Rspo2$^{-/-}$, and VAChT-RSPO2/Rspo2$^{-/-}$ mice at E18.5."

"A comparison of Rspo2$^{-/-}$ mice against MCK-RSPO2/Rspo2$^{-/-}$ and VAChT-RSPO2/Rspo2$^{-/-}$ mice revealed that SMN-derived Rspo2 contributes to branching and outgrowth of SMN axons, whereas muscle-derived Rspo2 has no substantial effect on SMN axons."

"We next investigated AChR clustering at the NMJs of the right hemi-diaphragms at high-magnification in wild-type, Rspo2$^{-/-}$, MCK-RSPO2/Rspo2$^{-/-}$, and VAChT-RSPO2/Rspo2$^{-/-}$ mice at E18.5."

"To differentiate the effects of muscle-derived Rspo2 and SMN-derived Rspo2 on the NMJ, we examined gene expression levels in the diaphragm and spinal cords of wild-type, Rspo2$^{-/-}$, MCK-RSPO2/Rspo2$^{-/-}$, and VAChT-RSPO2/Rspo2$^{-/-}$ mice at E18.5."

Discussion

"In three of the four categories (i, ii, and iv), SMN-derived Rspo2 mitigated aberrant NMJ features in Rspo2$^{-/-}$ mice more than muscle-derived Rspo2."

"SMN-derived Rspo2 improved the NMJ phenotypes more than muscle-derived Rspo2."

"Only SMN-derived Rspo2 over-corrected the bandwidth of AChR clusters, which was determined by branching of motor axons."

"The reason why only muscle-derived Rspo2, but not SMN-derived Rspo2, normalized three ultrastructural features remains unknown, but muscle-derived and SMN-derived Rspo2 may have different roles during normal development of the NMJ."

"Individual NMJ features that were mitigated by either muscle-derived or SMN-derived Rspo2 are further addressed in detail."

"Muscle-derived Rspo2 had no effect on this broadening, but SMN-derived Rspo2 made the bandwidth of AChR clusters even narrower than that in wild-type mice."

"We demonstrate that SMN-derived Rspo2 is required for AChR clustering at the motor endplate and for the NMJ formation, and muscle-derived Rspo2 also plays a substantial role in ultrastructural formation of the NMJ."

Materials and Methods

"For Rspo2 staining, sections were incubated with rabbit polyclonal anti-FLAG antibody (1:100, PM020, MBL) at 4 °C overnight and were then incubated with FITC-conjugated goat anti-rabbit IgG secondary antibody (1:100, FI1000, Vector Laboratories) at 4 °C for 2 h in a humidified chamber."

"For microscopic examination of the whole-mount diaphragms, embryos were sacrificed on E18.5, and the diaphragms were dissected and fixed in 2% paraformaldehyde (PFA) in 0.1 M phosphate buffer (pH 7.4) at room temperature (26 °C) for 2 h. After the removal of connective tissue, the whole-mount diaphragm was permeabilized with 0.5% Triton X-100 in 0.1 M phosphate buffer (pH 7.4) for 10 min and then incubated in a blocking buffer containing 3% bovine serum albumin (BSA) and 5% normal goat serum (NGS) at 4 °C overnight."

Statistical Analyses

"Data of multiple groups were analyzed by one-way ANOVA and post-hoc Tukey adjustments using Prism 5.0 (GraphPad) software."

"P values of 0.05 or less were considered statistically significant."

Neuromuscular Synaptic Patterning Requires the Function of Skeletal Muscle Dihydropyridine Receptors [72]

This is a machine-generated summary of:

Chen, Fujun; Liu, Yun; Sugiura, Yoshie; Allen, Paul D; Gregg, Ronald G; Lin, Weichun: Neuromuscular synaptic patterning requires the function of skeletal muscle dihydropyridine receptors [72]

Published in: Nature Neuroscience (2011)

Link to original: https://doi.org/10.1038/nn.2792

If you want to cite the papers, please refer to the original.

For technical reasons we could not place the page where the original quote is coming from.

Abstract-Summary "Chen and colleagues show that a functional skeletal muscle dihydropyridine receptor (DHPR) is required for muscle pre-patterning during neuromuscular junction development."

Main

"Imaging in zebrafish embryos demonstrates that AChR clusters not only are prepatterned in developing muscles, but also are actively incorporated into newly formed NMJs [73, 74]."

"These findings challenge the long-held neurocentric dogma and support the opposing view that postsynaptic muscle intrinsically defines its central region for the ingrowing nerves, which subsequently fine-tune the pre-patterned AChRs so that neuromuscular synapses are generated along a narrow endplate band of the muscle [75–77]."

"This raises the possibility that muscle electrical activity, contractile activity or both are involved in pre-patterning the muscle during NMJ development."

"These defects specifically resulted from the loss of DHPR function in the skeletal muscle, as reintroducing Cacnb1 into Cacnb1$^{-/-}$ muscles genetically rescued the muscle pre-patterning defects and restored the normal patterning and formation of the NMJ."

"Our findings suggest that a functioning skeletal muscle DHPR is specifically required for establishing muscle pre-patterning during NMJ formation."

Results

"These results indicate that muscle-specific expression of Cacnb1 is sufficient to rescue the patterning defects of the NMJ in Cacnb1$^{-/-}$ mice and that the muscle DHPR, but not the neuronal DHPR, is required for the establishment of the normal pattern of the NMJ."

"Using quantitative real-time RT-PCR, we found that Chrna1 expression was significantly increased in E14.5 Cacnb1$^{-/-}$ muscles compared with wild-type muscles (relative expression levels of Chrna1: wild type, 1.00 ± 0.20, n = 8 mice; Cacnb1$^{-/-}$, 2.82 ± 0.25, n = 8 mice; P = 0.007)."

"We therefore measured Musk expression levels in Cacnb1$^{-/-}$ muscles and found that Musk expression was significantly increased in E14.5 Cacnb1$^{-/-}$ muscles compared with E14.5 wild-type muscles (relative expression levels of Musk: wild type, 1.00 ± 0.23, n = 8 mice; Cacnb1$^{-/-}$, 2.12 ± 0.21, n = 8 mice; P = 0.006)."

Discussion

"In the absence of muscle DHPR function, AChR gene and Musk expression is upregulated and AChR and MuSK protein is broadly distributed, resulting in pre-patterning defects."

"In the absence of Cacnb1, L-type Ca^{2+} currents are absent and this negative regulation is lost, allowing increased AChR gene and Musk expression and a disruption of muscle pre-patterning [78, 79]."

"Our data indicate that the underlying mechanism that leads to an increase in AChR gene and Musk expression in Cacnb1$^{-/-}$ muscles is attributed to the loss of L-type Ca^{2+} channel activity, as application of L-type Ca^{2+} channel agonist and antagonists in wild-type muscles led to a substantial decrease and increase of AChR gene and Musk expression, respectively."

"These results suggest that the mechanism by which electrical activity suppresses AChR gene and Musk expression is that it must activate the muscle DHPRs, the L-type Ca^{2+} channel in skeletal muscles."

Methods

"Whole mounts of diaphragm and intercostal muscles were fixed in 2% paraformaldehyde (wt/vol, PFA) in 0.1 M phosphate buffer (pH 7.3) overnight at 4 °C."

"For electron microscopy, E18.5 diaphragm and intercostal muscles were immersion-fixed in freshly prepared 1% glutaraldehyde (wt/vol) and 4% PFA in 0.1 M phosphate buffer (pH 7.4)."

"Ribcages including diaphragm and intercostal muscles were fixed in freshly prepared 4% PFA in 0.1 M phosphate buffer at 4 °C overnight and then dehydrated through a series of methanol solutions (25, 50, 75 and 100%)."

"After hybridization, the samples were washed with TBS containing 1% Tween-20, blocked with 5% goat serum in dilution buffer, and incubated with alkaline phosphatase–conjugated antibody to digoxigenin (1:1,000, Boehringer Mannheim) overnight at 4 °C, and hybridization signals were detected in a staining solution containing 100 mM Tris (pH 9.5), 0.4 mg ml^{-1} nitro blue tetrazolium chloride (NBT), 0.19 mg ml^{-1} 5-bromo-4-chloro-3-indolyl phosphate (BCIP), 100 mM NaCl and 50 mM $MgCl_2$."

Bibliography

1. Arnold A-S, Gill J, Christe M, Ruiz R, McGuirk S, St-Pierre J, Tabares L, Handschin C (2014) Morphological and functional remodelling of the neuromuscular junction by skeletal muscle PGC-1α. Nat Commun 5. https://doi.org/10.1038/ncomms4569
2. Yumoto N, Kim N, Burden SJ (2012) Lrp4 is a retrograde signal for presynaptic differentiation at neuromuscular synapses. Nature 489:438–442
3. Handschin C et al (2007) PGC-1alpha regulates the neuromuscular junction program and ameliorates Duchenne muscular dystrophy. Genes Dev 21:770–783
4. Chakkalakal JV, Nishimune H, Ruas JL, Spiegelman BM, Sanes JR (2010) Retrograde influence of muscle fibers on their innervation revealed by a novel marker for slow motoneurons. Development 137:3489–3499
5. Karolczak J, Sobczak M, Majewski Ł, Yeghiazaryan M, Jakubiec-Puka A, Ehler E, Sławińska U, Wilczyński GM, Rędowicz MJ (2012) Myosin VI in skeletal muscle: its localization in the sarcoplasmic reticulum, neuromuscular junction and muscle nuclei. Histochem Cell Biol 139:873–885. https://doi.org/10.1007/s00418-012-1070-9
6. Mohiddin SA, Ahmed ZM, Griffith AJ, Tripodi D, Friedman TB, Fananapazir L, Morell RJ (2004) Novel association of hypertrophic cardiomyopathy, sensorineural deafness, and a mutation in unconventional myosin VI (MYO6). J Med Genet 41:309–314
7. Spudich G, Chibalina MV, Au JS, Arden SD, Buss F, Kendrick-Jones J (2007) Myosin VI targeting to clathrin-coated structures and dimerization is mediated by binding to disabled-2 and PtdIns(4,5)P2. Nat Cell Biol 9:176–183
8. Suzuki M, McHugh J, Tork C, Shelley B, Hayes A, Bellantuono I, Aebischer P, Svendsen CN (2008) Direct muscle delivery of GDNF with human mesenchymal stem cells improves motor neuron survival and function in a rat model of familial ALS. Mol Ther 16:2002–2010
9. Tomori K, Ohta Y, Nishizawa T, Tamaki H, Takekura H (2010) Low-intensity electrical stimulation ameliorates disruption of transverse tubules and neuromuscular junctional architecture in denervated rat skeletal muscle fibers. J Muscle Res Cell Motil 31:195–205. https://doi.org/10.1007/s10974-010-9223-8
10. Nishizawa T, Yamashia S, McGrath KF, Tamaki H, Kasuga N, Takekura H (2006) Plasticity of neuromuscular junction architectures in rat slow and fast muscle fibers following temporary denervation and reinnervation processes. J Muscle Res Cell Motil 27:607–615
11. Pellegrino C, Franzini-Armstrong C (1969) Recent contribution of electron microscope to the study of normal and pathological muscle. In: Richter GW, Epstein MA (eds) International review of experimental pathology. Academic, New York, pp 139–226
12. Adams GR (2002) Exercise effects on muscle insulin signaling and action invited review: autocrine and/or paracrine insulin-like growth factor-I activity in skeletal muscle. J Appl Physiol 93:1159–1167
13. Bannow LI, Bonaterra GA, Bertoune M, Maus S, Schulz R, Weissmann N, Kraut S, Kinscherf R, Hildebrandt W (2022) Effect of chronic intermittent hypoxia (CIH) on neuromuscular junctions and mitochondria in slow- and fast-twitch skeletal muscles of mice—The role of iNOS. Skelet Mus 12:6. https://doi.org/10.1186/s13395-022-00288-7
14. Zhan G, Fenik P, Pratico D et al (2005) Inducible nitric oxide synthase in long-term intermittent hypoxia: hypersomnolence and brain injury. Am J Respirat Crit Care Med 171(12):1414–1420
15. Kim K (2015) Interaction between HSP 70 and iNOS in skeletal muscle injury and repair. J Exerc Rehabil 11(5):240–243
16. Greenberg H, Ye X, Wilson D et al (2006) Chronic intermittent hypoxia activates nuclear factor-kappaB in cardiovascular tissues in vivo. Biochem Biophys Res Commun 343(2):591–596

17. Kostrominova TY, Pasyk KA, van Remmen H et al (2007) Adaptive changes in structure of skeletal muscles from adult Sod1 homozygous knockout mice. Cell Tissue Res 327(3):595–605
18. Sidlauskaite E, Gibson JW, Megson IL et al (2018) Mitochondrial ROS cause motor deficits induced by synaptic inactivity: implications for synapse pruning. Redox Biol 16:344–351
19. Maffiuletti NA, Roig M, Karatzanos E, Nanas S (2013) Neuromuscular electrical stimulation for preventing skeletal-muscle weakness and wasting in critically ill patients: a systematic review. BMC Med 11. https://doi.org/10.1186/1741-7015-11-137
20. Latronico N, Bolton CF (2011) Critical illness polyneuropathy and myopathy: a major cause of muscle weakness and paralysis. Lancet Neurol 10:931–941. https://doi.org/10.1016/S1474-4422(11)70178-8
21. de Jonghe B, Lacherade JC, Sharshar T, Outin H (2009) Intensive care unit-acquired weakness: risk factors and prevention. Crit Care Med 37:S309–S315
22. Gerovasili V, Tripodaki E, Karatzanos E, Pitsolis T, Markaki V, Zervakis D, Routsi C, Roussos C, Nanas S (2009) Short-term systemic effect of electrical muscle stimulation in critically iII patients. Chest 136:1249–1256. https://doi.org/10.1378/chest.08-2888
23. Gerovasili V, Stefanidis K, Vitzilaios K, Karatzanos E, Politis P, Koroneos A, Chatzimichail A, Routsi C, Roussos C, Nanas S (2009) Electrical muscle stimulation preserves the muscle mass of critically ill patients: a randomized study. Critical Care 13:R161. https://doi.org/10.1186/cc8123
24. Truong AD, Fan E, Brower RG, Needham DM (2009) Bench-to-bedside review: mobilizing patients in the intensive care unit–from pathophysiology to clinical trials. Critical Care 13:216-10.1186/cc7885
25. Laghi F, Jubran A (2011) Treating the septic muscle with electrical stimulations. Crit Care Med 39:585–586. https://doi.org/10.1097/CCM.0b013e31820e2f6f
26. Maddocks M, Gao W, Higginson IJ, Wilcock A (2011) Neuromuscular electrical stimulation for muscle weakness in adults with advanced disease. Coch Datab Syst Rev 11:CD009419
27. Abdellaoui A, Prefaut C, Gouzi F, Couillard A, Coisy-Quivy M, Hugon G, Molinari N, Lafontaine T, Jonquet O, Laoudj-Chenivesse D (2011) Skeletal muscle effects of electrostimulation after COPD exacerbation: a pilot study. Euro Respirat J 38:781–788. https://doi.org/10.1183/09031936.00167110
28. Zanotti E, Felicetti G, Maini M, Fracchia C (2003) Peripheral muscle strength training in bed-bound patients with COPD receiving mechanical ventilation: effect of electrical stimulation. Chest 124:292–296. https://doi.org/10.1378/chest.124.1.292
29. Karatzanos E, Gerovasili V, Zervakis D, Tripodaki ES, Apostolou K, Vasileiadis I, Papadopoulos E, Mitsiou G, Tsimpouki D, Routsi C (2012) Electrical muscle stimulation: an effective form of exercise and early mobilization to preserve muscle strength in critically ill patients. Crit Care Res Pract 2012:432752
30. Routsi C, Gerovasili V, Vasileiadis I, Karatzanos E, Pitsolis T, Tripodaki E, Markaki V, Zervakis D, Nanas S (2010) Electrical muscle stimulation prevents critical illness polyneuromyopathy: a randomized parallel intervention trial. Critical Care 14:R74-10.1186/cc8987
31. Gandevia SC (2001) Spinal and supraspinal factors in human muscle fatigue. Phys Rev 81:1725–1789
32. Hortobagyi T, Maffiuletti NA (2011) Neural adaptations to electrical stimulation strength training. Eur J Appl Physiol 111:2439–2449. https://doi.org/10.1007/s00421-011-2012-2
33. Yamaoka N, Shimizu K, Imaizumi Y, Ito T, Okada Y, Honda H (2019) Open-chamber co-culture microdevices for single-cell analysis of skeletal muscle myotubes and motor neurons with neuromuscular junctions. BioChip J 13:127–132. https://doi.org/10.1007/s13206-018-3202-3
34. He XP, Yang F, Xie ZP, Lu B (2000) Intracellular Ca^{2+} and Ca^{2+}/calmodulin-dependent kinase II mediate acute potentiation of neurotransmitter release by neurotrophin-3. J Cell Biol 149:783–791
35. Das M, Rumsey JW, Bhargava N, Stancescu M, Hickman JJ (2010) A defined long-term in vitro tissue engineered model of neuromuscular junctions. Biomaterials 31:4880–4888

36. Ionescu A, Zahavi EE, Gradus T, Ben-Yaakov K, Perlson E (2016) Compartmental microfluidic system for studying muscle-neuron communication and neuromuscular junction maintenance. Eur J Cell Biol 95:69–88
37. Tong Z, Seira O, Casas C, Reginensi D, Homs-Corbera A, Samitier J, Antonio Del Rio J (2014) Engineering a functional neuro-muscular junction model in a chip. RSC Adv 4:54788–54797
38. Shimizu K, Genma R, Gotou Y, Nagasaka S, Honda H (2017) Three-dimensional culture model of skeletal muscle tissue with atrophy induced by dexamethasone. Bioengineering 4
39. Greising SM, Call JA, Lund TC, Blazar BR, Tolar J, Lowe DA (2011) Skeletal muscle contractile function and neuromuscular performance in Zmpste24$^{-/-}$ mice, a murine model of human progeria. GeroScience 34:805–819. https://doi.org/10.1007/s11357-011-9281-x
40. Baltgalvis KA, Call JA, Nikas JB, Lowe DA (2009) Effects of prednisolone on skeletal muscle contractility in mdx mice. Muscle Nerve 40:443–454
41. Garlich MW, Baltgalvis KA, Call JA, Dorsey LL, Lowe DA (2010) Plantarflexion contracture in the mdx mouse. Am J Phys Med Rehabil 89(12):976–985
42. Moran AL, Warren GL, Lowe DA (2005) Soleus and EDL muscle contractility across the lifespan of female C57BL/6 mice. Experim Gerontol 40:966–975
43. Moran AL, Warren GL, Lowe DA (2006) Removal of ovarian hormones from mature mice detrimentally affects muscle contractile function and myosin structural distribution. J Appl Physiol 100:548–559
44. Cisternas P, Henriquez JP, Brandan E, Inestrosa NC (2013) Wnt signaling in skeletal muscle dynamics: myogenesis, neuromuscular synapse and fibrosis. Mol Neurobiol 49:574–589. https://doi.org/10.1007/s12035-013-8540-5
45. Burden SJ (1998) The formation of neuromuscular synapses. Genes Dev 12(2):133–148
46. Fallon JR, Gelfman CE (1989) Agrin-related molecules are concentrated at acetylcholine receptor clusters in normal and aneural developing muscle. J Cell Biol 108(4):1527–1535
47. Polesskaya A, Seale P, Rudnicki MA (2003) Wnt signaling induces the myogenic specification of resident CD45+ adult stem cells during muscle regeneration. Cell 113(7):841–852
48. von Maltzahn J, Chang NC, Bentzinger CF, Rudnicki MA (2012) Wnt signaling in myogenesis. Trends Cell Biol 22(11):602–609
49. Brack AS, Conboy IM, Conboy MJ, Shen J, Rando TA (2008) A temporal switch from notch to Wnt signaling in muscle stem cells is necessary for normal adult myogenesis. Cell Stem Cell 2(1):50–59
50. Brack AS, Conboy MJ, Roy S, Lee M, Kuo CJ, Keller C, Rando TA (2007) Increased Wnt signaling during aging alters muscle stem cell fate and increases fibrosis. Science 317(5839):807–810
51. Wang HY, Liu T, Malbon CC (2006) Structure–function analysis of Frizzleds. Cell Signal 18(7):934–941
52. Niehrs C (2012) The complex world of WNT receptor signalling. Nat Rev Mol Cell Biol 13(12):767–779
53. Munsterberg AE, Kitajewski J, Bumcrot DA, McMahon AP, Lassar AB (1995) Combinatorial signaling by sonic hedgehog and Wnt family members induces myogenic bHLH gene expression in the somite. Genes Dev 9(23):2911–2922
54. Zhang B, Liang C, Bates R, Yin Y, Xiong WC, Mei L (2012) Wnt proteins regulate acetylcholine receptor clustering in muscle cells. Mol Brain 5:7
55. Bitterman PB, Henke CA (1991) Fibroproliferative disorders. Chest 99(3 Suppl):81S–84S
56. Wang Q, Usinger W, Nichols B, Gray J, Xu L, Seeley TW, Brenner M, Guo G, Zhang W, Oliver N, Lin A, Yeowell D (2011) Cooperative interaction of CTGF and TGF-beta in animal models of fibrotic disease. Fibrogen Tissue Rep 4(1):4
57. Brandan E, Cabello-Verrugio C, Vial C (2008) Novel regulatory mechanisms for the proteoglycans decorin and biglycan during muscle formation and muscular dystrophy. Matrix Biol Intl Soc Matrix Biol 27(8):700–708
58. Morales MG, Cabello-Verrugio C, Santander C, Cabrera D, Goldschmeding R, Brandan E (2011) CTGF/CCN-2 over-expression can directly induce features of skeletal muscle dystrophy. J Pathol 225(4):490–501

59. Cabello-Verrugio C, Santander C, Cofre C, Acuna MJ, Melo F, Brandan E (2012) The internal region leucine-rich repeat 6 of decorin interacts with low density lipoprotein receptor-related protein-1, modulates transforming growth factor (TGF)-beta-dependent signaling, and inhibits TGF-beta-dependent fibrotic response in skeletal muscles. J Biol Chem 287(9):6773–6787
60. Zhu J, Li Y, Shen W, Qiao C, Ambrosio F, Lavasani M, Nozaki M, Branca MF, Huard J (2007) Relationships between transforming growth factor-beta1, myostatin, and decorin: implications for skeletal muscle fibrosis. J Biol Chem 282(35):25852–25863
61. Chen S, McLean S, Carter DE, Leask A (2007) The gene expression profile induced by Wnt 3a in NIH 3T3 fibroblasts. J Cell Commun Signal 1(3–4):175–183
62. Li J, Ito M, Ohkawara B, Masuda A, Ohno K (2018) Differential effects of spinal motor neuron-derived and skeletal muscle-derived Rspo2 on acetylcholine receptor clustering at the neuromuscular junction. Sci Rep 8. https://doi.org/10.1038/s41598-018-31949-7
63. Liu Y et al (2008) Essential roles of the acetylcholine receptor gamma-subunit in neuromuscular synaptic patterning. Development 135:1957–1967
64. Sanes JR, Lichtman JW (1999) Development of the vertebrate neuromuscular junction. Annu Rev Neurosci 22:389–442
65. Fischbach GD, Rosen KM (1997) ARIA: a neuromuscular junction neuregulin. Annu Rev Neurosci 20:429–458
66. Carraway KL, Burden SJ (1995) Neuregulins and their receptors. Curr Opin Neurobiol 5:606–612
67. Falls DL, Rosen KM, Corfas G, Lane WS, Fischbach GD (1993) ARIA, a protein that stimulates acetylcholine receptor synthesis, is a member of the neu ligand family. Cell 72:801–815
68. An MC et al (2010) Acetylcholine negatively regulates development of the neuromuscular junction through distinct cellular mechanisms. Proc Natl Acad Sci U S A 107:10702–10707
69. Lee YI et al (2016) Neuregulin1 displayed on motor axons regulates terminal Schwann cell-mediated synapse elimination at developing neuromuscular junctions. Proc Natl Acad Sci U S A 113:E479–E487
70. Wang J, Song F, Loeb JA (2017) Neuregulin1 fine-tunes pre-, post-, and perisynaptic neuromuscular junction development. Develop Dynam 246:368–380
71. Nakashima H et al (2016) R-spondin 2 promotes acetylcholine receptor clustering at the neuromuscular junction via Lgr5. Sci Rep 6:28512
72. Chen F, Liu Y, Sugiura Y, Allen PD, Gregg RG, Lin W (2011) Neuromuscular synaptic patterning requires the function of skeletal muscle dihydropyridine receptors. Nat Neurosci 14:570–577. https://doi.org/10.1038/nn.2792
73. Flanagan-Steet H, Fox MA, Meyer D, Sanes JR (2005) Neuromuscular synapses can form in vivo by incorporation of initially aneural postsynaptic specializations. Development 132:4471–4481
74. Panzer JA, Song Y, Balice-Gordon RJ (2006) In vivo imaging of preferential motor axon outgrowth to and synaptogenesis at prepatterned acetylcholine receptor clusters in embryonic zebrafish skeletal muscle. J Neurosci 26:934–947
75. Lin W et al (2001) Distinct roles of nerve and muscle in postsynaptic differentiation of the neuromuscular synapse. Nature 410:1057–1064
76. Misgeld T, Kummer TT, Lichtman JW, Sanes JR (2005) Agrin promotes synaptic differentiation by counteracting an inhibitory effect of neurotransmitter. Proc Natl Acad Sci U S A 102:11088–11093
77. Lin W et al (2005) Neurotransmitter acetylcholine negatively regulates neuromuscular synapse formation by a Cdk5-dependent mechanism. Neuron 46:569–579
78. Gregg RG et al (1996) Absence of the beta subunit (cchb1) of the skeletal muscle dihydropyridine receptor alters expression of the alpha 1 subunit and eliminates excitation-contraction coupling. Proc Natl Acad Sci USA 93:13961–13966
79. Strube C, Beurg M, Powers PA, Gregg RG, Coronado R (1996) Reduced Ca2+ current, charge movement, and absence of Ca2+ transients in skeletal muscle deficient in dihydropyridine receptor beta 1 subunit. Biophys J 71:2531–2543

80. Deschenes MR, Tenny KA, Wilson MH (2006) Increased and decreased activity elicits specific morphological adaptations of the neuromuscular junction. Neuroscience 137:1277–1283
81. Lømo T, Slater CR (1978) Control of acetylcholine sensitivity and synapse formation by muscle activity. J Physiol 275:391–402
82. Brown MC, Holland RL (1979) A central role for denervated tissues in causing nerve sprouting. Nature 282:724–726
83. Rodriguez PO, Setten M, Maskin LP, Bonelli I, Vidomlansky SR, Attie S, Frosiani SL, Kozima S, Valentini R (2012) Muscle weakness in septic patients requiring mechanical ventilation: protective effect of transcutaneous neuromuscular electrical stimulation. J Crit Care 27:e311–e318
84. Bergo MO, Gavino B, Ross J, Schmidt WK, Hong C, Kendall LV, Mohr A, Meta M, Genant H, Jiang Y, Wisner ER, Van Bruggen N, Carano RA, Michaelis S, Griffey SM, Young SG (2002) Zmpste24 deficiency in mice causes spontaneous bone fractures, muscle weakness, and a prelamin A processing defect. Proc Natl Acad Sci U S A 99:13049–13054

Skeletal Muscle Receptors

4

Christopher Myers

4.1 Introduction by the Editor

This amazing chapter takes a deep dive into the realm of skeletal muscle receptors, particularly emphasizing the Vitamin D receptor (VDR) and the Ryanodine receptor (RyR). The crucial roles these receptors play in skeletal muscle physiology have recently been the subject of numerous studies, and this chapter strives to consolidate these new insights.

One significant area of focus in this chapter is the role of the VDR, the primary mediator of the biological effects of Vitamin D in human physiology. The VDR is highly expressed in skeletal muscles, where it modulates muscle growth and function. Recent research has explored the intriguing link between VDR, muscle mass, and fracture risk in patients. Studies have demonstrated that serum vitamin D levels correlate with muscle mass and strength, and deficiencies in vitamin D or mutations in the VDR gene can lead to muscle weakness and an increased risk of fractures, especially in the distal forearm.

An integral part of this discussion is the role of VDR agonists in promoting muscle health and potentially reducing the risk of fractures. We'll explore the latest studies on how these VDR activators might enhance muscle mass and function, especially in patients with Vitamin D deficiency.

The chapter also highlights the RyR, a calcium release channel on the sarcoplasmic reticulum of skeletal muscle. Changes in RyR function have been linked to muscle fatigue, and recent findings suggest that RyR dysfunction may contribute to muscle disorders. This chapter looks at new research on the SPRY domains and their responsibility to RYR development. This discussion provides a new dimension to understanding intracellular calcium dynamics and muscle contraction. In

C. Myers (✉)
3 Sports Science, 3 Sports Science LLC, Akron, OH, United States of America
e-mail: chris@3sportsscience.com

C. Myers (ed.), *Skeletal Muscle Physiology*,
https://doi.org/10.1007/978-3-031-47065-3_4

conjunction with exploring individual receptors, the chapter investigates the complex interplay between these receptors, particularly the VDR and RyR, in muscle physiology. Emerging research suggests a nuanced crosstalk between these receptors, potentially affecting muscle health and disease.

In summary, this chapter explores the novel research shedding light on the roles of the VDR and RyR in skeletal muscle and other crucial skeletal muscle receptors. Through this exploration, the chapter highlights the significance of these receptors in muscle function, the potential impact of their dysregulation on patient health, and the prospective therapeutic possibilities presented by receptor agonists. The exciting discoveries in this field are redefining our understanding of skeletal muscle and opening new avenues for maintaining muscle health and treating muscle disorders.

In the dynamic realm of skeletal muscle physiology, the neuromuscular junction (NMJ) emerges as a focal point of interaction between neural impulses and muscular responses, orchestrating the delicate ballet of muscle contraction and relaxation. Anchoring this intricate dance are the molecular maestros: β-arrestin 1, the CB1 cannabinoid receptor, p43—a truncated thyroid hormone receptor α, and the Vitamin D Receptor (VDR), each playing a pivotal role in modulating skeletal muscle hypertrophy, contractility, lipid composition, metabolism, and autophagy.

β-arrestin 1, a multifunctional adaptor protein, is essential in regulating β2-adrenergic receptor-mediated skeletal muscle growth and contractility. Activating the β2-adrenergic receptor through clenbuterol administration highlights β-arrestin 1's pivotal role as a critical mediator in this process, introducing a new dimension to anabolic skeletal muscle growth. New research highlighted in this chapter demonstrates this signaling pathway bolsters the contractile force within the extensor digitorum longus muscle and fine-tunes calcium signaling within isolated flexor digitorum brevis fibers. Such modulation could emphasize the critical function of β-arrestin 1 in steering these physiological outcomes. Researchers have identified that clenbuterol's interaction with the β2-adrenergic receptor sets off a series of biochemical events leading to enhanced muscle performance. β-arrestin 1 emerges as a crucial intermediary in this signaling cascade, orchestrating the response to clenbuterol stimulation. This interaction between clenbuterol and the β2-adrenergic receptor, facilitated by β-arrestin 1, signifies a groundbreaking advancement in our understanding of muscle physiology. The increased contractility observed in the extensor digitorum longus muscle post-clenbuterol administration underscores the adaptability and responsiveness of skeletal muscle to β-arrestin one mediated signaling.

Furthermore, the research highlighted in this chapter suggests the adjustments in calcium signaling within the flexor digitorum brevis fibers illustrate the intricate regulatory mechanisms at play. This discovery of β-arrestin 1's involvement offers new insights into muscle growth and functionality. Consequently, β-arrestin 1's role in these processes illuminates potential therapeutic targets for enhancing muscle strength and addressing muscular disorders. The understanding of β-arrestin 1's roles is rudimentary at best. More research is needed to understand the underlying mechanisms at play.

The CB1 cannabinoid receptor, traditionally recognized for its roles in neural tissues, extends its influence on the regulation of autophagy in skeletal muscle. The manipulation of this receptor in the tibialis anterior skeletal muscle of mice delineates a complex interplay between cannabinoid signaling and insulin pathways, mainly through the AKT-GSK axis. The conclusions from this research suggest this crosstalk highlights the receptor's role in maintaining muscle homeostasis and presents a novel target for modulating muscle metabolism in conditions such as obesity and diabetes mellitus.

P43, a truncated variant of the thyroid hormone receptor α, intricately weaves its effects on the lipid composition and metabolism within skeletal muscle fibers. The expression of p43 in muscle tissue significantly impacts the muscle's lipid profile, affecting triglyceride levels, monounsaturated fatty acids synthesis, and the unsaturation index of fatty acids. The animal model research highlighted in this chapter highlights this modulation of lipid metabolism by p43, emphasizing its potential in regulating muscle energy utilization and oxidative capacity, which could be pivotal for muscle endurance and performance in humans.

The expression of the VDR in skeletal muscle establishes itself as a pivotal factor in determining muscle mass and its functional capabilities. This aspect is especially evident in conditions such as sarcopenia and muscle repair after an injury. The complex interplay between vitamin D signaling and the expression of VDR within muscle cells profoundly influences the size of muscle fibers. Additionally, VDR affects the transcriptional regulation of vital myogenic and metabolic genes essential for muscle growth and maintenance. Researchers have noted a distinct variation in VDR expression when comparing sarcopenic individuals to their non-sarcopenic counterparts, as noted in this chapter. This variation directly correlates with differences in muscle cross-sectional area and the expression of metabolic genes, highlighting the significant role of vitamin D and its receptor in both muscle health and disease. The differential expression of VDR underscores the hormone's critical involvement in muscle physiology, offering insights into potential therapeutic targets for sarcopenia and muscle recovery strategies.

Furthermore, the relationship between VDR expression and muscle fiber size suggests a link between vitamin D pathways and muscle structural integrity. In sarcopenic patients, research demonstrated the reduced expression of VDR points to a possible disruption in vitamin D-mediated muscle maintenance mechanisms. Conversely, higher levels of VDR expression indicate an optimal functioning of vitamin D signaling pathways essential for muscle preservation in individuals without sarcopenia. This discovery opens new avenues for exploring vitamin D's therapeutic potential in mitigating muscle loss and promoting muscle repair. Therefore, VDR's role in skeletal muscle represents a critical nexus in understanding and addressing muscle-related conditions, emphasizing the importance of maintaining a balanced vitamin D signaling system for muscle health.

These molecular players—β-arrestin 1, the CB1 receptor, p43, and VDR—form a complex network that governs the multifaceted aspects of skeletal muscle function. From hypertrophy and contractility to lipid metabolism, autophagy, and the hormonal regulation of muscle mass, elucidating these pathways offers profound insights into muscle biology. Moreover, the therapeutic potential of targeting these molecules opens new avenues for treating muscle-related diseases, offering hope for interventions that could enhance muscle performance, combat sarcopenia, and facilitate recovery from musculoskeletal injuries. Exploring these molecular dimensions enriches our understanding of muscle physiology and underscores the potential for innovative strategies to optimize muscle health and function.

serum vitamin, index, muscle mass, agonist, receptor vdr, woman, forearm, ryr, adenosine

4.2 Machine Generated Summaries

Disclaimer: The summaries in this chapter were generated from Springer Nature publications using extractive AI auto-summarization: An extraction-based summarizer aims to identify the most important sentences of a text using an algorithm and uses those original sentences to create the auto-summary (unlike generative AI). As the constituted sentences are machine selected, they may not fully reflect the body of the work, so we strongly advise that the original content is read and cited. The auto generated summaries were curated by the editor to meet Springer Nature publication standards. To cite this content, please refer to the original papers.

Machine generated keywords: vdr, vitamin, fracture, vitamin receptor, distal, patient, serum vitamin, index, muscle mass, agonist, receptor vdr, woman, forearm, ryr, adenosine

β-Arrestin 1 Regulates β2-Adrenergic Receptor-Mediated Skeletal Muscle Hypertrophy and Contractility [1]

This is a machine-generated summary of:

Kim, Jihee; Grotegut, Chad A.; Wisler, James W.; Li, Tianyu; Mao, Lan; Chen, Minyong; Chen, Wei; Rosenberg, Paul B.; Rockman, Howard A.; Lefkowitz, Robert J.: β-arrestin 1 regulates β2-adrenergic receptor-mediated skeletal muscle hypertrophy and contractility [1]

Published in: Skeletal Muscle (2018)

Link to original: https://doi.org/10.1186/s13395-018-0184-8

If you want to cite the papers, please refer to the original.

For technical reasons we could not place the page where the original quote is coming from.

Abstract-Summary "An important action of β_2AR stimulation on skeletal muscle is anabolic growth, which has led to the use of agonists such as clenbuterol by athletes to enhance muscle performance."

"While previous work has demonstrated that β_2ARs can engage distinct signaling and functional cascades mediated by either G proteins or the multifunctional adaptor protein, β-arrestin, the precise role of β-arrestin in skeletal muscle physiology is not known."

"We tested the hypothesis that agonist activation of the β_2AR by clenbuterol would engage β-arrestin as a key transducer of anabolic skeletal muscle growth."

"The contractile force of isolated extensor digitorum longus muscle (EDL) and calcium signaling in isolated flexor digitorum brevis (FDB) fibers were examined from the wild-type (WT) and β-arrestin 1 knockout mice (βarr1KO) followed by chronic administration of clenbuterol (1 mg/kg/d)."

"We performed a targeted phosphoproteomic analysis on clenbuterol stimulated primary cultured myoblasts from WT and βarr1KO mice."

"Chronic administration of clenbuterol to WT mice enhanced the contractile force of EDL muscle and calcium signaling in isolated FDB fibers."

"While clenbuterol-induced hypertrophic responses were observed in WT mice, this response was abrogated in mice lacking β-arrestin 1."

"We have identified a previously unappreciated role for β-arrestin 1 in mediating β_2AR-stimulated skeletal muscle growth and strength."

Backgrounds

"B-adrenergic receptors (βARs) play important roles in cardiovascular, pulmonary, and skeletal muscle physiology [2, 3]."

"The hypertrophic effect of β_2AR-mediated signaling in skeletal muscle is associated with enhanced gene transcription and protein translation that lead to increased expression of skeletal muscle contractile proteins and decreased myofibrillar proteolysis [4]."

"Based on the regulatory roles of β-arrestins in GPCR signaling, we hypothesized that the enhancement of skeletal muscle function by β_2AR stimulation might be mediated by β-arrestin 1, which is the dominant β-arrestin isoform in skeletal muscle."

"We investigated the contractile properties of skeletal muscle fibers after chronic administration of the selective β_2AR agonist clenbuterol in wild-type (WT) and β-arrestin 1 knockout (βarr1KO) mice."

"By assessing skeletal muscle fiber size, muscle composition, and contractile properties, we identified a fundamental role for β-arrestin 1 in β_2AR-mediated skeletal muscle hypertrophy and further elucidated the molecular mechanism for the action of clenbuterol in isolated myoblasts."

Methods

"ES cell targeting and generation of chimeric mice with arrb1 loxP allele were performed by the Duke Transgenic Mouse Facility."

"Positive ES clones were injected into mouse blastocysts to produce chimeric mice, which were then crossed with C57Bl/6J mice (The Jackson Laboratory, Bar Harbor, ME, USA) to allow germline transmission and produce heterozygote mice harboring arrb1 loxP/FRT-Neo-FRT allele."

"Excision of the FRT-Neo-FRT cassette done by crossing the heterozygote mice with a transgenic mouse expressing FLPe recombinase (B6SJLTg (ACTFLPe) 9205Dym/J; The Jackson Laboratory, Bar Harbor, ME, USA) led to the establishment of heterozygote mice harboring arrb1-loxP allele (arrb1wt/flox)."

"The stimulus-response curve was generated by applying such stimuli at a range of stimulus durations (100 ms to 5 s), with a single test stimulus applied every 50 s. Curves were compared using a two-way repeated measures ANOVA with Sidak's multiple comparison test."

Results

"These data indicate that β-arrestin 1 is essential in mediating the β_2AR-stimulated enhancement in contractile properties of skeletal muscle as assessed by twitch, tetanus, force-frequency, and fatigue parameters."

"Due to these distinct characteristics, we examined the role of β-arrestin 1 in the β_2AR-mediated hypertrophic response of these three muscles after chronic administration of clenbuterol (1 mg/kg/day) or vehicle for 4 weeks."

"We observed similar clenbuterol-induced percentage increases in IIb fibers in all tissue types in the βarr1KO mice as well suggesting β-arrestin 1 does not play a significant role in clenbuterol-mediated increases in IIb muscle fibers."

"These data show that β-arrestin 1 differentially modulates muscle cell growth in response to β_2AR stimulation with little effect on fiber type remodeling as assessed by fiber composition."

"To address the potential cellular mechanisms for β-arrestin 1-mediated signaling in skeletal muscle, we performed a targeted phosphoproteomic analysis on clenbuterol stimulated primary cultured myoblasts from WT and βarr1KO mice."

Discussion

"The increase in skeletal muscle fiber size with chronic clenbuterol-induced β_2AR stimulation was β-arrestin 1-dependent indicating that β-arrestin 1 is a critical signaling molecule regulating skeletal muscle growth and function downstream of the β_2AR."

"Our findings support the concept that enhancing β-arrestin 1-dependent, β_2AR-mediated signaling pathways may be useful in stimulating muscle growth in conditions of muscle wasting including the elderly and patients suffering from chronic illness such as cancer or heart failure."

"Our data show that clenbuterol stimulated Ca^{2+} transients from βarr1KO fibers were markedly blunted indicating a dependence on β-arrestin 1 signaling for excitation-contraction (EC) coupling in skeletal muscle."

"Since the roles of β-arrestin 1 in regulating β_2AR signaling differ among pathological conditions, a better understanding of the mechanisms by which β_2AR activation leads to enhanced muscle performance may provide a platform to discover drugs that selectively activate β-arrestin 1 signaling as therapeutic candidates for pathological conditions associated with skeletal muscle wasting."

Conclusions

"We have identified a previously unappreciated role for β-arrestin 1 in mediating β_2AR-stimulated skeletal muscle growth and strength."

"We propose these findings could have important implications in the design of future pharmacologic agents aimed at reversing pathological conditions associated with skeletal muscle wasting."

Ubiquitous SPRY Domains and Their Role in the Skeletal Type Ryanodine Receptor [5]

This is a machine-generated summary of:

Tae, HanShen; Casarotto, Marco G.; Dulhunty, Angela Fay: Ubiquitous SPRY domains and their role in the skeletal type ryanodine receptor [5]

Published in: European Biophysics Journal (2009)

Link to original: https://doi.org/10.1007/s00249-009-0455-8

If you want to cite the papers, please refer to the original.

For technical reasons we could not place the page where the original quote is coming from.

Abstract-Summary "We recently identified the second of three SPRY domains in the skeletal muscle ryanodine receptor type 1 (RyR1) as a potential binding partner in the RyR1 ion channel for the recombinant II–III loop of the skeletal muscle dihydropyridine receptor, for a scorpion toxin, Imperatoxin A and for an interdomain interaction within RyR1."

"We review the general characteristics of a range of SPRY domains and discuss evidence that the SPRY2 domain in RyR1 supports interactions with binding partners that contain a structural surface of aligned basic residues."

Introduction

"The membrane spanning $\alpha 1_S$ subunit of the DHPR is a voltage sensor, and it is thought to communicate the action potential to the RyR via a conformational coupling process between the intracellular II–III loop of the DHPR and RyR1."

"Despite extensive searches the specific binding residues on the RyR for the essential C region of the II–III loop have not been located."

"The SPRY domains that are highly conserved across RyR isoforms had not been assigned any specific function until we found that the second of the three SPRY domains (SPRY2) is an in vitro binding partner for the II–III loop from $\alpha 1_S$ subunit of the skeletal muscle dihydropryidine receptor (DHPR)."

"We found that the recombinant SPRY2 domain binds to the DHPR II–III loop through its N-terminal A region."

"We review the structure and function of SPRY domains in general and consider their function in RyR channels."

SPRY and B30.2 Domains

"The B30.2 domain contains a SPRY subdomain structure."

"The B30.2 domain consists of PRY and SPRY subdomains."

"B30.2 domains are ~170 residues long while SPRY2 domains contain ~100 residues."

"To the preferred C-terminal location of the B30.2 domain, SPRY domains have no positional preference and are found at both the N- and C-termini of the proteins as well as in central locations [6–9]."

"The similarities in the primary sequence of SPRY and B30.2 domains led to the proposal that they share a common evolutionary relationship."

"SPRY domains are highly conserved across all taxa whereas the B30.2 domains have so far only been discovered in vertebrate species with an adaptive immune system."

"SPRY domains are more ancient than B30.2 domains which evolved recently and incorporated the PRY domain into the N-terminal end of the SPRY domain."

Disease-Causing Point Mutations in the SPRY Domain

"The functional importance of SPRY domains is indicated by diseases caused by point mutations in their sequences."

"Most of the mutations are located in the SPRY domain and are thought to alter pyrin protein function."

"The Opitz syndrome (OS) is the result of mutations in the MID1 gene, which encodes a 72-kDa MID1 protein with several conserved functional domains."

"Mutations causing OS are located in the C-terminal SPRY domain of the MID1 protein [10–12]."

"It is proposed that the OS mutations dissociate the MID1 protein from the microtubule complex and remove its regulatory role on ventral midline development [13–16]."

The Structure of the B30.2/SPRY Domain

"The differences between the unstructured regions endow unique properties on SPRY domains in different SPRY proteins which allow interactions with specific protein binding partners."

"The structures of the B30.2/SPRY domain from two members of the SSB protein family and TRIM21 are described below."

"The SSB-2 SPRY domain has been suggested to interact directly with c-MET, which has diverse biological roles."

"Four members of the SOCS box (SSB) protein family (SSB-1 to -4) contain a central conserved SPRY domain and a C-terminal SOCS box motif."

"The SSB-2 SPRY domain is thought to interact directly with c-MET, which has diverse biological roles."

"GUSTAVUS also belongs to the family of SPRY domain-containing SOCS box proteins [17–19], and is the only SSB protein found in D. melanogaster [20]."

"Several studies suggest that SPRY domains function as protein-interaction modules."

"The B30.2/SPRY domain of all SSB proteins interacts with c-Met, a hepatocyte growth factor receptor [9]."

The SPRY Domain of the Skeletal RyR1

"Both RyR1 fragments contain the second repeat of the RyR1 SPRY domain (SPRY2) in residues Ser^{1085}-Val^{1208}; thus it is possible that the II–III loop interacts with RyR1 via this domain."

"Our preliminary data suggests that the ASI region and the A region of the II–III loop bind to the same acidic residues in the SPRY2 domain (Tae and others, unpublished observations)."

"Other preliminary data show that mutation of the residues in SPRY2 that bind to peptide A and the ASI region does not alter EC coupling in RyR1-null myotubes expressing the RyR1 with mutations in key acidic residues the SPRY2 domain (Wei and others, unpublished observations)."

"The lack of an effect of mutation of these residues on EC coupling suggests that if II–III loop and the ASI region do bind to the SPRY2 domain in vivo, this binding is not involved in the role of the two domains in EC coupling."

Conclusion

"The critical interaction interface of the SPRY domain is usually located on an unstructured peripheral loop."

"Our homology model of the SPRY2 domain in RyR1 predicts a high β-sheet content and the presence of a distinct acidic loop that is likely to be involved in its interaction with basic residues in other protein domains."

"This indicates that interactions involving the RyR1 SPRY2 domain are dependent on the charge and the alignment of positive residues in its binding partners."

"These findings provide evidence that SPRY2 domain functions as a protein-interaction module in the RyR and suggest a general role for SPRY domains in maintaining RyR dependent Ca^{2+} signalling."

The CB1 Cannabinoid Receptor Regulates Autophagy in the Tibialis Anterior Skeletal Muscle in Mice [21]

This is a machine-generated summary of:

Sepúlveda, Carlos; Rodríguez, Juan Manuel; Monsalves-Álvarez, Matías; Donoso-Barraza, Camila; Pino-de la Fuente, Francisco; Matías, Isabelle; Leste-Lasserre, Thierry; Zizzari, Philippe; Morselli, Eugenia; Cota, Daniela; Llanos, Miguel; Troncoso, Rodrigo: The CB1 cannabinoid receptor regulates autophagy in the tibialis anterior skeletal muscle in mice [21]

Published in: Biological Research (2023)

Link to original: https://doi.org/10.1186/s40659-023-00426-5

If you want to cite the papers, please refer to the original.

For technical reasons we could not place the page where the original quote is coming from.

Abstract-Summary "That the CB1 regulates signaling pathways that control autophagy, however, it is currently unknown whether the ECS could regulate autophagy in the skeletal muscle of obese mice."

"This study aimed to investigate the role of the CB1 in regulating autophagy in skeletal muscle."

"We found concomitant deregulation in the ECS and autophagy markers in high-fat diet-induced obesity."

"In obese CB1-KO mice, the autophagy-associated protein LC3 II does not accumulate when mTOR and AMPK phosphorylation levels do not change."

"Our results suggest that the CB1 regulates autophagy in the tibialis anterior skeletal muscle in both lean and obese mice."

Background

"Treatment with a CB1 agonist increases de novo fatty acid synthesis in the liver or isolated hepatocytes, thus contributing to diet-induced obesity and strongly suggesting a role in the progress of obesity and its comorbidities [22]."

"Although it is known that the CB1 participates in autophagy regulation in neurons, it is unclear if the CB1 can modulate autophagy by canonical pathways in vivo in the tibialis anterior muscle [23, 24]."

"Many studies have investigated the role of the ECS in regulating energy balance and metabolism in the nervous system and peripheral organs, however, very few have examined the physiological role in skeletal muscle and, specially, the control of relevant biological processes, such as autophagy [25–27]."

"This study aims to investigate the role of the CB1 in regulating autophagy in the tibialis anterior muscle."

"We show evidence that the CB1 regulates basal autophagy in the tibialis anterior muscle in normal and diet-induced obese mice."

Results

"To evaluate the effect of diet-induced obesity on autophagy markers, we then measured the protein levels of LC3 and p62/Sqstm1 (markers) in TA muscle."

"These results suggest that diet-induced obesity increases ECS activity and LC3 II protein levels."

"Since we observed concomitant deregulation in the ECS and basal autophagy markers, we wanted to determine if the deletion of the CB1 prevented LC3 II accumulation induced by a HFD in the TA."

"To investigate whether CB1 knockout prevented LC3 II accumulation induced by a high-fat diet, we determined autophagy-related proteins levels."

"Regarding LC3 proteins, JD-5037 produced an effect ($p = 0.0002$), suggesting that pharmacological blockade of CB1 prevents chloroquine-induced changes in autophagic flux."

"These results suggest that while genetic deletion of CB1 does not regulate autophagic flux, acute pharmacological inhibition of the CB1 with JD-5037 decreases LC3 II protein accumulation and autophagic flux."

Discussion

"We observed that the deletion of the CB1 in mice fed with a HFD prevented LC3 II accumulation, supporting the role of CB1 in favoring the increase of LC3 II observed in HFD."

"While genetic deletion of CB1 regulated autophagy, acute pharmacological inhibition of the CB1 with JD-5037 decreased LC3 II protein accumulation and autophagic flux."

"In agreement, Li and others showed that a HFD inhibits the production of autophagic lysosomal expression of autophagy-related genes (LC3 II, Becn1, and ATG5) and increases the accumulation of p62/Sqstm1 in the gastrocnemius muscle [28]."

"Mice treated only with a CB1 antagonist, JD-5037, reduced LC3 II basal levels."

"Our data suggest that CB1-KO prevents LC3 II accumulation, indicating restoration of the autophagy in mice fed a HFD."

"They performed an autophagic flux assay in human embryonic kidney cells, showing LC3 II accumulation, suggesting a possible role of the CB1 in autophagy regulation."

Conclusions

"Our results also indicate a regulatory role of the CB1 on fast muscle, such as the tibialis anterior in mice, reducing LC3 II accumulation induced by a HFD."

"Acute pharmacological inhibition of the CB1 reduces LC3 II accumulation induced by chloroquine, suggesting a reduction of autophagic flux."

"CB1 regulates autophagy in the tibialis anterior muscle in mice."

Methods

"Mice were fed with a control diet (CD, n = 6) containing 70% carbohydrates, 10% fat, and 20% protein in terms of kcals (Research Diet Inc. D12450J, New Brunswick, NJ, USA) and a high-fat diet (HFD, n = 6) containing 60% fat (90% lard, 10% soybean oil), 20% carbohydrates and 20% protein in terms of kcal (D12492, Research Diets, New Brunswick, NJ, USA)."

"For the second experiment, male C57BL/6 $CB1^{+/+}$ (WT) and $CB1^{-/-}$ (CB1-KO) littermate mice were maintained under standard conditions (food and tap water ad libitum; 12 h–12 h light–dark cycle)."

"Glycemia was measured in mice fasting for 6 h. Finally, the mice were sacrificed the next day after measuring the glycemia, and the TA was collected and frozen (−80 °C)."

"Mice were kept in fasting condition for 24 h. 4 h before sacrifice, mice were injected with 100 mg kg body weight of chloroquine or PBS."

A longitudinal Histologic Evaluation of Vitamin D Receptor Expression in the Skeletal Muscles of Patients with a Distal Radius Fracture [29]

This is a machine-generated summary of:

Shim, B. J.; Lee, M. H.; Lim, J. Y.; Gong, H. S.: A longitudinal histologic evaluation of vitamin D receptor expression in the skeletal muscles of patients with a distal radius fracture [29]

Published in: Osteoporosis International (2021)

Link to original: https://doi.org/10.1007/s00198-020-05809-y

If you want to cite the papers, please refer to the original.

For technical reasons we could not place the page where the original quote is coming from.

Abstract-Summary "We investigated the effect of vitamin D supplementation on the expression of muscle vitamin D receptor (VDR) and cross-sectional area (CSA) in patients with a distal radius fracture (DRF)."

"Significant increases in VDR expression and CSA were observed, especially in vitamin D-deficient patients."

"Vitamin D supplementation is known to enhance muscle mass and function, but whether the VDR is essential in this process remains unknown."

"We evaluated the change in VDR expression and CSA in the forearm muscles following vitamin D supplementation in patients with a DRF."

"The median VDR expression increased from 0.72 to 0.78 ($P = 0.002$), and the median CSA increased from 1290.0 to 1685.8 μm^2 ($P = 0.022$)."

"Significant increases in VDR expression and CSA were observed in vitamin D-deficient patients [25(OH)D] < 20 ng/mL, but not in vitamin D-non-deficient patients."

"Vitamin D supplementation may increase muscle VDR expression and CSA in patients with a DRF, especially in vitamin D-deficient patients."

"The increase in CSA without an increase in VDR expression in some patients indicates that the effect of vitamin D supplementation on muscle mass could be mediated by indirect effect of serum vitamin D restoration and by VDR."

Introduction

"VDR regulates the genomic and non-genomic pathways of vitamin D, influencing the muscle either directly or indirectly [30, 31]."

"Ceglia and others revealed that vitamin D supplementation increased both the intramyonuclear VDR concentration and muscle fiber size in elderly subjects with limited mobility and vitamin D deficiency [32]."

"These studies were limited to the vitamin D-deficient status, and the role of VDR in the increase in muscle mass remained uncertain."

"The purpose of this study was to evaluate the change in VDR expression and cross-sectional area (CSA) of the forearm flexor muscles after vitamin D supplementation in patients with a DRF."

Materials and Methods

"The patients who wanted hardware removal were enrolled in this study because the study needed follow-up muscle biopsy."

"Serum 25(OH)D levels were measured in all patients before the first and second surgeries via Diels–Alder derivatization using ultrahigh-performance liquid chromatography–tandem mass spectrometry (Waters Corp, Milford, MA, USA), which can be the most accurate and precise method."

"In a little more detail, the day before surgery, serum 25(OH)D laboratory test was performed together with the preoperative examination for an anesthesia."

"Vitamin D levels were measured again at the time of the second surgery."

"We compared the pre- and postoperative parameters that included serum vitamin D levels, VDR expression, muscle CSA, and grip strength using the Wilcoxon signed-rank tests."

"The participants were classified according to the serum 25OHD level (cutoff value of 20 ng/mL) , and we compared the vitamin D-deficient and non-deficient group using the Mann–Whitney U test and Wilcoxon signed-rank tests [33]."

Results

"The median serum vitamin D levels increased from 14.3 ng/mL (IQR, 11.7–25.2) to 32.1 ng/mL (IQR, 29.1–37.0 ng/mL) ($P = 0.001$)."

"VDR expression increased from 0.72 (IQR, 0.52–0.81) to 0.78 (IQR, 0.63–0.81); $P = 0.002$."

"The non-deficient group had a higher baseline VDR expression than the deficient group ($P = 0.032$)."

"In the deficient group, significant increases in VDR expression and muscle CSA were observed ($P = 0.005$ and $P = 0.023$, respectively)."

"VDR expression and CSA moved in the same direction in 13 patients, in which both increased in 11 patients and decreased in 2."

"The changes in VDR expression and CSA moved in opposite directions in the remaining five patients."

Discussion

"We evaluated the changes in VDR expression and CSA of the forearm flexor muscles after vitamin D supplementation in patients with a DRF."

"We found that vitamin D supplementation has a significant positive effect on the forearm muscle VDR expression and CSA, especially in vitamin D-deficient patients."

"Both the VDR expression and CSA significantly increased with vitamin D supplementation in patients with a DRF."

"The increases in VDR expression and muscle CSA were significant in vitamin D-deficient patients, whereas these changes were not significant in vitamin D-non-deficient patients."

"Our study suggests that vitamin D supplementation is not effective for increasing the muscle size or VDR expression in vitamin D-non-deficient patients."

"Vitamin D may have affected the muscles both directly via VDR and indirectly via other pathways."

Conclusion

"The vitamin D supplementation may increase the forearm muscle VDR expression and CSA in patients with a DRF, especially in vitamin D-deficient patients."

"Although the changes in VDR expression and CSA were in the same direction in most patients, the directionality was opposite in some patients, suggesting that the effect of vitamin D supplementation on muscle mass is mediated not only by VDR but also by an indirect effect of serum vitamin D restoration."

Skeletal Muscle Expression of p43, a Truncated Thyroid Hormone Receptor α, Affects Lipid Composition and Metabolism [34]

This is a machine-generated summary of:

Casas, François; Fouret, Gilles; Lecomte, Jérome; Cortade, Fabienne; Pessemesse, Laurence; Blanchet, Emilie; Wrutniak-Cabello, Chantal; Coudray, Charles; Feillet-Coudray, Christine: Skeletal muscle expression of p43, a truncated thyroid hormone receptor α, affects lipid composition and metabolism [34]

Published in: Journal of Bioenergetics and Biomembranes (2018)

Link to original: https://doi.org/10.1007/s10863-018-9743-2

If you want to cite the papers, please refer to the original.

For technical reasons we could not place the page where the original quote is coming from.

Abstract-Summary "Using mouse models overexpressing p43 in skeletal muscle (p43-Tg) or lacking p43 (p43−/−), we have investigated the lipid composition in quadriceps muscle and in mitochondria."

"We reported in the quadriceps muscle of p43−/− mice, a fall in triglycerides, an inhibition of monounsaturated fatty acids (MUFA) synthesis, an increase in elongase index and an decrease in desaturase index."

"In the quadriceps muscle of p43-Tg mice, MUFA content was decreased whereas the unsaturation index was increased."

"In quadriceps mitochondria of p43-Tg mice, we found an increase of linoleic acid level and unsaturation index."

"We showed that cardiolipin content, a key phospholipid for mitochondrial function, remained unchanged both in quadriceps muscle and in its mitochondria whatever the mice genotype."

"This study shows that muscle lipid content and fatty acid profile are strongly affected in skeletal muscle by p43 levels."

Introduction

"Thyroid hormone (TH) is a major regulator of metabolism and mitochondrial function [35, 36]."

"Hypothyroid state in rat induced a decrease in myocardial CL content associated with a reduction in cytochrome c oxidase activity [37]."

"Hyperthyroid state in rats increases CL content and cytochrome c oxidase activity [38]."

"We demonstrated that the specific overexpression of p43 in skeletal muscle (p43-Tg) increases mitochondrial respiration and induces a shift in the metabolic and contractile features of muscle fibers toward a slower and more oxidative phenotype [39]."

"P43 depletion in mice (p43−/−) reduces mitochondrial respiratory chain activities and induces a shift toward a faster and more glycolytic muscle fiber phenotype [40]."

"This set of data indicates that p43 regulates the metabolic and contractile features of myofibers through the modulation of mitochondrial activity."

Materials and Methods

"The lipids Folch extracts of whole quadriceps muscle and of quadriceps muscle mitochondria were applicated on silica gel 60 HPTLC plates (250 μm, 20*10 cm, Merck, Germany) pretreated with 1.5% w/v boric acid in ethanol (100%), was automatically performed on a 4 mm band width using a CAMAG ATS4 apparatus (Muttenz, Switzerland) [41]."

"The lipids Folch extracts providing from whole quadriceps muscle and from quadriceps muscle mitochondria were applicated on silica gel 60 HPTLC plates (250 μm, 20*10 cm, Merck, Germany) pretreated with 2.3% w/v boric acid in ethanol (100%), was automatically performed on a 4 mm band width using a CAMAG ATS4 apparatus (Muttenz, Switzerland) [42]."

Results

"Of these modifications, the unsaturation index was significantly increased in p43−/− animals by comparison to WT mice (+10%)."

"The unsaturation index was significantly increased in p43-Tg mice by comparison to WT animals (+8%)."

"In mitochondria from p43−/− mice, fatty acid profile was barely modified by comparison to WT animals."

"Δ9-desaturase [16:1n-7/16:0] and elongase indices [18:1n-7/16:1n-7] were modified (−23% and +24%, respectively) by comparison to WT mice."

Discussion

"This fall of TG (−62%) observed in the skeletal muscle of p43−/− mice is consistent with our previous results reporting that p43 depletion in mice induces a reduction in mitochondrial respiratory chain activities and a shift toward a faster and more glycolytic muscle fiber phenotype [40]."

"In mice lacking p43, the fall in TG, the inhibition of MUFA synthesis, the increase in elongase index and the decrease in desaturase index observed in the quadriceps muscle, are fully consistent with our previous works."

"In mice overexpressing p43 specifically in skeletal muscle, the increase of linoleic acid and unsaturation index found at the mitochondrial level are consistent with our previous studies demonstrating that p43 overexpressing in mice increase respiratory chain activities."

The Vitamin D Receptor Expression in Skeletal Muscle of Women with Distal Radius Fracture [43]

This is a machine-generated summary of:

Kim, Kahyun; Gong, Hyun Sik; Lim, Jae-Young; Kim, Jong Hee; Baek, Goo Hyun: The vitamin D receptor expression in skeletal muscle of women with distal radius fracture [43]

Published in: Archives of Osteoporosis (2018)

Link to original: https://doi.org/10.1007/s11657-018-0442-8

If you want to cite the papers, please refer to the original.

For technical reasons we could not place the page where the original quote is coming from.

Abstract-Summary "We evaluated the vitamin D receptor (VDR) expression in the forearm flexor muscle of women with distal radius fracture."

"High VDR expression was associated with low appendicular lean mass index."

"We aimed to evaluate the relationship between the VDR expression in the muscle cell and the muscle mass in women with a distal radius fracture (DRF)."

"The clinical parameters including grip strength, gait speed, body mass index (BMI), bone mineral density (BMD), and serum vitamin D levels were compared between patients grouped by appendicular lean mass index and were correlated with the VDR expression."

"Patients with a low appendicular lean mass index had significantly lower muscle CSA ($p = 0.037$), but a higher VDR expression ($p = 0.045$) than those with higher indices."

"VDR expression was negatively correlated with BMI ($r = -0.417$, $p = 0.004$) and appendicular lean mass index ($r = -0.316$, $p = 0.044$)."

"DRF patients with low appendicular lean mass index presented high VDR expression and low CSA in forearm muscle cells."

Introduction

"Muscle mass and strength reductions with decreased physical performance, as well as the consequent frailty in mature adults, are known by the term "sarcopenia."".

"Some studies suggest a positive association between serum vitamin D levels and muscle strength as well as physical performance in older adults [44, 45]."

"That individuals with a low vitamin D status improve in both muscle strength and performance through vitamin D supplements [46]."

"The relationship between VDR expressions and muscle mass has not been well studied."

"The purpose of this study was to evaluate the relationship between the VDR expression in the muscle cell and the muscle mass in women with a DRF."

Patients and Methods

"The participants were classified as sarcopenic if they had a low lean mass in addition to weakness (assessed by the grip strength) or slowness (classified by the gait speed)."

"Serum 25(OH)D (25-hydroxyvitamin D) levels were assessed in all patients preoperatively using Diels-Alder derivatization and ultrahigh-performance liquid chromatography-tandem mass spectrometry (Waters Corp, Milford, MA, USA), which constitute the gold standard for this measurement."

"The subjects were divided into two groups by the lean mass index with the cut-off value of 5.4 kg/m^2, following the AWGS definition."

"Patients with and without low muscle mass were compared regarding age, the body mass index (BMI), grip strength, gait speed, the SPPB score, serum vitamin D levels, the VDR expression, muscle CSA, and the femur neck bone mineral density (BMD), using independent sample t tests."

"Based on that study, we designed to determine differences of 1 standard deviation in VDR expression between patients with and without low muscle mass index."

Results

"Twelve patients (26%) of the total sample had an appendicular lean mass index lower than 5.4 kg/m^2."

"Only two out of the 12 patients with low muscle mass had weak grip strength and were defined as sarcopenic."

"The gait speed was greater than 0.8 m/s in all studied patients."

Discussion

"Further studies with older patients are necessary to ascertain clinical relevance of age on VDR expression with respect to progression of muscle mass decrement."

"Patients with a low appendicular muscle mass index had a higher VDR expression than those with a higher index, which was an unexpected finding."

"Mouse studies reported correlations between both vitamin D signaling and the VDR expression with grip strength, fiber quality, and myostatin expression which is an antagonist of myogenesis or muscle fiber development and growth [47–50]."

"One possible explanation could be that patients with a low lean mass may increase the VDR expression and maximize the use of vitamin D to compensate for reduced muscle mass."

"This study of patients with DRF found that those with low appendicular lean mass had high VDR expression and low CSA in the forearm's muscle cells."

The Crosstalk Between Adenosine A2B Receptor and Insulin Signalling in Rat Skeletal Muscle Cells [51]

This is a machine-generated summary of:

Habahbeh, Suna; Imraish, Amer; Zihlif, Malek: The crosstalk between adenosine A2B receptor and insulin signalling in rat skeletal muscle cells [51]

Published in: Biologia Futura (2020)

Link to original: https://doi.org/10.1007/s42977-020-00035-3

If you want to cite the papers, please refer to the original.

For technical reasons we could not place the page where the original quote is coming from.

Abstract-Summary "Diabetes mellitus (DM) is a group of metabolic diseases characterised by hyperglycaemia resulting from defects in insulin secretion, insulin action, or both."

"This study aims to find out if there is an association between AR2B and insulin signalling, especially the metabolic pathways (AKT-GSK)."

"Differentiated L6 cell rat muscle cells were treated with insulin, adenosine agonist NECA, selective AR2B antagonist PSB 603 and combinations between these reagents, the expression of AKT2, GSK3α, and GSK3β were measured by qPCR hydrolysis probe technique."

"Insulin increases AKT2, GSK3α and GSK3β mRNA expression, while AR2B antagonist inhibits AKT2 GSK3α and GSK3β mRNA expression and combining AR2B antagonist with insulin diminish insulin action and decrease AKT2 GSK3α and GSK3β mRNA expression, which means a strong relationship between AR2B and insulin action."

Introduction

"Accumulative evidence shows that the adenosine and its receptors play a crucial role in glucose homoeostasis, insulin resistance and pathophysiology of type 2 diabetes mellitus [52, 53]."

"Adenosine modulates a wide array of biological processes on many different tissues and organ systems, which suggests a potential effect in skeletal muscle regarding the glucose metabolism."

"Adenosine A2B receptor signalling pathway is a critical regulatory pathway in skeletal muscle that may be responsible for the modulation of muscle metabolism in response to adenosine [54, 55], as adenosine A2B receptor activation accumulated cAMP level in skeletal muscle cells [56]."

"Adenosine A2B receptors might modulate glucose metabolism in skeletal muscle tissue."

"Taken the metabolic functions of skeletal muscle tissue concerning of glucose metabolism [57], it is essential to evaluate the association between AR2B and insulin signalling particularly the metabolic pathways Protein kinase B (AKT)–Glycogen synthase kinase 3 (GSK3)."

Materials and Methods

"NECA and PSB 603 were obtained from Tocris Bioscience, UK; Trizol was brought from Invitrogen, USA."

"Quanti script reverse transcription kit was brought from Qiagen, USA."

"Dulbecco's Modified Eagle's medium (DMEM) high glucose was from Healthcare, USA."

"DMEM low glucose was sourced from Biowest, USA."

"Rat L6 skeletal muscle cell line was obtained from the American Type Culture Collection (ATCC CRL-1458, USA)."

"L6 cells were cultured in DMEM with high glucose supplemented with l-glutamine containing 10% heat-inactivated foetal bovine serum (FBS) and a cocktail of 100 μg/ml penicillin–streptomycin."

"Statistical analyses were performed using an unpaired, two-tailed Student's t test with F test to compare variances between two groups, by using the GraphPad Prism version 7.0 (GraphPad Software, San Diego, CA)."

Results

"To assess whether stimulation of adenosine A2 receptors could modulate AKT2, GSK3α, and GSK3β mRNA gene expression in rat L6 skeletal muscle cells, the effects of NECA (a non-selective adenosine receptor agonist) were relatively quantified using qRT-PCR."

"Starved L6 rat skeletal muscle cells were incubated with NECA (100 nM) for 2 h, and mRNA gene expression of above genes was subsequently quantified."

"Incubation of one week starved L6 skeletal muscle cells with PSB 603 (10 μM) inhibits AKT2, GSK3α and GSK3β mRNA gene expression (− 0.28-fold, v0.46-fold and − 0.57-fold, respectively, $F (2, 2) = 1.957$ $P = 0.67$, $F (2, 2) = 4.884$, $P = 0.34$, $F (2, 2) = 7.57$, $P = 0.233$, respectively), but these differences did not reach statistical significance."

Discussion

"Adenosine analogue modulates metabolic gene expression levels in cultures of rat skeletal muscle throughout 2 h. This effect is seemingly mediated by A2 receptors and involves the activation of genes encoding proteins, which have a crucial role in the regulation of insulin metabolism."

"It has been reported that the adenosine system modulation, in particular adenosine A2B receptors, regulates the mRNA expression of genes that modulate skeletal muscle glycolysis, insulin resistance/sensitivity, and glucose oxidation (Haddad 49)."

"Of greater interest, our findings that preexposure of skeletal muscle cells to the A2B receptor antagonist PSB 603 blocks the insulin-induced GSK3 pathway further confirm the role for these receptors in the role in the regulation of glycolysis, insulin resistance and metabolism."

Conclusion for Future Biology

"The results of this study indicate a strong relationship between A2B and insulin action through the metabolic pathways Protein kinase B (AKT)–Glycogen synthase kinase 3 (GSK)."

"The results may elect the AR2B agonist as a good candidate for development as an anti-diabetic drug to enhance insulin action."

"Such new anti-diabetic drug may be used in order to reduce insulin resistance in diabetic patients."

"To reach such point, the finding of this study has to be assured first in a diabetic animal model."

Altered Gene and Protein Expressions of Vitamin D Receptor in Skeletal Muscle in Sarcopenic Patients Who Sustained Distal Radius Fractures [58]

This is a machine-generated summary of:

Roh, Young Hak; Hong, Seok Woo; Chung, Seok Won; Lee, Yong-Soo: Altered gene and protein expressions of vitamin D receptor in skeletal muscle in sarcopenic patients who sustained distal radius fractures [58]

Published in: Journal of Bone and Mineral Metabolism (2019)

Link to original: https://doi.org/10.1007/s00774-019-00995-0

If you want to cite the papers, please refer to the original.

For technical reasons we could not place the page where the original quote is coming from.

Abstract-Summary "The purpose of this study was to compare VDR gene and protein expression in the forearm muscle between sarcopenic and non-sarcopenic individuals who have sustained distal radius fractures."

"Sarcopenic patients showed a significantly lower level of gene expression for VDR and myogenin, but a greater level of gene expression for myostatin than the controls according to qRT-PCR analysis."

"The density of VDR protein expressions was 2.1 times greater, while that of myostatin was 2.6 times lower, in the control group than in the sarcopenic group according to Western blot analysis."

"On immunohistochemical analysis, the density of the cells expressing VDR was significantly decreased in the sarcopenic patients."

"Sarcopenic patients who sustained distal radius fractures presented lower vitamin D receptor gene and protein expression in skeletal muscles compared to non-sarcopenic individuals."

Introduction

"Aging is associated with a progressive decline of muscle mass, muscle quality, and muscle strength, which is a condition known as sarcopenia [59]."

"Some observational studies have demonstrated an association between vitamin D level and muscle strength/physical performance in older adults [44, 45]."

"Recent studies show the presence of vitamin D receptor (VDR) in skeletal muscle cells, showing contradictory effects of serum vitamin D on the expression of intramuscular VDR [60–64]."

"Despite the possible association between VDR expression and muscle mass change, the change of VDR gene or protein expressions in sarcopenic patients has not been well studied."

"The purpose of this study was to compare VDR gene and protein expression in the muscle between sarcopenic and non-sarcopenic individuals who sustained distal radius fractures."

Materials and Methods

"The mRNA expression levels of VDR and some myokine of interest that may be associated with muscle mass change (myogenin and myostatin) were evaluated using real-time quantitative reverse transcription PCR (qRT-PCR) analyses."

"The tissue samples from the isolated pronator quadratus muscles were immediately frozen at −80 °C until RNA extraction."

"The real-time Cycler 480 Multiwell Plate 96 contained 0.5 mM of each primer, 0.25 mM of the probe, 10-mL FastStart Essential DNA Probes Master (Roche Diagnostics), and 2 mL of DNA template, in a final reaction volume of 20 mL. The thermal cycling conditions on the Light-Cycler 480 System were as follows: enzyme activation, 95 °C for 10 min; 45 cycles of amplification, 95 °C for 10 s; and 60 °C for 30 s. Each quantitative real-time PCR analysis was performed in triplicate for both target genes and internal control, and the gene expression levels were reported as the relative ratio to the internal control of the β-actin gene."

Results

"The mRNA expression levels of VDR were more significantly decreased in the sarcopenic group than in the control group (2.3-fold, $p = .012$)."

"The mRNA expression of myogenin was significantly decreased (2.1-fold, $p = .017$), but that of myostatin was increased (2.4-fold, $p = .010$) in the sarcopenic group compared to the control group."

"The mean optical density VDR was 2.1 times greater in the control group than in the sarcopenic group ($p = .018$)."

Discussion

"Despite the presence of the vitamin D receptor (VDR) in skeletal muscle cells, the relationship between VDR expressions and muscle mass or function has not been well studied."

"Several previous studies investigating the effect of VDR expression on muscle mass suggested an incremental effect of VDR on muscle fibers [48, 63]."

"The previous study showing negative association between VDR expression and muscle mass did not evaluate gene expression for VDR, and it was supposed that patients with a low lean mass increased the VDR expression and maximized the use of vitamin D to compensate for reduced muscle mass and strength [65]."

"It can be speculated that a decreased VDR expression in skeletal muscle leads lower muscle mass and function in this study."

"Sarcopenic patients had decreased VDR gene and protein expression in their skeletal muscles."

Bibliography

1. Kim J, Grotegut CA, Wisler JW, Li T, Mao L, Chen M, Chen W, Rosenberg PB, Rockman HA, Lefkowitz RJ (2018) β-arrestin 1 regulates β2-adrenergic receptor-mediated skeletal muscle hypertrophy and contractility. Skelet Muscle 8:39. https://doi.org/10.1186/s13395-018-0184-8
2. Tomiyama H, Yamashina A (2014) Beta-blockers in the management of hypertension and/or chronic kidney disease. Int J Hypertens 2014:919256
3. Ryall JG, Lynch GS (2008) The potential and the pitfalls of beta-adrenoceptor agonists for the management of skeletal muscle wasting. Pharmacol Ther 120:219–232
4. Goncalves DA, Silveira WA, Lira EC, Graca FA, Paula-Gomes S, Zanon NM et al (2012) Clenbuterol suppresses proteasomal and lysosomal proteolysis and atrophy-related genes in denervated rat soleus muscles independently of Akt. Am J Physiol Endocrinol Metab 302:E123–E133
5. Tae HS, Casarotto MG, Dulhunty AF (2009) Ubiquitous SPRY domains and their role in the skeletal type ryanodine receptor. Eur Biophys J 39:51–59. https://doi.org/10.1007/s00249-009-0455-8
6. Godbout R, Packer M, Katyal S, Bleoo S (2002) Cloning and expression analysis of the chicken DEAD box gene DDX1. Biochim Biophys Acta 1574:63–71
7. Ponting C, Schultz J, Bork P (1997) SPRY domains in ryanodine receptors (Ca(2 +)-release channels). Trends Biochem Sci 22:193–194
8. Shtifman A, Ward CW, Yamamoto T, Wang J, Olbinski B, Valdivia HH, Ikemoto N, Schneider MF (2002) Interdomain interactions within ryanodine receptors regulate Ca2+ spark frequency in skeletal muscle. J Gen Physio 119:15–32
9. Wang D, Li Z, Messing EM, Wu G (2005) The SPRY domain-containing SOCS box protein 1 (SSB-1) interacts with MET and enhances the hepatocyte growth factor-induced Erk-Elk-1-serum response element pathway. J Biol Chem 280:16393–16401
10. Cox TC, Allen LR, Cox LL, Hopwood B, Goodwin B, Haan E, Suthers GK (2000) New mutations in MID1 provide support for loss of function as the cause of X-linked Opitz syndrome. Hum Mol Genet 9:2553–2562. https://doi.org/10.1093/hmg/9.17.2553
11. De Falco F, Cainarca S, Andolfi G, Ferrentino R, Berti C, Rodriguez Criado G, Rittinger O, Dennis N, Odent S, Rastogi A, Liebelt J, Chitayat D, Winter R, Jawanda H, Ballabio A, Franco B, Meroni G (2003) X-linked Opitz syndrome: novel mutations in the MID1 gene and redefinition of the clinical spectrum. Am J Med Genet A 120A:222–228

12. Mnayer L, Khuri S, Merheby HA, Meroni G, Elsas LJ (2006) A structure-function study of MID1 mutations associated with a mild Opitz phenotype. Mol Genet Metab 87:198–203
13. Aranda-Orgilles B, Aigner J, Kunath M, Lurz R, Schneider R, Schweiger S (2008a) Active transport of the ubiquitin ligase MID1 along the microtubules is regulated by protein phosphatase 2A. PLoS One 3:e3507. https://doi.org/10.1371/journal.pone.0003507
14. Aranda-Orgilles B, Trockenbacher A, Winter J, Aigner J, Kohler A, Jastrzebska E, Stahl J, Muller EC, Otto A, Wanker EE, Schneider R, Schweiger S (2008b) The Opitz syndrome gene product MID1 assembles a microtubule-associated ribonucleoprotein complex. Hum Genet 123:163–176. https://doi.org/10.1007/s00439-007-0456-6
15. Cainarca S, Messali S, Ballabio A, Meroni G (1999) Functional characterization of the Opitz syndrome gene product (midin): evidence for homodimerization and association with microtubules throughout the cell cycle. Hum Mol Genet 8:1387–1396. https://doi.org/10.1093/hmg/8.8.1387
16. Schweiger S, Foerster J, Lehmann T, Suckow V, Muller YA, Walter G, Davies T, Porter H, van Bokhoven H, Lunt PW, Traub P, Ropers HH (1999) The Opitz syndrome gene product, MID1, associates with microtubules. Proc Natl Acad Sci USA 96:2794–2799
17. Hilton DJ, Richardson RT, Alexander WS, Viney EM, Willson TA, Sprigg NS, Starr R, Nicholson SE, Metcalf D, Nicola NA (1998) Twenty proteins containing a C-terminal SOCS box form five structural classes. Proc Natl Acad Sci USA 95:114–119
18. Kile BT, Schulman BA, Alexander WS, Nicola NA, Martin HM, Hilton DJ (2002) The SOCS box: a tale of destruction and degradation. Trends Biochem Sci 27:235–241
19. Krebs DL, Hilton DJ (2000) SOCS: physiological suppressors of cytokine signaling. J Cell Sci 113(Pt 16):2813–2819
20. Styhler S, Nakamura A, Lasko P (2002) VASA localization requires the SPRY-domain and SOCS-box containing protein, GUSTAVUS. Dev Cell 3:865–876
21. Sepúlveda C, Rodríguez JM, Monsalves-Álvarez M, Donoso-Barraza C, Pino-de la Fuente F, Matías I, Leste-Lasserre T, Zizzari P, Morselli E, Cota D, Llanos M, Troncoso R (2023) The CB1 cannabinoid receptor regulates autophagy in the tibialis anterior skeletal muscle in mice. Biol Res 56:14. https://doi.org/10.1186/s40659-023-00426-5
22. Osei-Hyiaman D, DePetrillo M, Pacher P, Liu J, Radaeva S, Bátkai S et al (2005) Endocannabinoid activation at hepatic CB1 receptors stimulates fatty acid synthesis and contributes to diet-induced obesity. J Clin Invest 115:1298–1305
23. Kataoka K, Bilkei-Gorzo A, Nozaki C, Togo A, Nakamura K, Ohta K et al (2020) Age-dependent alteration in mitochondrial dynamics and autophagy in hippocampal neuron of cannabinoid CB1 receptor-deficient mice. Brain Res Bull 160:40–49. https://doi.org/10.1016/j.brainresbull.2020.03.014
24. Lin B, Gao Y, Li Z, Zhang Z, Lin X, Gao J (2020) Cannabidiol alleviates hemorrhagic shock-induced neural apoptosis in rats by inducing autophagy through activation of the PI3K/AKT pathway. Fundam Clin Pharmacol 34:640–649
25. Matias I, Gatta-Cherifi B, Cota D (2012) Obesity and the endocannabinoid system: circulating endocannabinoids and obesity. Curr Obes Rep 1:229–235
26. Matias I, Gonthier M-P, Orlando P, Martiadis V, De Petrocellis L, Cervino C et al (2006) Regulation, function, and dysregulation of endocannabinoids in models of adipose and beta-pancreatic cells and in obesity and hyperglycemia. J Clin Endocrinol Metab 91:3171–3180
27. Cota D (2007) CB1 receptors: emerging evidence for central and peripheral mechanisms that regulate energy balance, metabolism, and cardiovascular health. Diabetes Metab Res Rev 23:507–517
28. Li J, Kanasaki M, Xu L, Kitada M, Nagao K, Adachi Y et al (2018) A ketogenic amino acid rich diet benefits mitochondrial homeostasis by altering the AKT/4EBP1 and autophagy signaling pathways in the gastrocnemius and soleus. Biochim Biophys Acta 1862:1547–1555
29. Shim BJ, Lee MH, Lim JY, Gong HS (2021) A longitudinal histologic evaluation of vitamin D receptor expression in the skeletal muscles of patients with a distal radius fracture. Osteoporos Int 32:1387–1393. https://doi.org/10.1007/s00198-020-05809-y

30. Lee SM, Meyer MB, Benkusky NA, O'Brien CA, Pike JW (2018) The impact of VDR expression and regulation in vivo. J Steroid Biochem Mol Biol 177:36–45. https://doi.org/10.1016/j.jsbmb.2017.06.002
31. de la Puente YM, Collado Yurrita L, Ciudad Cabanas MJ, Cuadrado Cenzual MA (2020) Role of vitamin D in athletes and their performance: current concepts and new trends. Nutrients 12(2):759. https://doi.org/10.3390/nu12020579
32. Ceglia L, Niramitmahapanya S, da Silva MM, Rivas DA, Harris SS, Bischoff-Ferrari H, Fielding RA, Dawson-Hughes B (2013) A randomized study on the effect of vitamin D(3) supplementation on skeletal muscle morphology and vitamin D receptor concentration in older women. J Clin Endocrinol Metab 98:E1927–E1935. https://doi.org/10.1210/jc.2013-2820
33. Kai Y, Devendra KA (2014) Vitamin D and inflammatory disease. J Inflamm Res 7:69–87. https://doi.org/10.2147/JIR.S63898
34. Casas F, Fouret G, Lecomte J, Cortade F, Pessemesse L, Blanchet E, Wrutniak-Cabello C, Coudray C, Feillet-Coudray C (2018) Skeletal muscle expression of p43, a truncated thyroid hormone receptor α, affects lipid composition and metabolism. J Bioenerg Biomembr 50:71–79. https://doi.org/10.1007/s10863-018-9743-2
35. Davis PJ, Goglia F, Leonard JL (2016) Nongenomic actions of thyroid hormone. Nat Rev Endocrinol 12:111–121
36. Lanni A, Moreno M, Goglia F (2016) Mitochondrial actions of thyroid hormone. Comprehen Physiol 6:1591–1607
37. Paradies G, Ruggiero FM, Dinoi P, Petrosillo G, Quagliariello E (1993) Decreased cytochrome oxidase activity and changes in phospholipids in heart mitochondria from hypothyroid rats. Arch Biochem Biophys 307:91–95
38. Paradies G, Ruggiero FM, Petrosillo G, Quagliariello E (1994) Enhanced cytochrome oxidase activity and modification of lipids in heart mitochondria from hyperthyroid rats. Biochim Biophys Acta 1225:165–170
39. Casas F et al (2008) Overexpression of the mitochondrial T3 receptor p43 induces a shift in skeletal muscle fiber types. PLoS One 3:e2501
40. Pessemesse L et al (2012) Depletion of the p43 mitochondrial T3 receptor in mice affects skeletal muscle development and activity. FASEB J 26:748–756
41. Baati N et al (2017) Myostatin deficiency is associated with lipidomic abnormalities in skeletal muscles. Biochim Biophys Acta 1862:1044–1055
42. Fouret G et al (2015) The mitochondrial-targeted antioxidant, MitoQ, increases liver mitochondrial cardiolipin content in obesogenic diet-fed rats. Biochim Biophys Acta 1847:1025–1035
43. Kim K, Gong HS, Lim J-Y, Kim JH, Baek GH (2018) The vitamin D receptor expression in skeletal muscle of women with distal radius fracture. Arch Osteoporos 13. https://doi.org/10.1007/s11657-018-0442-8
44. Visser M, Deeg DJ, Lips P, Longitudinal Aging Study A (2003) Low vitamin D and high parathyroid hormone levels as determinants of loss of muscle strength and muscle mass (sarcopenia): The Longitudinal Aging Study Amsterdam. J Clin Endocrinol Metab 88(12):5766–5772. https://doi.org/10.1210/jc.2003-030604
45. Snijder MB, van Schoor NM, Pluijm SM, van Dam RM, Visser M, Lips P (2006) Vitamin D status in relation to one-year risk of recurrent falling in older men and women. J Clin Endocrinol Metab 91(8):2980–2985. https://doi.org/10.1210/jc.2006-0510
46. Ceglia L, Harris SS (2013) Vitamin D and its role in skeletal muscle. Calcif Tissue Int 92(2):151–162. https://doi.org/10.1007/s00223-012-9645-y
47. Girgis CM, Clifton-Bligh RJ, Hamrick MW, Holick MF, Gunton JE (2013) The roles of vitamin D in skeletal muscle: form, function, and metabolism. Endocr Rev 34(1):33–83. https://doi.org/10.1210/er.2012-1012
48. Girgis CM, Cha KM, Houweling PJ, Rao R, Mokbel N, Lin M, Clifton-Bligh RJ, Gunton JE (2015) Vitamin D receptor ablation and vitamin D deficiency result in reduced grip strength, altered muscle fibers, and increased myostatin in mice. Calcif Tissue Int 97(6):602–610. https://doi.org/10.1007/s00223-015-0054-x

49. Garcia LA, King KK, Ferrini MG, Norris KC, Artaza JN (2011) 1, 25(OH)2vitamin D3 stimulates myogenic differentiation by inhibiting cell proliferation and modulating the expression of promyogenic growth factors and myostatin in C2C12 skeletal muscle cells. Endocrinology 152(8):2976–2986
50. Perez-Siles G, Grant A, Ellis M, Ly C, Kidambi A, Khalil M, Llanos RM, La Fontaine S, Strickland AV, Züchner S (2016) Characterizing the molecular phenotype of an Atp7a T985I conditional knock in mouse model for X-linked distal hereditary motor neuropathy (dHMNX). Metallomics 8(9):981–992
51. Habahbeh S, Imraish A, Zihlif M (2020) The crosstalk between adenosine A2B receptor and insulin signalling in rat skeletal muscle cells. Biologia Futura 71:283–288. https://doi.org/10.1007/s42977-020-00035-3
52. Novitskiy SV et al (2008) Adenosine receptors in regulation of dendritic cell differentiation and function. Blood 112:1822–1831
53. Wojcik M, Zieleniak A, Mac-Marcjanek K, Wozniak LA, Cypryk K (2014) The elevated gene expression level of the A2B adenosine receptor is associated with hyperglycemia in women with gestational diabetes mellitus. Diabetes Metab Res Rev 30:42–53
54. Haddad M (2015) Adenosine receptors machinery and purinergic receptors in rat primary skeletal muscle cells. Biomed Pharmacol J 7:383–398
55. Haddad M (2016) The impact of adenosine A2B receptors modulation on nuclear receptors (NR4A) gene expression. Biomed Pharmacol J 9:177–185
56. Lynge J, Schulte G, Nordsborg N, Fredholm BB, Hellsten Y (2003) Adenosine A 2B receptors modulate cAMP levels and induce CREB but not ERK1/2 and p38 phosphorylation in rat skeletal muscle cells. Biochem Biophys Res Commun 307:180–187
57. Turcotte LP, Fisher JS (2008) Skeletal muscle insulin resistance: roles of fatty acid metabolism and exercise. Phys Ther 88:1279–1296
58. Roh YH, Hong SW, Chung SW, Lee Y-S (2019) Altered gene and protein expressions of vitamin D receptor in skeletal muscle in sarcopenic patients who sustained distal radius fractures. J Bone Miner Metab 37:920–927. https://doi.org/10.1007/s00774-019-00995-0
59. Candow DG, Chilibeck PD (2005) Differences in size, strength, and power of upper and lower body muscle groups in young and older men. J Gerontol A Biol Sci Med Sci 60:148–156
60. Bouillon R, Gielen E, Vanderschueren D (2014) Vitamin D receptor and vitamin D action in muscle. Endocrinology 155:3210–3213
61. Girgis CM, Mokbel N, Cha KM, Houweling PJ, Abboud M, Fraser DR, Mason RS, Clifton-Bligh RJ, Gunton JE (2014) The vitamin D receptor (VDR) is expressed in skeletal muscle of male mice and modulates 25-hydroxyvitamin D (25OHD) uptake in myofibers. Endocrinology 155:3227–3237
62. Pike JW (2014) Expression of the vitamin D receptor in skeletal muscle: are we there yet? Endocrinology 155:3214–3218
63. Pojednic RM, Ceglia L, Olsson K, Gustafsson T, Lichtenstein AH, Dawson-Hughes B, Fielding RA (2015) Effects of 1,25-dihydroxyvitamin D3 and vitamin D3 on the expression of the vitamin d receptor in human skeletal muscle cells. Calcif Tissue Int 96:256–263. https://doi.org/10.1007/s00223-014-9932-x
64. Bischoff-Ferrari H, Borchers M, Gudat F, Dürmüller U, Stähelin H, Dick W (2004) Vitamin D receptor expression in human muscle tissue decreases with age. J Bone Miner Res 19(2):265–269
65. Kim K, Gong HS, Lim JY, Kim JH, Baek GH (2018) The vitamin D receptor expression in skeletal muscle of women with distal radius fracture. Arch Osteoporos 13:24. https://doi.org/10.1007/s11657-018-0442-8
66. Oláh T, Bodnár D, Tóth A, Vincze J, Fodor J, Reischl B et al (2016) Cannabinoid signalling inhibits sarcoplasmic Ca2+ release and regulates excitation-contraction coupling in mammalian skeletal muscle. J Physiol 594:7381–7398

Neuromuscular Signal Transmission

5

Christopher Myers

5.1 Introduction by the Editor

The intricate architecture of skeletal muscle, characterized by myriad fibers with distinct functional and metabolic properties, plays a pivotal role in orchestrating force generation and endurance capabilities in response to diverse demands. Among the different fibers, type 1 and type 2 fibers emerge as the protagonists, each endowed with unique attributes for varying requirements of muscle performance. Type 1 fibers, often heralded as the marathoners of muscle fibers, exhibit a remarkable proficiency in sustained aerobic activities, courtesy of their abundant mitochondrial content and propensity for oxidative metabolism. Conversely, type 2 fibers, the sprinters of the muscular realm, are tailored for rapid, high-intensity bursts of activity, relying predominantly on glycolytic pathways for their energetic needs. This dichotomy in functional specialization is underpinned by a molecular framework that tailors each fiber type to its specific role. This framework includes the nuanced expression and regulation of mechanosensors and signaling molecules pivotal for muscle adaptation and homeostasis.

An essential and well-studied neurosignaling pathway is the nitric oxide (NO) pathway. NO is primarily produced by neuronal NO synthases (nNOSμ) and is pivotal for mitochondrial bioenergetics, muscle development, and overall function. Despite the acknowledged importance, the impact of NO/nNOSμ interaction with mitochondria on muscle function remained uncharted until recent investigations utilizing a mouse model deficient in nNOSμ unveiled critical insights. These studies highlighted in this chapter revealed that nNOSμ absence leads to significant alterations in mitochondrial bioenergetics, network remodeling, and a surge in mitochondrial unfolded protein response (UPRmt) and autophagy, coupled with compromised

C. Myers (✉)
3 Sports Science, 3 Sports Science LLC, Akron, OH, United States of America
e-mail: chris@3sportsscience.com

C. Myers (ed.), *Skeletal Muscle Physiology*,
https://doi.org/10.1007/978-3-031-47065-3_5

muscle development during early perinatal growth. Additionally, nNOSμ deficiency manifested in reduced muscle force, diminished resistance to fatigue, and heightened muscle damage post-exercise. The findings collectively underscore the necessity of nNOSμ/NO in regulating essential homeostatic mechanisms within skeletal muscle, pointing to mitochondrial bioenergetics and network remodeling, UPRmt, and autophagy as crucial elements. This line of research suggests that nNOSμ-dependent impairments in muscle fiber growth culminate in a notable deficit in muscle performance, offering new perspectives on the molecular underpinnings of muscle function and disease.

Costameres, subcellular structures, are the structural linchpins of this molecular framework. They are critical mediators of force transduction and mechanical signaling within skeletal muscle fibers. Costameres are nestled beneath the sarcolemma, forging a mechanical and signaling nexus between the extracellular matrix and the contractile apparatus of the muscle fiber. Through their interaction with integrins and other anchoring proteins, new research highlighted in this chapter suggests costameres bolster the structural integrity of muscle fibers during the rigors of contraction and initiate a cascade of signaling events that modulate muscle adaptation to mechanical stresses. The composition and organization of costameres are finely tuned to the distinctive demands placed on different muscle fiber types, suggesting a fiber type-specific orchestration of mechanical signaling pathways that govern muscle adaptation and resilience.

In the face of mechanical loading or unloading, skeletal muscle fibers exhibit remarkable plasticity, adjusting their structural and metabolic attributes to accommodate the altered mechanical environment. The research highlighted in this chapter demonstrates that changes in the expression and activity of essential elements of the costameric complex partially mediate the adaptive response. These elements translate mechanical cues into molecular signals that stimulate muscle remodeling. However, the precise nature, which is still not fully understood, of these adaptive changes appears to be differentially modulated in type 1 and type 2 fibers, reflecting their distinct roles and mechanical properties. Mechanical loading, for instance, enhances the expression of specific costameric proteins and signaling molecules in type 1 fibers, reinforcing their oxidative capacity and endurance potential. In contrast, type 2 fibers may respond to similar stimuli with adaptations that favor rapid force production and glycolytic metabolism, aligning with their role in high-intensity, anaerobic activities.

The complex architecture of skeletal muscle, consisting of diverse fibers with unique functional and metabolic characteristics, is crucial for generating force and endurance in response to varied demands. Type 1 and type 2 fibers, each with distinct attributes, play central roles in muscle performance. Type 1 fibers, known for their endurance capabilities, are rich in mitochondria and favor oxidative metabolism. In contrast, type 2 fibers excel in rapid, high-intensity activities through glycolytic pathways. This functional specialization is supported by a molecular framework that adapts each fiber type to its specific function. It includes the expression and regulation of mechanosensors and signaling molecules crucial for muscle adaptation. The neuro signaling NO pathway, produced mainly by nNOSμ, is vital for mitochondrial function, muscle development, and function. Recent studies using nNOSμ-deficient mouse models have shown that the absence of nNOSμ affects mitochondrial bioenergetics and network remodeling, increasing UPRmt and autophagy, leading to compromised muscle development and function. These findings emphasize the essential role of nNOSμ/NO in regulating fundamental mechanisms in skeletal muscle, highlighting the impact on muscle performance. Costameres, pivotal in this molecular framework, mediate force transduction and mechanical signaling, adjusting muscle structure and metabolism to mechanical changes. This adaptive response, evidenced by changes in costameric complex expressions, underlines the differential modulation in type 1 and type 2 fibers, showcasing the fiber type-specific orchestration of mechanical signaling pathways that guide muscle adaptation and resilience.

The intricate architecture of skeletal muscle, characterized by myriad fibers with distinct functional and metabolic properties, plays a pivotal role in orchestrating force generation and endurance capabilities in response to diverse demands. Among the different fibers, type 1 and type 2 fibers emerge as the protagonists, each endowed with unique attributes for varying requirements of muscle performance. Type 1 fibers, often heralded as the marathoners of muscle fibers, exhibit a remarkable proficiency in sustained aerobic activities, courtesy of their abundant mitochondrial content and propensity for oxidative metabolism. Conversely, type 2 fibers, the sprinters of the muscular realm, are tailored for rapid, high-intensity bursts of activity, relying predominantly on glycolytic pathways for their energetic needs. This dichotomy in functional specialization is underpinned by a molecular framework that tailors each fiber type to its specific role. This framework includes the nuanced expression and regulation of mechanosensors and signaling molecules pivotal for muscle adaptation and homeostasis. One of the first mechanisms explored in this chapter is the eccentric:concentric strength ratio.

The eccentric: concentric strength ratio of human skeletal muscle, which compares the force produced during muscle lengthening (eccentric) to that during muscle shortening (concentric), is crucial in understanding muscle function and adaptation across various conditions. This chapter explores a comprehensive meta-analysis of existing research that reveals that several factors, including sex, age, joint action, and movement velocity, influence this ratio. Generally, muscles exhibit greater strength during eccentric contractions compared to concentric ones. However, this analysis shows that the magnitude of this difference can vary, with factors such as older age and slower velocities tending to enhance the eccentric: concentric ratio. Interestingly, sex differences are also evident, with women sometimes showing a smaller disparity between eccentric and concentric strength than men. However, this can vary based on the specific muscle group and type of joint action involved. The joint being examined also plays a significant role, with certain joints demonstrating more pronounced differences in this strength ratio. Understanding these nuanced influences on the eccentric: concentric strength ratio is essential for designing more effective training and rehabilitation programs that cater to the specific needs of individuals based on these variables.

An essential and well-studied neurosignaling pathway is the nitric oxide (NO) pathway. NO is primarily produced by neuronal NO synthases (nNOSμ) and is pivotal for mitochondrial bioenergetics, muscle development, and overall function. Despite the acknowledged importance, the impact of NO/nNOSμ interaction with mitochondria on muscle function remained uncharted until recent investigations utilizing a mouse model deficient in nNOSμ unveiled critical insights. These studies highlighted in this chapter revealed that nNOSμ absence leads to significant alterations in mitochondrial bioenergetics, network remodeling, and a surge in mitochondrial unfolded protein response (UPRmt) and autophagy, coupled with compromised muscle development during early perinatal growth. Additionally, nNOSμ deficiency manifested in reduced muscle force, diminished resistance to fatigue, and heightened muscle damage post-exercise. The findings collectively underscore the necessity of nNOSμ/NO in regulating essential homeostatic mechanisms within skeletal muscle, pointing to mitochondrial bioenergetics and network remodeling, UPRmt, and autophagy as crucial elements. This line of research suggests that nNOSμ-dependent impairments in muscle fiber growth culminate in a notable deficit in muscle performance, offering new perspectives on the molecular underpinnings of muscle function and disease.

Costameres, subcellular structures, are the structural linchpins of this molecular framework. They are critical mediators of force transduction and mechanical signaling within skeletal muscle fibers. Costameres are nestled beneath the sarcolemma, forging a mechanical and signaling nexus between the extracellular matrix and the contractile apparatus of the muscle fiber. Through their interaction with integrins and other anchoring proteins, new research highlighted in this chapter suggests costameres bolster the structural integrity of muscle fibers during the rigors of contraction and initiate a cascade of signaling events that modulate muscle adaptation to mechanical stresses. The composition and organization of costameres are finely tuned to the distinctive demands placed on different muscle fiber types, suggesting a fiber type-specific orchestration of mechanical signaling pathways that govern muscle adaptation and resilience.

In the face of mechanical loading or unloading, skeletal muscle fibers exhibit remarkable plasticity, adjusting their structural and metabolic attributes to accommodate the altered mechanical environment. The research highlighted in this chapter demonstrates that changes in the expression and activity of essential elements of the costameric complex partially mediate the adaptive response. These elements translate mechanical cues into molecular signals that stimulate muscle remodeling. However, the precise nature, which is still not fully understood, of these adaptive changes appears to be differentially modulated in type 1 and type 2 fibers, reflecting their distinct roles and mechanical properties. Mechanical loading, for instance, enhances the expression of specific costameric proteins and signaling molecules in type 1 fibers, reinforcing their oxidative capacity and endurance potential. In contrast, type 2 fibers may respond to similar stimuli with adaptations that favor rapid force production and glycolytic metabolism, aligning with their role in high-intensity, anaerobic activities.

The complex architecture of skeletal muscle, consisting of diverse fibers with unique functional and metabolic characteristics, is crucial for generating force and endurance in response to varied demands. Type 1 and type 2 fibers, each with distinct attributes, play central roles in muscle performance. Type 1 fibers, known for their endurance capabilities, are rich in mitochondria and favor oxidative metabolism. In contrast, type 2 fibers excel in rapid, high-intensity activities through glycolytic pathways. This functional specialization is supported by a molecular framework that adapts each fiber type to its specific function. It includes the expression and regulation of mechanosensors and signaling molecules crucial for muscle adaptation. The neuro signaling NO pathway, produced mainly by nNOSμ, is vital for mitochondrial function, muscle development, and function. Recent studies using nNOSμ-deficient mouse models have shown that the absence of nNOSμ affects mitochondrial bioenergetics and network remodeling, increasing UPRmt and autophagy, leading to compromised muscle development and function. These findings emphasize the essential role of nNOSμ/NO in regulating fundamental mechanisms in skeletal muscle, highlighting the impact on muscle performance. Costameres, pivotal in this molecular framework, mediate force transduction and mechanical signaling, adjusting muscle structure and metabolism to mechanical changes. This adaptive response, evidenced by changes in costameric complex expressions, underlines the differential modulation in type 1 and type 2 fibers, showcasing the fiber type-specific orchestration of mechanical signaling pathways that guide muscle adaptation and resilience.

5.2 Machine Generated Summaries

Disclaimer: The summaries in this chapter were generated from Springer Nature publications using extractive AI auto-summarization: An extraction-based summarizer aims to identify the most important sentences of a text using an algorithm and uses those original sentences to create the auto-summary (unlike generative AI). As the constituted sentences are machine selected, they may not fully reflect the body of the work, so we strongly advise that the original content is read and cited. The auto generated summaries were curated by the editor to meet Springer Nature publication standards. To cite this content, please refer to the original papers.

Machine generated keywords: mitochondrial, insulin, insulin signal, vesicle, extracellular, glucose, autophagy, type fiber, adaptation, sclerosis, overexpression, autophagic, exercise, regeneration, impact.

Deficient Nitric Oxide Signalling Impairs Skeletal Muscle Growth and Performance: Involvement of Mitochondrial Dysregulation [1]

This is a machine-generated summary of:

De Palma, Clara; Morisi, Federica; Pambianco, Sarah; Assi, Emma; Touvier, Thierry; Russo, Stefania; Perrotta, Cristiana; Romanello, Vanina; Carnio, Silvia; Cappello, Valentina; Pellegrino, Paolo; Moscheni, Claudia; Bassi, Maria Teresa; Sandri, Marco; Cervia, Davide; Clementi, Emilio: Deficient nitric oxide signalling impairs skeletal muscle growth and performance: involvement of mitochondrial dysregulation [1]

Published in: Skeletal Muscle (2014)

Link to original: https://doi.org/10.1186/s13395-014-0022-6

If you want to cite the papers, please refer to the original.

For technical reasons we could not place the page where the original quote is coming from.

Abstract-Summary "Nitric oxide (NO), generated in skeletal muscle mostly by the neuronal NO synthases (nNOSμ), has profound effects on both mitochondrial bioenergetics and muscle development and function."

"How the effects of NO/nNOSμ on mitochondria impact on muscle function, however, has not been investigated yet."

"We have examined the relationship between the NO system, mitochondrial structure/activity and skeletal muscle phenotype/growth/functions using a mouse model in which nNOSμ is absent."

"We show that nNOSμ deficiency in mouse skeletal muscle leads to altered mitochondrial bioenergetics and network remodelling, and increased mitochondrial unfolded protein response (UPR^{mt}) and autophagy."

"The absence of nNOSμ is also accompanied by an altered mitochondrial homeostasis in myogenic precursor cells with a decrease in the number of myonuclei per fibre and impaired muscle development at early stages of perinatal growth."

"nNOSμ deficiency was also accompanied by functional changes in muscle with reduced muscle force, decreased resistance to fatigue and increased degeneration/damage post-exercise."

"Our results indicate that nNOSμ/NO is required to regulate key homeostatic mechanisms in skeletal muscle, namely mitochondrial bioenergetics and network remodelling, UPR^{mt} and autophagy."

"These events are likely associated with nNOSμ-dependent impairments of muscle fibre growth resulting in a deficit of muscle performance."

Background

"NO has an important role in regulating skeletal muscle physiological activity, including excitation-contraction coupling, muscle force generation, auto-regulation of blood flow, calcium homeostasis, metabolism and bioenergetics [2–4]."

"The role of NO in regulating oxidative phosphorylation and mitochondrial biogenesis in skeletal muscle physiology has been established [5, 6]."

"We have examined the relationship between the NO system, mitochondrial structure/activity and skeletal muscle phenotype/growth/functions using a mouse model in which nNOSμ is absent (NOS1-/-)."

"Our results indicate that the deficit in NO signalling leads in skeletal muscle to alterations in mitochondrial morphology, bioenergetics and network remodelling, accompanied by defective autophagy and the induction of a UPR^{mt} response."

"These events, while not severely altering the overall resting skeletal muscle structure, are associated with modifications in the Akt-mammalian target of rapamycin (mTOR) pathway and Akt-FoxO3-mitochondrial E3 ubiquitin protein ligase 1 (Mul-1) axis and are sufficient to dysregulate skeletal muscle growth and exercise performance."

Methods

"Tibialis anterior and diaphragm muscles were dissected, trimmed clean of visible fat and connective tissue, minced with scissors and digested in ATP medium, containing 50 mM Tris-HCl (pH 7.4), 100 mM KCl, 5 mM $MgCl_2$, 1.8 mM ATP, 1 mM ethylenediaminetetraacetic acid (EDTA), and 0.1% collagenase type V for 10 minutes at 37°C under strong agitation."

"Satellite cells and muscle tissue samples were homogenised, and RNA was extracted using the TRIzol protocol (Invitrogen-Life Technologies, Monza, Italy)."

"Tibialis anterior muscles were dissected and fixed for one hour in a solution containing 4% paraformaldehyde and 0.5% glutaraldehyde in 0.1 M cacodylate buffer, pH 7.4, immobilised on a Nunc Sylgard coated Petri dish (ThermoFisher Scientific, Waltham, MA, USA) to prevent muscular contraction as previously described [7]."

Results

"Mitochondrial function in skeletal muscles of adult NOS1-/- mice, that is, at P120, was dissected and compared with that of the respective age-matched wild-type littermattes (control)."

"We investigated whether the latent mitochondrial dysfunction observed in muscles of NOS1-/- mice affected the muscle bioenergetic parameters."

"The CII-linked respiratory capacity in tibialis anterior of NOS1-/- mice was lower than that in control muscle fibres, while no difference was observed in the diaphragm."

"To identify the changes in mitochondrial network morphology, tibialis anterior muscles of P120 NOS1-/- and wild-type control mice were imaged using pDsRed2-Mito, a mitochondrially targeted red fluorescent protein, by in situ two-photon confocal microscopy [8–10]."

"The enhanced autophagy in the absence of nNOSμ in skeletal muscle was confirmed by Western blot analysis."

"Enhanced activity of FoxO transcription factors has also been associated with disruption of mitochondrial function and organisation leading to impaired skeletal muscle function and development [8]."

Discussion

"The fact that NOS1-/- mice in our in vivo experiments exhibited a deficit in forward pulling tension and resistance to fatigue during a forced exercise indicates that nNOSμ is important to maintain skeletal muscle strength and the animal's ability to perform in repetitive exercise training."

"We raise the possibility that mitochondrial dysfunction, UPR^{mt} and autophagy are functionally related to each other and promoted by a single event, that is, the deficit in NO signalling, thus suggesting that the association of altered mitochondrial homeostasis and muscle phenotype/performance in NOS1-/- mice is not coincidental."

"The experiments we carried-out in myogenic precursor cells and NOS1-/- mice during critical stages of muscle development are consistent with an association of altered mitochondrial homeostasis and muscle phenotype/performance and provide an indication of the mechanism responsible for the impaired fibre growth resulting in a deficit of muscle performance."

Conclusions

"Muscle exercise performance is a complex physiological process that can occur by many different mechanisms and NO has long been described to be relevant among them [2]."

"Our study now suggests that the relevance of NO also resides in the fact that it regulates key homeostatic mechanisms in skeletal muscle, namely mitochondrial bioenergetics and network remodelling, UPR^{mt} and autophagy."

"Although NOS1-/- mice do not display the overt features of myopathies, such as muscle degeneration, reactive regeneration and replacement of muscle with fibro-adipous tissue [11, 12], we clearly show that alterations of the NO system significantly impair muscle fibre growth, thus resulting in a deficit of muscle force and the ability to sustain prolonged exercise."

"This aspect may explain why NO deficiency contributes to muscle impairment in degenerative disease of the muscle, such as muscular dystrophies."

Authors' Information

"EA, TT, VR, SC and VC are post-doctoral research fellows."

"EC is a Professor of Pharmacology and the Head of the Pharmacology group."

The Eccentric:Concentric Strength Ratio of Human Skeletal Muscle In Vivo: Meta-analysis of the Influences of Sex, Age, Joint Action, and Velocity [13]

This is a machine-generated summary of:

Nuzzo, James L.; Pinto, Matheus D.; Nosaka, Kazunori; Steele, James: The Eccentric:Concentric Strength Ratio of Human Skeletal Muscle In Vivo: Meta-analysis of the Influences of Sex, Age, Joint Action, and Velocity [13].

Published in: Sports Medicine (2023)

Link to original: https://doi.org/10.1007/s40279-023-01851-y

If you want to cite the papers, please refer to the original.

For technical reasons we could not place the page where the original quote is coming from.

Abstract-Summary "Knowledge of the ECC:CON strength ratio is incomplete and might inform resistance exercise prescriptions."

"Our purposes were to determine the magnitude of the ECC:CON ratio of human skeletal muscle in vivo and explore if sex, age, joint actions/exercises, and movement velocity impact it."

"It was possible to analyse 1516 ECC:CON ratios, aggregated from 12,546 individuals who made up 564 groups in 335 of the identified studies."

"The overall main model estimate for the ECC:CON ratio was 1.41 (95% credible interval [CI] 1.38–1.44)."

"The ECC:CON ratio was slightly less in men (1.38 [CI 1.34–1.41]) than women (1.47 [CI 1.43–1.51]), and greater in older adults (1.62 [CI 1.57–1.68]) than younger adults (1.39 [CI 1.36–1.42])."

"The ratio was similar between grouped upper-body (1.42 [CI 1.38–1.46]) and lower-body joint actions/exercises (1.40 [CI 1.37–1.44])."

"Heterogeneity in the ratio existed across joint actions/exercises, with point estimates ranging from 1.32 to 2.61."

"The ECC:CON ratio was most greatly impacted by movement velocity, with a 0.20% increase in the ratio for every 1°/s increase in velocity."

"The ECC:CON ratio is greatly affected by movement velocity and to lesser extents age and sex."

Introduction

"Across the six exercises that they assessed, the ECC:CON strength ratio ranged from 1.57 to 2.87 in women and 1.30 to 1.51 in men [14]."

"Practitioners [15] and researchers [16–25] prescribe a wide range of relative loads for accentuated ECC and ECC-only resistance exercise (usually between 105 and 150% of the CON 1RM), and there is no consensus on the magnitude of ECC overload that should be prescribed and whether factors such as muscle group, sex, and age should be considered."

"Meta-analysis of the magnitude of the ECC:CON strength ratio, and the factors that impact it, could help to inform and optimize delivery of ECC overload for specific exercises and populations, particularly as exercise technology continues to evolve to make accentuated ECC exercise safer and more feasible."

"We examined the extent to which sex, age, joint actions/exercises, and movement velocity impact the ECC:CON strength ratio of human skeletal muscle in vivo."

Methods

"The data extracted from papers included sample size, number of study groups, study type (non-training or training study), sex, age group, joint actions/exercises, movement velocity, and means and standard deviations (SDs) of the ECC:CON strength ratios or the ECC and CON strength values."

"As the included studies often had multiple groups/conditions, and reported multiple strength measures within these, the data had a nested structure."

"A main model included all ratios reported for all groups in each study."

"This exploratory model included velocity and age (grand mean centred) as a fixed effect to adjust for the fact that some joints only had low numbers of effects at specific velocities or were from studies in one age group, and we anticipated that both velocity and age would impact the ECC:CON ratio."

Results

"Not all identified studies were included in the meta-analyses because effect sizes could not be calculated when only mean ratios without variances were reported."

"The studies identified included 12,582 (12,546 included in analyses) participants from 575 (564 included in analyses) separate study groups, with a median sample size of 15 (range 2–734)."

"Some studies reported that participants were either competitive or recreational athletes, with 25% and 5% of the extracted ratios coming from either population, respectively."

"Due to the sex and age differences in ECC:CON strength ratios, we also examined the standardized mean differences and log ratio of means for ECC and CON separately between these groups for studies that included either both men and women or both younger and older adults."

"The difference between younger and older adults was larger for CON than ECC strength, favouring younger adults, and thus leading to the lower ECC:CON strength ratios among younger than older adults."

Discussion

"The purpose of this meta-analysis was to determine the magnitude of the ECC:CON strength ratio of human skeletal muscle in vivo and explore if sex, age, joint actions/exercises, and movement velocity impact it."

"No difference in the ECC:CON strength ratio was observed between upper-body and lower-body joint actions/exercises generally speaking, but exploratory analysis suggested heterogeneity in the ECC:CON strength ratio across specific joint actions/exercises."

"In the current analysis, muscle strength measurements acquired from joint actions/exercises about the wrist, elbow, and shoulder were combined into one upper-body ECC:CON strength ratio."

"Such machines might then be used to establish ECC:CON muscle strength ratios for various joint actions/exercises."

"A notable limitation of the current meta-analysis is that the vast majority of studies reported ECC and CON strength from isokinetic dynamometers, whereas most strength and conditioning coaches use isoinertial equipment when prescribing ECC resistance exercise to athletes [15, 26, 27]."

Conclusion

"We report a main model estimate for the ECC:CON strength ratio of 1.41."

"The ratio does not differ between upper- and lower-body muscles generally speaking, but an exploratory analysis indicated that there is likely heterogeneity in ratios across different joint actions/exercises."

"Further systematic study will be necessary to identify more precise estimates of exercise-specific ECC:CON strength ratios."

"Exercise practitioners can use the ECC:CON ratios from the current analysis to guide prescriptions of ECC overload to healthy individuals."

Multi-omics Analysis of Sarcospan Overexpression in mdx Skeletal Muscle Reveals Compensatory Remodeling of Cytoskeleton-Matrix Interactions That Promote Mechanotransduction Pathways [28]

This is a machine-generated summary of:

McCourt, Jackie L.; Stearns-Reider, Kristen M.; Mamsa, Hafsa; Kannan, Pranav; Afsharinia, Mohammad Hossein; Shu, Cynthia; Gibbs, Elizabeth M.; Shin, Kara M.; Kurmangaliyev, Yerbol Z.; Schmitt, Lauren R.; Hansen, Kirk C.; Crosbie, Rachelle H.: Multi-omics analysis of sarcospan overexpression in mdx skeletal muscle reveals compensatory remodeling of cytoskeleton-matrix interactions that promote mechanotransduction pathways [28]

Published in: Skeletal Muscle (2023)

Link to original: https://doi.org/10.1186/s13395-022-00311-x

Abstract-Summary "The dystrophin-glycoprotein complex (DGC) is a critical adhesion complex of the muscle cell membrane, providing a mechanical link between the extracellular matrix (ECM) and the cortical cytoskeleton that stabilizes the sarcolemma during repeated muscle contractions."

"Overexpression of SSPN in the skeletal muscle of mdx mice (murine model of DMD) restores muscle fiber attachment to the ECM in part through an associated increase in utrophin and integrin adhesion complexes at the cell membrane, protecting the muscle from contraction-induced injury."

"We utilized transcriptomic and ECM protein-optimized proteomics data sets from wild-type, mdx, and mdx transgenic (mdx^{TG}) skeletal muscle tissues to identify pathways and proteins driving the compensatory action of SSPN overexpression."

"The tibialis anterior and quadriceps muscles were isolated from wild-type, mdx, and mdx^{TG} mice and subjected to bulk RNA-Seq and global proteomics analysis using methods to enhance capture of ECM proteins."

"We identified signaling pathways that may contribute to the rescue phenotype, most notably cytoskeleton and ECM organization pathways."

"Using ingenuity pathway analysis, we identified upstream regulators that are computationally predicted to drive compensatory changes, revealing a possible mechanism of SSPN rescue through a rewiring of cell-ECM bidirectional communication."

"Our findings indicate that SSPN overexpression rescues dystrophin deficiency partially through mechanotransduction signaling cascades mediated through components of the ECM and the cortical cytoskeleton."

Background

"The dystrophin-deficient mdx mouse is the most widely used mouse model for DMD as it exhibits much of the pathology observed in patient skeletal muscle, albeit much milder, including elevated creatine kinase levels, increased levels of degeneration and regeneration, fibrosis, and reduced grip strength and whole-body tension [29, 30]."

"While many therapeutic genes that improve sarcolemma instability, decrease fibrosis, and increase force in mdx muscle have been identified [31–38], there are gaps in understanding whether their protective mechanisms affect overlapping molecular processes and signaling pathways."

"SSPN increases SG protein levels and restores integrity of the SG-SSPN subcomplex at the sarcolemma in mdx skeletal and cardiac muscles [39–44]."

"We found that SSPN overexpression in mdx muscle does not restore all transcripts and proteins to wild-type levels but ameliorates muscle pathology through compensatory changes in ECM and cytoskeletal composition, including key signaling molecules associated with regulation of cytoskeleton organization and mechanotransduction."

Methods

"Data acquisition was performed using the instrument supplied Xcalibur (Thermo Fisher Scientific, San Jose, CA, USA) software in positive ion mode."

"Quantification of immunofluorescence analysis was performed on 20× images using ImageJ software (NIH, version 1.50i) for collagen II, collagen V, and β-spectrin by line scan analysis and Sox9 by measuring the integrated intensity in 4–12 images per genotype."

"Lines were drawn along or perpendicular to positive-stained areas, and peak intensity values were plotted denoted as "max" measurement on ImageJ. Gene ontology (GO) enrichment analysis for RNA sequencing and proteomic data sets was performed using the PANTHER classification system online user interface [45] and the statistical overrepresentation test."

"Differentially expressed transcripts and proteins between mdx and mdx^{TG} were pooled as input for integrated gene set enrichment analysis using Database for Annotation, Visualization and Integrated Discovery (DAVID, version 2021)."

Results

"Our goal was to interrogate the effects of SSPN overexpression on gene and protein expression in mdx muscle relative to wild-type and mdx (non-transgenic) controls."

"Gene expression analysis was performed using traditional poly-A-enriched RNA sequencing of tibialis anterior muscles isolated from 12-week-old wild-type, mdx, and mdx^{TG} mice."

"From our proteomics analysis of wild-type, mdx, and mdx^{TG} quadriceps muscle, we identified a total of 1679 proteins in all three genotypes."

"Spectrin is a key linker protein in the response to mechanical stimuli, coupling changes in cellular tension to downstream signaling networks [46, 47]."

"To identify candidate networks and upstream regulators that drive compensatory changes in mdx^{TG} muscle, we analyzed the transcriptomic and proteomic data sets using the ingenuity pathway analysis (IPA) application."

"There are compensatory and restorative changes in several factors that regulate Rho GTPase activity including the GTPase activating proteins (Arhgap 6, 22, and 44), all of which are increased in mdx^{TG} muscle."

Discussion

"Collagen is the most abundant protein in the ECM, and its high tensile strength allows transmission of forces generated by skeletal muscle fibers."

"Nghiem and colleagues [48] identified spectrin as a candidate protein in a study of the cranial sartorius muscle that is spared from dystrophic pathology through compensatory hypertrophy via myostatin signaling."

"The upregulation of collagens and scaffolding cytoskeletal components such as spectrin suggests that SSPN overexpression and the increase of adhesion complexes at the membrane may induce a rewiring of signaling between the ECM and the cytoskeleton to stabilize the sarcolemma during contraction and enhance mechanotransduction in order to improve force transmission."

"Our findings of compensatory changes in muscle of other cartilage-related proteins including collagen II and collagen XI are novel and suggest a possible new mechanism of improved force transmission in dystrophic muscle."

Intermolecular and Dynamic Investigation of the Mechanism of Action of Reldesemtiv on Fast Skeletal Muscle Troponin Complex Toward the Treatment of Impaired Muscle Function [49]

This is a machine-generated summary of:

Issahaku, Abdul Rashid; Ibrahim, Mahmoud A. A.; Mukelabai, Namutula; Soliman, Mahmoud E. S.: Intermolecular And Dynamic Investigation of The

Mechanism of Action of Reldesemtiv on Fast Skeletal Muscle Troponin Complex Toward the Treatment of Impaired Muscle Function [49]

Published in: The Protein Journal (2023)
Link to original: https://doi.org/10.1007/s10930-023-10091-y

If you want to cite the papers, please refer to the original.

For technical reasons we could not place the page where the original quote is coming from.

Abstract-Summary "This has spurred the search for small molecules which could activate fast skeletal muscle troponin complex as a means to increase muscle strength."

"Discovered small molecules have however been punctuated by off-target and side effects leading to the development of the second-generation small molecule, Reldesemtiv."

"We investigated the impact of Reldesemtiv binding to the fast skeletal troponin complex and the molecular determinants that condition the therapeutic prowess of Redesemtiv through computational techniques."

"It was revealed that Reldesemtiv binding possibly potentiates troponin C compacting characterized by reduced exposure to solvent molecules which could favor the slow release of calcium ions and the resultant sensitization of the subunit to calcium."

"These findings present useful insights which could lay the foundation for the development of fast skeletal muscle small molecule activators with high specificity and potency."

Introduction

"Direct activation of skeletal muscle has therefore become a therapeutic target to improve the physical performance in patients with compromised muscle function [50]."

"Fast skeletal muscle fibers vary from cardiac and slow muscles in the uniqueness of its troponin and myosin components [51]."

"Sequence homology between fast and slow muscle troponins C, T, and I is about 50% offering an opportunity to target fast skeletal muscle therapeutically [52]."

"Tirasemtiv selectively interacts with fast skeletal muscle troponin complex inducing an increase affinity for calcium and amplifies the muscle cell response to neural input hence increasing force production at submaximal muscle stimulation frequencies [50, 53, 54]."

"It is our hope that insights unraveled herein will aid in the design of selective small molecule activators with improved therapeutic prowess and less side effects on fast skeletal muscle."

Methodology

"The troponin complex was prepared using UCSF Chimera 1.13.1 by deleting all non-standard residues and the protein saved in pdb format [55]."

"Calcium ions which were co-crystalised with the structure and are reported to play a critical role in Troponin C function was removed because they did not conform to standard AMBER parameterization."

"Two systems were created: the unbound troponin complex system with the labile calcium ions (apo) and the Reldesemtiv bound troponin (Ca^{2+}) complex system."

"Additional hydrogens were then added and the systems neutralized by adding Na^{+} and Cl^{-} counter ions and the systems solvated using the LEap module incorporated in AMBER18 [56]."

"The systems were suspended in bulk solvent of Transferable Intermolecular Potential with 3 Points (TIP3P) water box size of 10 Å. Before MD simulation production, the systems were minimized and relaxed, heated and then equilibrated."

Results and Discussion

"The second-generation small molecule Reldesemtiv binds to the N-terminal of the troponin C components thus slowing the rate of calcium release and therefore sensitizing the muscle to calcium."

"We began the analysis by estimating the RMSD of the C-α atoms of the 159 residues of the troponin C component of the troponin complex to determine the stability of the protein upon Reldesemtiv binding during the simulation period."

"As observed, the Reldesemtiv-troponin C complex showed average RMSD value of 2.73 ± 0.37 Å compared to the unbound troponin C which presented 2.82 ± 0.33 Å. These values also express confidence in the simulation process suggesting the systems converged during the 150 ns simulation since they fell below 3.0 Å. The RMSF estimation which is informative on the flexibility of the residues, revealed Reldesemtiv induced an increase in troponin C residual fluctuation compared to the unbound form."

Conclusion

"In our attempt to unravel the atomistic and molecular sights into the therapeutic activity of Reldesemtiv on the fast skeletal troponin complex, we docked the small molecule into the binding pocket of the troponin C subunit of the complex and then subjected the complex system to 150 ns MD simulation."

"C-a atoms analyses of the troponin C subunit of the complex revealed the binding of Reldesemtiv could have induced stability of the troponin C complex in a dense and compact state with reduced solvent accessibility surface area."

"Further analyses suggest Reldesemtiv induced more conformational changes on the distant C-terminal than the N-terminal where it binds taking into account the residues scale difference."

"These findings therefore provide insights into the binding and impacts of Reldesemtiv on the troponin complex and would therefore be useful in the design of small molecule activators with high specificity and potency."

Molecular Regulation of Skeletal Muscle Mitochondrial Biogenesis Following Blood Flow-Restricted Aerobic Exercise: A Call to Action [57]

This is a machine-generated summary of:

Preobrazenski, Nicholas; Islam, Hashim; Gurd, Brendon J.: Molecular regulation of skeletal muscle mitochondrial biogenesis following blood flow-restricted aerobic exercise: a call to action [57]

Published in: European Journal of Applied Physiology (2021)

Link to original: https://doi.org/10.1007/s00421-021-04669-6

If you want to cite the papers, please refer to the original.

For technical reasons we could not place the page where the original quote is coming from.

Abstract-Summary "Little is known about the acute responses within skeletal muscle following BFR aerobic exercise (AE)."

"Although preliminary evidence suggests chronic BFR AE may augment certain training adaptations in skeletal muscle mitochondria more than non-BFR AE, the underlying mechanisms are poorly understood."

"We summarise the acute BFR AE literature examining mitochondrial biogenic signalling pathways and provide insight into mechanisms linked to skeletal muscle remodelling following BFR AE."

"We focus on signalling pathways potentially contributing to augmented peroxisome proliferator-activated receptor gamma coactivator 1-alpha (PGC-1α) mRNA following work-rate-matched BFR AE compared with non-BFR AE."

Introduction

"Most BFR exercise studies to date demonstrate increased muscular strength and/or hypertrophy following resistance training (RT) at workloads as low as 20% of 1-repetition maximum [58]."

"Although BFR may limit prolonged exercise at moderate-to-high intensities [59, 60] and can be painful [61–63], the lower intensity used in BFR RT makes it an appealing and increasingly popular exercise modality for older adults [64], untrained individuals [65], rehabilitation patients [66], and athletes [67, 68] to augment or achieve similar adaptations to non-BFR RT."

"Our current understanding of the molecular mechanisms underlying mitochondrial adaptations following BFR AE are incomplete."

"The purpose of this review is to (1) review the BFR AE literature to provide deeper mechanistic insight into the molecular pathway(s) underlying mitochondrial adaptations following BFR AE and (2) evaluate the quality of the acute BFR AE literature using the Cochrane Collaboration's risk of bias assessment tool [69]."

BFR AE Background

"Cuff-based BFR may increase hypoxia-mediated metabolite production and cause venous pooling-mediated decreases in metabolite clearance."

"Both metabolite accumulation and venous pooling may contribute to augmented skeletal muscle adaptations following cuff-based BFR AE [70]."

"Similar to cuff-induced BFR, lower body positive pressure (LBPP) induces hypoxia within exercising muscle by decreasing blood flow [71]."

"Unlike cuff-based BFR, LBPP increases venous return thereby eliminating the metabolite accumulation and venous pooling associated with cuff-based BFR [72]."

"LBPP and placing exercising limbs above the heart (i.e. a gravity-based model), therefore, represent models to study the impact of BFR-induced hypoxia on signalling and muscular adaptations in isolation (i.e. in the absence of venous pooling and cellular swelling)."

"Comparing cuff-based with non-cuff-based BFR may also allow researchers to separate oxygen-dependent mechanisms (via local hypoxia) from cellular swelling-induced signalling (via venous pooling)."

Acute Blood Flow-Restricted Aerobic Exercise and PGC-1α

"BFR AE can provide insight into in vivo PGC-1α regulation because BFR AE may modulate energetic stress (i.e. [AMP]:[ATP] ratio) in insolation (see below for details), as evidenced by BFR AE activating AMPK and increasing PGC-1α mRNA more than non-BFR AE in human skeletal muscle [73–75]."

"AMPK activation and PGC-1α mRNA increases in an intensity-dependent manner following non-BFR AE [76]."

"These findings [73–75] present enhanced AMPK activation as the most likely mechanism underlying augmented PGC-1α mRNA following work rate-matched BFR AE."

"P38 MAPK activation and PGC-1α mRNA both increase in human skeletal muscle following non-BFR AE [77, 78]."

"Augmented p38 MAPK activation and PGC-1α mRNA, however, have not been observed following BFR AE."

"Skeletal muscle PGC-1α mRNA often [73–75, 79], but not always [79], increases more following BFR applied during AE compared with non-BFR AE."

Chronic BFR Aerobic Exercise and Mitochondrial Biogenesis

"If BFR induces consistently greater activation of the AMPK–PGC-1α pathway during BFR AE, then repeated bouts of BFR AE (i.e. aerobic training) may also augment changes in mitochondrial content [80]."

"Citrate synthase (CS) activity, a validated biomarker of skeletal muscle mitochondrial content in humans [81], is the only marker of mitochondrial content examined within the BFR AE literature to date."

"Only two chronic BFR AE studies from the early 1990s have examined changes in skeletal muscle mitochondrial content, and both reported significantly higher CS activities in the BFR leg compared with the non-BFR leg following 4 weeks of one-legged training with LBPP [82, 83]."

"The impact of chronic BFR AE on mitochondrial content in skeletal muscle requires further investigation."

"Because training-induced changes in mitochondrial function can dissociate from changes in mitochondrial content [84], measuring mitochondrial respiration may provide additional information into mitochondrial adaptations following chronic BFR AE."

Methodological Considerations for Future Work

"To evaluate the quality of existing acute BFR AE studies, two reviewers (NP and HI) independently assessed relevant BFR AE studies based on what was reported in published manuscripts using the Cochrane Collaboration's risk of bias assessment tool [69]."

"Studies that reported an adequate methodology (e.g. blinding outcome assessors to protect against detection bias) for protecting against a given source of bias were judged as having a "low" risk of bias."

"Studies were judged as having a "high" risk of reporting bias if they did not report publicly registering their trial or reporting their methods in a public database/registry."

"We generally observed an unclear risk of bias across acute BFR AE studies."

"The unclear risk of bias in BFR AE studies paired with the generally poor chronic BFR AE study quality reported in Bennett and Slattery [85] demonstrates a need to use adequately powered, rigorously designed studies to examine BFR AE."

Conclusion

"The AMPK-PGC-1α axis is emerging as an important mechanistic pathway underlying acute mitochondrial biogenic responses to BFR applied during acute AE."

"Augmented increases in PGC-1α mRNA following work rate-matched BFR AE concur only with augmented AMPK signalling using cuff-, gravity-, and pressure chamber-based BFR models."

"Although BFR aerobic training appears to increase mitochondrial content more than work-rate matched non-BFR aerobic training, this observation requires replication."

Secretome of Adipose-Derived Mesenchymal Stem Cells Promotes Skeletal Muscle Regeneration Through Synergistic Action of Extracellular Vesicle Cargo and Soluble Proteins [86]

This is a machine-generated summary of:

Mitchell, Robert; Mellows, Ben; Sheard, Jonathan; Antonioli, Manuela; Kretz, Oliver; Chambers, David; Zeuner, Marie-Theres; Tomkins, James E.; Denecke, Bernd; Musante, Luca; Joch, Barbara; Debacq-Chainiaux, Florence; Holthofer, Harry; Ray, Steve; Huber, Tobias B.; Dengjel, Joern; De Coppi, Paolo; Widera, Darius; Patel, Ketan: Secretome of adipose-derived mesenchymal stem cells promotes skeletal muscle regeneration through synergistic action of extracellular vesicle cargo and soluble proteins [86].

Published in: Stem Cell Research & Therapy (2019)

Link to original: https://doi.org/10.1186/s13287-019-1213-1

If you want to cite the papers, please refer to the original.

For technical reasons we could not place the page where the original quote is coming from.

Abstract-Summary "Various in vitro models were used to evaluate the effects of the secretome on cellular processes that promote tissue regeneration."

"A cardiotoxin-induced skeletal muscle injury model was used to test the regenerative effects of the whole secretome or isolated extracellular vesicle fraction in vivo."

"We have demonstrated that the secretome from adipose-derived mesenchymal stem cells shows robust effects on cellular processes that promote tissue regeneration."

"We assessed the efficacy of the total secretome compared with the extracellular vesicle fraction on a number of assays that inform on tissue regeneration and demonstrate that both fractions affect different aspects of the process in vitro and in vivo."

"Our in vitro, in vivo, and bioinformatic results show that factors that promote regeneration are distributed both within extracellular vesicles and the soluble fraction of the secretome."

"Our study implies that extracellular vesicles and soluble molecules within ADSC secretome act in a synergistic manner to promote muscle generation."

Background

"We reported that the secretome from human amniotic fluid stem cells (AFSCs) has anti-inflammatory properties, promotes stem cell proliferation, protects against cellular senescence and facilitates tissue regeneration in a cardiotoxin (CTX)-induced model of muscle degeneration [87]."

"We wanted to investigate if the application of the same isolation method would produce an ADSC secretome that has similar regenerative potential and if the EV fraction was solely responsible for these effects."

"We detected significant differences between the impact of the whole secretome and the EV fraction on the muscle stem cells during regeneration."

"Our comparison of ADSC and AFSC secretomes revealed 108 mutually exclusive protein hits in the EV fractions of both secretomes and 50 proteins present in both soluble fractions."

"Our data implicates that the beneficial effects of the paracrine factors are a result of a synergistic action of the miRNA and protein cargo of the EV fraction and the soluble proteins within the total secretome."

Materials and Methods

"ADSC EV were labelled with fluorescent dye PKH67 (Sigma Aldrich MIDI67) by adding 40 μL of EV fraction (EV from 1×10^6 cells) to PKH67 dye solution followed by incubation for 5 min at room temperature before being ultracentrifuged at 200,000g for 18 h at 4 °C."

"C2C12 cells were seeded into 6-well cell culture plates (Sarstedt) at 3000 cells/cm^2 and allowed to adhere for 24 h. Two percent whole secretome (v/v) in growth media was added to the cells followed by incubation for 48 h and manual cell counting using a haemocytometer."

"Non-immunosuppressed mice were warmed for 10 min in a hot box at 40 °C prior to the intra-venous (IV) injection of 100 μL sterile 0.1 M PBS, 100 μL whole secretome or 100 μL EV (from 1×10^6 cells) through the lateral tail vein (5 mice per condition)."

Results

"We investigated the impact of the total ADSC secretome on these three hallmarks of regeneration."

"An acute CTX-induced muscle injury model was used to assess the impact of a systemic administration of either the whole ADSC secretome or the EV fraction on tissue regeneration."

"In order to comparatively quantify the effects of the total ADSC secretome and the EV fraction on muscle regeneration, we quantified the cross-sectional area (CSA) of newly formed muscle fibres identified by the expression of embryonic myosin heavy-chain (eMHC)."

"EVs were isolated from total secretome of three independent ADSC culture and the miRNA content of ADSC EV was analysed using a GeneChip® miRNA 4.0 array with the Affymetrix Expression Console™ software."

"We have taken a bioinformatic approach to compare molecular components of the ADSC secretome with the secretome of AFSCs that also promotes muscle regeneration [87]."

Discussion

"Although a number of groups have shown that the EV fraction of MSC secretomes support muscle regeneration [88, 89], the identity of the putative effective agents differs greatly."

"Due to the strong pro-angiogenic effects of the total ADSC secretome but not the EV fraction, we expected that the soluble protein fraction would contain all or at least some of these."

"Analysis of the miRNAs within the EV fraction of the ADSC secretome revealed a presence of many well-characterised anti-inflammatory miRNAs, including the miR-let7 family (let7a, b, c and e)."

"Two of the top 50 miRNAs within the ADSC EVs, miR-23a and miR-23b, are known to be strongly pro-angiogenic [90]."

"This comparative analysis of the protein and miRNA components of both stem cell types reveals distinct profiles despite similar biological effects in muscle regeneration."

Conclusions

"Our data suggest that soluble and EV-associated factors promote muscle regeneration both in vitro and in vivo in a highly synergistic manner."

"This is supported by our findings showing that both fractions affect different aspects of tissue regeneration after muscle injury."

"Our comparative analysis of ADSC and AFSC secretomes indicates that different molecular factors might mediate similar beneficial biological outcomes."

Profiling of Ob/Ob Mice Skeletal Muscle Exosome-Like Vesicles Demonstrates Combined Action of miRNAs, Proteins and Lipids to Modulate Lipid Homeostasis in Recipient Cells [91]

This is a machine-generated summary of:

Jalabert, Audrey; Reininger, Laura; Berger, Emmanuelle; Coute, Yohann; Meugnier, Emmanuelle; Forterre, Alexis; Errazuriz-Cerda, Elizabeth; Geloen, Alain; Aouadi, Myriam; Bouzakri, Karim; Rieusset, Jennifer; Rome, Sophie: Profiling of ob/ob mice skeletal muscle exosome-like vesicles demonstrates combined action of miRNAs, proteins and lipids to modulate lipid homeostasis in recipient cells [91].

Published in: Scientific Reports (2021)

Link to original: https://doi.org/10.1038/s41598-021-00983-3

If you want to cite the papers, please refer to the original.

For technical reasons we could not place the page where the original quote is coming from.

Abstract-Summary "We have determined the lipid, protein and miRNA composition of skeletal muscle (SkM)-released extracellular vesicles (ELVs) from Ob/ob (OB) vs wild-type (WT) mice."

"The results showed that atrophic insulin-resistant OB-SkM released less ELVs than WT-SkM, highlighted by a RAB35 decrease and an increase in intramuscular cholesterol content."

"Proteomic analyses of OB-ELVs revealed a group of 37 proteins functionally connected, involved in lipid oxidation and with catalytic activities."

"The most significant function targeted by the 7 miRNAs altered in OB-ELVs was lipid metabolism."

"In agreement, OB-ELVs induced lipid storage in recipient adipocytes and increased lipid up-take and fatty acid oxidation in recipient muscle cells."

"OB-ELVs altered insulin-sensitivity and induced atrophy in muscle cells, reproducing the phenotype of the releasing OB muscles."

Introduction

"Previous studies have demonstrated that skeletal muscle releases ELVs (SkM-ELVs), in addition to myokines, and have suggested that SkM-ELVs would act as deleterious signals during the development of insulin-resistance (IR) associated to obesity-induced diabetes [92]."

"It was demonstrated that SkM insulin-resistance induced by a diet enriched in palm oil triggered the release of a new population of SkM-ELVs enriched in palmitate, able to spread the deleterious effect of this lipid between muscle cells [93]."

"The same population of SkM-ELVs could also transfer specific miRNAs into recipient pancreatic beta cells to induce their proliferation, suggesting that muscle-released ELVs might contribute to adaptations in beta cell mass occurring during IR development associated with obesity in mice [94]."

"For other cell types like the adipocytes, it was demonstrated that obesity affected adipocyte-released ELV content which exported a high proportion of obesity and IR-related proteins/lipids/RNAs from adipocytes able to modify the metabolic responses of other insulin-sensitive target cells [95]."

Material and Methods

"We used 100 μg of protein-containing SkM-ELVs for total RNA extraction."

"For SkM-ELV miRNA profiling, 33 ng of total RNA were used for multiplex reverse transcription (RT)."

"SkM-ELV proteins were resuspended in Laemmli buffer, stacked in the top of a SDS-PAGE gel (NuPAGE 4–12%, ThermoFisher Scientific) and stained with Coomassie blue (R250, Bio-Rad) before being in-gel digested using modified trypsin (Promega, sequencing grade) as previously described [96]."

"For quantitative comparison of WT-ELV and OB-ELV proteomes, we used a beta-binomial test specifically developed to test the significance of differential protein abundances expressed in SC."

"Heatmap showing the significantly enriched miRNAs in SkM-ELVs and muscle from OB vs mice were created by using ClustVis (https://biit.cs.ut.ee/clustvis/)."

"PANTHER (http://pantherdb.org) was used to analyze the set of proteins contained in SkM-ELVs (G.O functions, cellular pathways, and reactome), and clustering was performed with MORPHEUS (https://software.broadinstitute.org/morpheus/)."

Results

"These data demonstrated for the first time a strong selection for lipid export in SkM-ELVs, which is altered in obese SkM. As specific lipids and proteins were differentially enriched between OB-ELVs and WT-ELVs, we wondered whether obesity also affected the concentrations of miRNAs."

"This data demonstrated that obesity affected SkM-ELV miRNAs content but that this altered miRNA signature does not reflect intramuscular miRNA alterations, in OB mice."

"As the functions of the SkM-ELV miRNAs released from muscle explants had never been studied before, we predicted their target genes and performed functional enrichment analyses to determine which pathways they affected in the recipient cells."

"As this function was targeted by miRNAs depleted (miR-340-5p) or less abundant in OB-ELVs vs WT-ELVs (i.e.; miR-1a-3p, miR-101a-3p, miR-101b-3p), this data suggested that fatty acid metabolism would be less repressed in recipient cells for SkM-ELVs in obese animals than in healthy animals."

Discussion

"Considering that in OB mice the first relevant target cells of SkM-ELVs would be the muscle itself and adipocytes, we have treated these 2 cell types with OB-ELVs and WT-ELVs released from SkM explants of OB and lean mice."

"We determined the consequence of obesity on the lipid composition of SkM-ELVs and report, for the first time, significant differences between OB and lean mice."

"Considering that SkM-ELVs are from endosomal origin and are formed inside MBVs, the OB-ELV lipidomic data also indicate that obesity affects MVB lipid composition."

"As depleted OB-ELV miRNAs targeted genes involved in lipid metabolism, this function would be less repressed by SkM-ELVs in obese mice."

"Through the modulation of its miRNA population released in SkM-ELVs, SkM can both increase lipid metabolism and concomittantly alter mice muscle mass though the release of 'atrophic miRNAs' in OB-ELVs."

Conclusions

"These data demonstrated that SkM-ELVs have a very specific composition which matches only partially with the compositions of the ELV-releasing muscle in terms of lipids and miRNAs."

"We highlight an unexpected function of OB-ELVs in the control of lipids accumulation in adipocytes which might be very important for the control of the muscle mass/adipose tissue ratio in the body."

"Recently a possible cross-talk between adipocytes and muscle cells has been described, through adipocyte-derived ELVs; i.e., miR-27a released from adipocytes of DIO obese mice was associated with TAG accumulation and IR in recipient muscle cells [97]."

"We demonstrate that SkM is also able to modulate the adipose tissue function through the release of SkM-ELVs."

"Like myokines and adipokines, the exchange of ELVs between adipose tissues and SkM represents a new way of communication that has to be taken into account to develop new therapeutic strategies."

CYB5R3 Overexpression Preserves Skeletal Muscle Mitochondria and Autophagic Signaling in Aged Transgenic Mice [98]

This is a machine-generated summary of:

López-Bellón, Sara; Rodríguez-López, Sandra; González-Reyes, José A.; Burón, M. Isabel; de Cabo, Rafael; Villalba, José M.: CYB5R3 overexpression preserves skeletal muscle mitochondria and autophagic signaling in aged transgenic mice [98]

Published in: GeroScience (2022)

Link to original: https://doi.org/10.1007/s11357-022-00574-8

If you want to cite the papers, please refer to the original.

For technical reasons we could not place the page where the original quote is coming from.

Abstract-Summary "The aim of our study was to understand how CYB5R3 overexpression targets key pathways that modulate the rate of aging in skeletal muscle, a postmitotic tissue with a greater contribution to resting energy expenditure."

"Mitochondrial function, autophagy and mitophagy markers were evaluated in mouse hind limb skeletal muscles from young-adult (7 months old) and old (24 months old) males of wild-type and CYB5R3-overexpressing genotypes."

"CYB5R3, which was efficiently overexpressed and targeted to skeletal muscle mitochondria regardless of age, increased the abundance of complexes I, II, and IV in old mice and prevented the age-related decrease of complexes I, III, IV, and V and the mitofusin MFN-2."

"ATP was significantly decreased by aging, which was prevented by CYB5R3 overexpression."

"Both aging and CYB5R3 overexpression upregulated SIRT3 and the mitochondrial fission markers FIS1 and DRP-1, although with different outcomes on mitochondrial ultrastructure: old wild-type mice exhibited mitochondrial fragmentation whereas CYB5R3 overexpression increased mitochondrial size in old transgenic mice concomitant with an improvement of autophagic recycling."

Introduction

"One of the principal effects of aging on skeletal muscle is sarcopenia, whose onset is related with a decrease in mitochondrial content and functionality, highlighting the importance of the studies focused on this organelle [99]."

"To gain further insights into the participation of CYB5R3 in the regulation of those pathways that influence the rate of aging, we generated mice overexpressing CYB5R3 (TG mice), which displayed enhanced protection against induced cancer, increased insulin sensitivity, less oxidative damage, and extended longevity [100]."

"The aim of this work was to study how CYB5R3 overexpression affects mitochondrial morphology and function as well as autophagy in skeletal muscle from young-adult and aged TG mice."

"Our results demonstrate that, consistent with its prolongevity effects in mice, CYB5R3 overexpression protects mitochondria from aging-associated biochemical and structural alterations and modulates autophagic signaling in skeletal muscle."

Methods

"Hind limb skeletal muscle samples from six animals per group were homogenized in radioimmunoprecipitation assay (RIPA) buffer (50 mM Tris–HCl pH 8; 150 mM NaCl; 0.5% deoxycholate; 0.1% SDS; 1% Triton X-100; 1 mM DTT; 1 mM phenylmethylsulfonyl fluoride (PMSF); 10 μg/mL each of chymostatin, leupeptin, antipain, and pepstatin A (CLAP); and phosphatase inhibitor cocktails 2 and 3 (Sigma-Aldrich) diluted at 1/100) for 30 s using a high-performance dispersing instrument (Ultra-Turrax T25, IKA, Staufen, Germany)."

"Hind limb skeletal muscles clean of fat and connective tissue were homogenized during 30 s at 4 °C in ice-cold buffer containing 20 mM Tris–HCl pH 7.6, 40 mM KCl, 0.2 M sucrose, 1 mM PMSF, 10 mM EDTA, 1 mM DTT, 20 μg/μL CLAP, and phosphatase inhibitor cocktails 2 and 3 (Sigma-Aldrich) diluted at 1/100, using an electric tissue disrupter (Ultra-Turrax T25, IKA, Staufen, Germany)."

Results

"We measured the amounts of CYB5R3 polypeptide not only in whole extracts, but also in enriched mitochondrial fractions that had been isolated from hind limb skeletal muscle of both WT and TG mice at two age points: 7 months (young-adult; Y-WT and Y-TG groups, respectively) and 24 months (old; O-WT and O-TG groups, respectively)."

"We next studied the effect of aging and/or CYB5R3 overexpression on protein markers of mitochondrial dynamics."

"The changes in protein markers of mitochondrial dynamics are consistent with a prevalence of fission in O-WT and in Y-TG, without a further increase of this process in O-TG mice."

"In WF cross-sections, we found that CYB5R3 overexpression led to a remarkable decrease of mean mitochondrial size in Y-TG compared with Y-WT mice."

"This analysis uncovered a more subtle effect of CYB5R3 overexpression increasing the number of smaller mitochondrial profiles in Y-TG compared with Y-WT, as we had also encountered in WF (see above)."

Discussion

"Representative protein markers of mitochondrial complexes I, III, IV, and V decreased with aging in skeletal muscle from WT mice."

"While the aging-related decline of mitochondrial biogenesis markers and coenzyme Q was not prevented in TG mice, CYB5R3 overexpression did avoid the decrease in mitochondrial complexes observed in aged WT mice (see above), which could contribute to maintain the functionality of this organelle with advanced age."

"In our model, the abundance of the cleaved (mitochondrial) isoform of SIRT-3 was significantly increased in hind-limb skeletal muscle from aged WT mice, and a trend towards an increase due to CYB5R3 overexpression was also observed in young-adult mice, followed by the stabilization of SIRT-3 abundance in old TG mice."

"A lack of MFN-2 decrease with aging in TG mice reinforces the importance of CYB5R3 expression to prevent aging-associated mitochondrial dysfunction in skeletal muscle."

Elucidating the Contribution of Skeletal Muscle Ion Channels to Amyotrophic Lateral Sclerosis in Search of New Therapeutic Options [101]

This is a machine-generated summary of:

Camerino, Giulia Maria; Fonzino, Adriano; Conte, Elena; De Bellis, Michela; Mele, Antonietta; Liantonio, Antonella; Tricarico, Domenico; Tarantino, Nancy; Dobrowolny, Gabriella; Musarò, Antonio; Desaphy, Jean-Francois; De Luca, Annamaria; Pierno, Sabata: Elucidating the Contribution of Skeletal Muscle Ion Channels to Amyotrophic Lateral Sclerosis in search of new therapeutic options [101]

Published in: Scientific Reports (2019)

Link to original: https://doi.org/10.1038/s41598-019-39676-3

If you want to cite the papers, please refer to the original.

For technical reasons we could not place the page where the original quote is coming from.

Abstract-Summary "We investigated the role of the most important ion channels in skeletal muscle of an ALS animal model (MLC/SOD1^{G93A}) carrying a mutated SOD1 exclusively in this tissue, avoiding motor-neuron involvement."

"Ion channels are fundamental proteins for muscle function, and also to sustain neuromuscular junction and nerve integrity."

"The expression of ClC-1 chloride channel, present only in skeletal muscle, was reduced."

"The functional characterization confirmed the reduction of ClC-1 activity, leading to hyperexcitability and impaired relaxation."

"The increased expression of ion channel coupled AMPA-receptor may contribute to sustained depolarization and functional impairment."

"These findings show that ion channel function impairment in skeletal muscle may lead to motor-neuron increased vulnerability, and opens the possibility to investigate on new compounds as promising therapy."

Introduction

"The very earliest manifestation of disease in both the SOD1^{G93A} mouse model (carrying G93A mutation in SOD1 protein) and ALS patients occurs at the NMJ, where significant levels of denervation can be observed before the onset of motor neuron degeneration [102]."

"The ClC-1 chloride channel is typically expressed in skeletal muscle and controls resting membrane potential and excitability [103, 104]."

"Hereditary loss-of-function mutations in the ClC-1 channel are responsible for Myotonia Congenita, a disease characterized by impairment of muscle excitability and relaxation [105]."

"There are no reports on ClC-1 channel involvement in ALS, a channel exclusively expressed in skeletal muscle."

"In the light of these considerations, we focused our study on ion channel expression and function in skeletal muscles of transgenic SOD1^{G93A} mice and in muscle specific MLC/SOD1^{G93A} mice (carrying the SOD1^{G93A} mutant gene under the transcriptional control of Myosin Light Chain, MLC, a muscle-specific promoter) to elucidate the mechanisms and to find potential therapeutic targets in ALS."

Results

"The quantitative Real-Time PCR analysis have been used to detect skeletal muscle alterations due to SOD1 mutation in both SOD1^{G93A} and MLC/SOD1^{G93A} mice with respect to age-matched and strain-matched wild-type (WT)."

"Principal Component Analysis (PCA) and Linear Discriminant Analysis (LDA) The gene expression dataset was composed of 27 columns representing the expression level of genes normalized to β-Actin (housekeeping gene) and 30 rows representing the analyzed TA muscles."

"In this case, the clusters segregated muscles belonging to ubiquitously $SOD1^{G93A}$ into ($MLC/SOD1^{G93A}$) and WT, highlighting that the major part of the variance is due to variation in gene expression level occurring in $SOD1^{G93A}$ mice."

"Since the mRNA level of ClC-1 gene was not significantly changed in the muscle of $MLC/SOD1^{G93A}$ mice, it seems that post-transcriptional events are involved, i.e. an alteration of translation or a rapid degradation of the protein."

"The restCa level was not significantly modified in skeletal muscle of $MLC/SOD1^{G93A}$ mice."

Discussion

"Mstn expression was decreased also in $MLC/SOD1^{G93A}$ mice, the animal model expressing the SOD1 mutant gene selectively in skeletal muscle [106], again suggesting the early involvement of skeletal muscle in the pathogenesis of ALS."

"The ClC-1 protein expression and resting gCl were significantly reduced also in muscles of $MLC/SOD1^{G93A}$ mice, likely through a mechanism involving Reactive Oxygen Species (ROS) production."

"$MLC/SOD1^{G93A}$ muscles were characterized by a minor number of transcriptional modifications, and the identified most discriminant genes involved in the onset of damage were PKC-theta, that control muscle ClC-1 activity, AMPAR2, the ionotropic glutamate receptor involved in synaptic plasticity, and irisin, a recently-discovered hormone secreted by skeletal muscle and able to communicate in a paracrine and endocrine manner with nervous system [107]."

"The results indicate that muscle hyperexcitability is due to ClC-1 expression reduction and increased activity of PKC-theta with consequent gCl reduction in $MLC/SOD1^{G93A}$ mice."

Material and Methods

"The ATP-sensitive potassium (KATP) channel currents were recorded immediately after excision during voltage steps from 0 to −60 mV with 150 mM KCl on both sides of the membrane patches in the absence (control) or the presence of ATP in the muscle bath [108]."

"Multivariate analysis of gene expression dataset has been performed following the PCA-LDA analysis [109, 110]."

"For each animal 27 genes have been analyzed as normalized values ratio with house-keeping gene β-Actin (HK: Actb; target genes: ClC1, PKC Theta, PKC Alpha, Ryr1, Serca1, Serca2, CN, Bk, Kir6.2, Sur1, Surb2a, Sur2b, Nav1.4, Nmdar1, Ampar2, Murf1, Irisin, Notch1, TauT, Ampk, Ngf, Mstn, Achr1, Mhc2b, Mhc2x, Mhc2a, Mhc1)."

"3 vectors of LDA coefficients and a matrix containing the PCA loading weights of each genes have been obtained."

"The three discriminant lists of genes that separate the classes have been obtained multiplying the PCA loadings matrix with each transposed vector of LDA coefficients."

Unacylated Ghrelin Restores Insulin and Autophagic Signaling in Skeletal Muscle of Diabetic Mice [111]

This is a machine-generated summary of:

Tam, Bjorn T.; Pei, Xiao M.; Yung, Benjamin Y.; Yip, Shea P.; Chan, Lawrence W.; Wong, Cesar S.; Siu, Parco M.: Unacylated ghrelin restores insulin and autophagic signaling in skeletal muscle of diabetic mice [111]

Published in: Pflügers Archiv—European Journal of Physiology (2015)
Link to original: https://doi.org/10.1007/s00424-015-1721-5

If you want to cite the papers, please refer to the original.

For technical reasons we could not place the page where the original quote is coming from.

Abstract-Summary "Impairment of insulin signaling in skeletal muscle detrimentally affects insulin-stimulated disposal of glucose."

"Restoration of insulin signaling in skeletal muscle is important as muscle is one of the major sites for disposal of blood glucose."

"We examined the effects of UnAG on restoring the impaired insulin signaling in skeletal muscle of db/db diabetic mice."

"Our results demonstrated that UnAG effectively restored the impaired insulin signaling in diabetic muscle."

"Our data indicated that UnAG normalized the suppressed autophagic signaling in diabetic muscle."

"Our findings illustrated that UnAG restored the impaired insulin and autophagic signaling in skeletal muscle of diabetic mice, which are valuable to understand the underlying mechanisms of the anti-diabetic action of UnAG at peripheral skeletal muscle level."

Introduction

"Defects at different levels of insulin signaling pathway in skeletal muscle lead to insulin resistance, in which muscle cells do not respond adequately to normal insulin concentration and is a key feature of type 2 diabetes [112]."

"A highly regulated cellular process known as autophagy has been linked to insulin resistance in skeletal muscle."

"Inactivated or insufficient autophagy in skeletal muscle, liver, and adipose tissue has been shown to result in (or associate with) defective insulin signaling [113–115]."

"Another loss-of-function study has reported that autophagy-related protein 7 (Atg7) knockdown leads to enhanced endoplasmic reticulum stress and defective insulin signaling in liver tissue of mice."

"Skeletal muscle-specific autophagy deficiency has been demonstrated to protect mice from diet-induced obesity and insulin resistance [116]."

"The effects of UnAG on skeletal muscle insulin signaling pathway and autophagic signaling pathway remain largely unknown."

Methods

"Non-diabetic db/+ mice (heterozygous allelic deficient for the leptin receptor gene) were used as the non-diabetic control because db/db mice and db/+ mice have similar genetic background except that db/db mice exhibit abnormal blood glucose level and show the type 2 diabetic phenotype."

"Mice assigned to UnAG treatment were exposed to intraperitoneal injection of unacylated ghrelin (Tocris Bioscience, USA) twice daily for ten consecutive days."

"Sections were then incubated with rabbit IgG anti-Glut 4 (#1404, Millipore, 1:400) and mouse IgG anti-dystrophin (D505, Vector Laboratories 1:50) in 1 % BSA in PBST in a humidified chamber overnight at 4 °C."

"A commercially available plasma membrane protein extraction kit (#k268-50, BioVision, Mountain View, CA, USA) was used to extract plasma membrane protein from skeletal muscle."

Results

"To further assess the beneficial effects of UnAG on intracellular insulin signaling in muscle of db/db mice, we measured the enzymatic activity of PDH in muscle."

"Besides the alterations in insulin signaling and enzymatic activity, UnAG also tended to ameliorate the fasting blood glucose concentration of db/db mice."

"UnAG was found to upregulate the protein abundances of Beclin-1 and Atg5 by 215 and 140 %, respectively, in muscle of UnAG-treated db/db mice when compared to db/db control mice."

"Skeletal muscle autophagy was impaired in db/db control mice as indicated by the 79 % decrease in LC3 ratio and 164 % increase in p62 accumulation in db/db diabetic muscle relative to db/+ non-diabetic muscle."

"UnAG restored the suppressed autophagy in muscle of UnAG-treated db/db mice by increasing LC3 ratio and reducing p62 accumulation by 638 and 67 %, respectively, when compared to db/db control mice."

Discussion

"We demonstrated that unacylated ghrelin restored the impaired insulin signaling and re-activated the suppressed autophagic signaling in skeletal muscle of db/db diabetic mice."

"UnAG reduced the protein tyrosine nitration and restored insulin signaling in diabetic skeletal muscle making it a potential drug to treat diabetes."

"We have demonstrated that UnAG can restore the impaired insulin signaling in skeletal muscle of db/db diabetic mice."

"We demonstrated that insulin signaling, PDH activity, and fasting blood glucose were impaired in skeletal muscle of db/db diabetic mice."

"UnAG could reduce phosphorylation of IRS-1 at Ser307 and enhance activation of several downstream kinases such as PDK1 and Akt in skeletal muscle of db/db mice."

"Our data illustrated that UnAG partially restored the dysregulated insulin signaling in diabetic skeletal muscle by modulating and resetting the phosphorylation status of IRS-1, PDK1, PKCζ, Akt, and PRAS40."

Conclusion

"UnAG is effective in normalizing autophagy and partially restoring the impaired insulin signaling in skeletal muscle of diabetic mice."

"Coincidentally, the re-activation of autophagy restored the impaired IRS/Akt/mTOR signaling in diabetic muscle."

"Although UnAG is only an unacylated product of ghrelin, it clearly possesses beneficial cellular effects on muscle insulin signaling in diabetic muscle."

Insulin Signaling and Skeletal Muscle Atrophy and Autophagy in Transition Dairy Cows Either Overfed Energy or Fed a Controlled Energy Diet Prepartum [117]

This is a machine-generated summary of:

Mann, S.; Abuelo, A.; Nydam, D. V.; Leal Yepes, F. A.; Overton, T. R.; Wakshlag, J. J.: Insulin signaling and skeletal muscle atrophy and autophagy in transition dairy cows either overfed energy or fed a controlled energy diet prepartum [117]

Published in: Journal of Comparative Physiology B (2016)

Link to original: https://doi.org/10.1007/s00360-016-0969-1

If you want to cite the papers, please refer to the original.

For technical reasons we could not place the page where the original quote is coming from.

Abstract-Summary "This is important in transition dairy cows experiencing negative energy and protein balance postpartum."

"Overconsumption of energy during late pregnancy affects resting glucose and insulin concentrations peripartum and increases the risk for hyperketonemia postpartum, but the effects on muscle tissue are not fully understood."

"Our objective was to study peripartal skeletal muscle insulin signaling as well as muscle accretion and atrophy in cows with excess energy consumption prepartum."

"Skeletal muscle biopsies were obtained 28 and 10 days prepartum, as well as 4 and 21 days postpartum from 24 Holstein cows."

"Excess energy consumption in late pregnancy did not lead to changes in insulin-dependent molecular regulation of muscle accretion or atrophy compared with the controlled energy group."

"This study showed that in addition to the proteasome pathway of muscle atrophy, macroautophagy is upregulated in postpartum negative energy and protein balance regardless of dietary energy strategy prepartum and was higher in cows overfed energy throughout the study period."

Introduction

"Insulin has a central role in regulating skeletal muscle hypertrophy and atrophy in all mammals via the AKT and mammalian target of rapamycin (mTORC1) signaling pathway, and through E3 ubiquitin ligases, such as atrogin-1 and muscle RING-finger protein-1 (muRF1) [118, 119]."

"Hyperketonemic animals have a more pronounced drop in insulin concentrations in the first weeks after calving [120] and could therefore experience a change in the rate and regulation of muscle atrophy and muscle specific insulin signaling compared with cows that maintain nutrient balance more adequately."

"The regulation of the mTORC1 pathway during the precarious energy and protein balance in postpartum dairy cows is of particular interest."

"The objective of our study was to describe changes in insulin signaling and regulation of muscle hypertrophy and atrophy in vivo during the peripartal period in dairy cattle either overfed energy or fed a controlled energy diet prepartum."

Materials and Methods

"Approximately 50 mg of frozen muscle tissue from each sample were transferred to 500 μL of Trizol reagent (Life Technologies, Thermo Fisher Scientific, Waltham, MA), placed in the pre-chilled adapter of a tissue disruptor (TissueLyser LT, Qiagen, Hilden, Germany) and homogenized at maximum oscillation for 3 min with sterile 5 mm stainless steel beads."

"Repeated ultrasonographic measurements of the diameter of M. longisssimus dorsi, dry matter intake pre- and postpartum, energy balance and MP balance estimates, proteasome activity and baseline gene and protein expression before glucose infusion, as well as the change after infusion were analyzed using separate models of repeated measures ANOVA (Proc MIXED, SAS, v. 9.3, Cary, NC) with the fixed effects group, time and including a group × time interaction, and including the dry-off measurement as a covariate."

Results

"During the first 3 weeks postpartum, predicted energy balance was negative for both groups due to the low energy intake and did not differ between the two groups (group H: 64.8 (59.9–69.8), group C: 70.5 (65.4–75.7), $p = 0.11$)."

"Muscle diameter was different for both groups with a more rapid increase in diameter prepartum as well as a more rapid decrease postpartum in group H compared with group C. In fact, the estimated daily loss of muscle diameter was greater on average in group H (0.35 [0.27–0.44] mm) compared with the average in group C (0.19 [0.11–0.27] mm, $p = 0.008$)."

"Differences in baseline concentrations over time were noted for the phosphorylation of AKT which was significantly reduced on d 4 postpartum compared to prepartum time-points in both groups (Time $p = 0.0002$)."

Discussion

"We found that the glucose-induced increase in circulating insulin concentration in vivo led to the activation of the AKT pathway in muscle tissue, and that the difference in resting insulin concentrations between groups did not translate into a difference in the baseline phosphorylation status of AKT."

"Decreased AKT-phosphorylation postpartum as a result of the decreased stimulation of the insulin pathway increased the gene expression of atrogin-1 in bovine muscle tissue samples from cows with and without high concentrations of BHB postpartum."

"This might be due to the fact that autophagy supports cell remodeling and proliferation [121, 122] and the higher rate of both accretion and mobilization of muscle tissue in this group as reflected by ultrasound measurements with the observed changes resulting in potentially larger losses of lean mass over time in group H. We did not detect any difference in the ratio of phosphorylated ERK kinase either over time, between groups or after glucose challenge."

Conclusions

"This study showed skeletal muscle changes due to parturition and lactation, particularly in the phosphorylation status of AKT at baseline and in response to an endogenous insulin surge."

"These changes were not driven by prepartum plane of nutrition or differences in lipid metabolism postpartum."

"We identified the potential role of macroautophagy in skeletal muscle proteolysis during negative energy balance in postpartum dairy cows in addition to the well characterized proteasome pathway of degradation."

"The changes in myostatin expression relative to the onset of lactation were pronounced, yet counterintuitive to myostatin's traditional regulatory role."

Proinflammatory NFkB Signalling Promotes Mitochondrial Dysfunction in Skeletal Muscle in Response to Cellular Fuel Overloading [123]

This is a machine-generated summary of:

Nisr, Raid B.; Shah, Dinesh S.; Ganley, Ian G.; Hundal, Harinder S.: Proinflammatory NFkB signalling promotes mitochondrial dysfunction in skeletal muscle in response to cellular fuel overloading [123]

Published in: Cellular and Molecular Life Sciences (2019)

Link to original: https://doi.org/10.1007/s00018-019-03148-8

If you want to cite the papers, please refer to the original.

For technical reasons we could not place the page where the original quote is coming from.

Abstract-Summary "Sustained nutrient (fuel) excess, as occurs during obesity and diabetes, has been linked to increased inflammation, impaired mitochondrial homeostasis, lipotoxicity, and insulin resistance in skeletal muscle."

"We examine the contribution that an increase in proinflammatory NFkB signalling makes towards regulation of mitochondrial bioenergetics, morphology, and dynamics and its impact upon insulin action in skeletal muscle cells subject to chronic fuel (glucose and palmitate) overloading."

"Pharmacological or genetic repression of NFkB activity ameliorated disturbances in mitochondrial respiratory function/morphology, attenuated loss of SDHA, ANT-1, UCP3, and MFN2 and mitigated the increase in ROS and the associated reduction in myotube insulin sensitivity."

"Our findings indicate that sustained oversupply of metabolic fuel to skeletal muscle cells induces heightened NFkB signalling and that this serves as a critical driver for disturbances in mitochondrial function and morphology, redox status, and insulin signalling."

Introduction

"There is mounting evidence that chronic activation of proinflammatory signalling in tissues such as skeletal muscle and adipose is a significant contributing factor in the development and progression of metabolic disorders such as insulin resistance, obesity and Type II diabetes [124, 125]."

"In support of this latter idea, signalling initiated by toll-like receptor-4 (TLR-4) and the tumor necrosis factor-α (TNF-α) receptor results in activation of the NFkB pathway, which has been linked to reduced mitochondrial respiration and suppressed activation of transcriptional regulators that promote mitochondrial biogenesis and the shift towards a muscle oxidative phenotype [126, 127]."

"We, and others, have also previously demonstrated that obesity in rodents and chronic oversupply of metabolic fuel to skeletal muscle cells in vitro is associated with an increase in proinflammatory NFkB signalling and insulin resistance [128–130]."

"We hypothesise that sustained oversupply of metabolic fuel will promote activation of proinflammatory NFkB signalling in muscle cells and that this contributes significantly to disturbances in mitochondrial biology that impact negatively upon myocellular insulin sensitivity."

Materials and Methods

"L6 muscle cells were cultured to myotubes as described previously in α-minimal essential media (αMEM) containing 2% (v/v) foetal bovine serum (FBS) and 1% (v/v) antibiotic/antimycotic solution (100 units/ml penicillin, 100 μg/ml streptomycin, and 250 ng/ml amphotericin B) at 37 °C with 5% CO_2 [131]."

"A mitochondrial-enriched membrane fraction was isolated from L6 myotubes using a mitochondria isolation kit (#89874, Thermo Fisher Scientific) as per manufacturer's instructions."

"Cell lysates, cytosolic, nuclear, or mitochondrial-enriched fractions (20 μg protein) from L6 myotubes and human LHCN-M2 myotubes were subjected to SDS/PAGE on 10% resolving gels and transferred onto nitrocellulose membranes (Millipore, Harts, UK), as described previously [131]."

"For determination of hydrogen peroxide (H_2O_2) under live cell conditions, L6 myotubes were incubated with 5 μM MitoPYI (a mitochondrial targeted H_2O_2 probe) and 1 μM deep red cell tracker at 37 °C in a 5% CO_2 incubator for 45 min."

Results

"To explore whether the increase in NFkB signalling seen in myotubes subjected to nutrient overload is a contributing factor to mitochondrial ROS production and respiratory dysfunction, we applied two distinct but complimentary strategies."

"Since inhibition of NFkB signalling suppresses the increase in cellular ROS associated with nutrient overload, we subsequently tested whether this might be linked to improved mitochondrial function."

"To assess whether the improved respiratory capacity that was seen upon inhibiting NFkB signalling in nutrient-overloaded myotubes was associated with changes in mitochondrial morphology or mitochondrial mass we subsequently performed live cell imaging of mitochondria stained with Mitotracker Green and also assessed cellular mitochondrial DNA content."

"To assess whether preserving the respiratory capacity and integrity of the mitochondrial network in nutrient overloaded myotubes by inhibition of NFkB signalling improves insulin action, we assessed insulin-stimulated Akt phosphorylation and glucose uptake as readouts."

Discussion

"Our results indicate that the sustained oversupply of both GLC and PA imposes a major substrate burden on mitochondria in L6 myotubes and that this promotes (1) a reduction in mitochondrial respiratory capacity, (2) reduced expression of key mitochondrial proteins, (3) increased generation of ROS and (4) increased mitochondrial fragmentation and mitophagy."

"It is noteworthy that the notion that mitochondrial expression/activity of ANT-1 may influence insulin sensitivity of muscle cells is supported by studies in cultured C2C12 myotubes, in which partial silencing of ANT-1 results in reduced cellular sensitivity to fatty acid-induced uncoupling and a significant reduction in insulin-stimulated glucose uptake [132]."

"Our findings indicate that muscle cells subjected to sustained oversupply of metabolic fuel exhibit an increase in proinflammatory NFkB signalling that is mechanistically linked to diminished mitochondrial function as evidenced by our ability to antagonise the marked decline in respiratory capacity, expression of key mitochondrial maker proteins and the increased fission and mitophagy by pharmacological or genetic repression of the IKK–NFkB axis."

Evidence for Skeletal Muscle Fiber Type-Specific Expressions of Mechanosensors [133]

This is a machine-generated summary of:

Mathes, Sebastian; Vanmunster, Mathias; Bloch, Wilhelm; Suhr, Frank: Evidence for skeletal muscle fiber type-specific expressions of mechanosensors [133]

Published in: Cellular and Molecular Life Sciences (2019)

Link to original: https://doi.org/10.1007/s00018-019-03026-3

If you want to cite the papers, please refer to the original.

For technical reasons we could not place the page where the original quote is coming from.

Abstract-Summary "Costameres physically link the muscle extracellular matrix to contractile and signaling 'hubs' inside muscle fibers mainly via integrins and are localized beneath sarcolemmas of muscle fibers."

"The fiber type-specific costamere architecture in muscles is unexplored."

"We hypothesized that fiber types differ in the expression of genes coding for costamere components."

"Overexpression of slow type-inducing NFATc1 in muscle precursor cells did not change Ilk or other costamere gene expressions."

"We demonstrate fiber type-specific costamere gene regulation upon mechanical loading and unloading conditions."

"Our data imply that costamere genes, such as Ilk, are involved in the control of muscle fiber characteristics."

"Further, they identify costameres as muscle fiber type-specific loading management 'hubs' and may explain adaptation differences of muscle fiber types to mechanical (un)loading."

Introduction

"Defined subcellular structures, known as costameres (Cstms), convert mechanical forces into biological signaling pathways triggering physiological adaptions of individual SkM fibers to mechanical loading [134]."

"Cstms function as 'hubs' to convert mechanical impacts reaching individual SkM fibers into biological signals to mediate SkM fiber adaptations to these impacts [134]."

"We hypothesized (1) that type 1 and type 2 fibers differ in their Cstms expressions under sedentary conditions, (2) that altered mechanical loading changes Cstms gene expressions in type 1 and type 2 fibers differentially and (3) that Ilk is involved in the control of defined Myh gene expressions in muscle precursor cells."

"Our data provide evidence that individual SkM fibers possess distinct molecular tools to manage mechanical (un)loading conditions and explain that type 1 and type 2 fibers adapt differentially to mechanical (un)loading due to fiber type-specific spatiotemporal assemblies of Cstms 'decision hubs' to govern downstream signals."

Methods

"Laser manipulation of tissue was performed through a graphical user interface that allowed selection for subsequent microdissection of fibers in a collection device."

"For subsequent RNA analysis, individual type 1 or type 2 fibers (comprising 2A, 2X and 2B fibers) corresponding to a total surface area of 3×10^6 µm^2 were laser-microdissected and collected in 20 µL GITC buffer to preserve RNA integrity."

"For subsequent protein analysis, fibers corresponding to a total surface area of 4×10^6 µm^2 were laser-microdissected and collected in 20 µL RIPA buffer and protease inhibitor."

"Complete volumes of cDNA reverse transcribed from laser-microdissected fibers were subjected to 17-plex PreAmp PCR."

"cDNA input was standardized given that complete cDNA prepared from a defined laser-microdissected fiber area of 3×10^6 µm^2 was used."

"Detailed information on tissue collection, immunofluorescence, ATPase staining, immunohistochemistry, RNA extraction, real-time RT-qPCR, RNA gel electrophoresis, cDNA synthesis, DNA gel electrophoresis, DNA extraction and sequencing, protein extraction, NuPAGE SDS-PAGE, western blot, Coomassie staining and buffers can be found in the supplements."

Results

"To further explore the protein levels observed by confocal localization studies, we stained soleus muscles from sedentary control mice with a specific Cstms antibodies and developed the respective signal by DAB, while we used immunofluorescence in parallel to stain type 1 SkM fibers (MyHC1) on the same section."

"With significant differences between fiber type-specific Cstms gene expression profiles under sedentary conditions, we next studied the influence of mechanical loading (WBV) vs. mechanical unloading (IML) on fiber type-specific Cstms expression, because mechanical cues trigger fiber type-specific adaptations via Cstms."

"IML significantly reduced Ilk gene expressions in both fiber types compared to WBV, but not compared to sedentary control."

"In line with Cstms genes, also mTor and p70s6kb1 gene expressions were significantly increased upon WBV in both fiber types, with significantly higher levels in type 1 compared to type 2 fibers."

Discussion

"We developed a multiplex tandem PCR approach coupled to laser microdissection-assisted enrichment of SkM fiber types to study Cstms protein and gene regulation in detail under basal sedentary as well as under conditions of increased (WBV) vs. reduced (IML) mechanical impacts."

"These data provide evidence that SkM Cstms assembly depends on externally applied stimuli and that the composition of Cstms can change rapidly in a fiber type-dependent manner upon defined stimuli to manage downstream adaptation processes."

"mTor gene expression was significantly increased upon WBV in both fiber types compared to control condition, whereas mTor showed higher expressions in type 1 compared to type 2 fibers."

"We found similar results for p70s6kb1 gene expressions, which were significantly increased in both fiber types upon WBV compared to control condition."

"We demonstrate that different mechanosensors and -amplifiers of the IRC/IPP Cstms show specific type 1 or type 2 fiber expressions under sedentary control and (un)loading conditions."

Bibliography

1. De Palma C, Morisi F, Pambianco S, Assi E, Touvier T, Russo S, Perrotta C, Romanello V, Carnio S, Cappello V, Pellegrino P, Moscheni C, Bassi MT, Sandri M, Cervia D, Clementi E (2014) Deficient nitric oxide signalling impairs skeletal muscle growth and performance: involvement of mitochondrial dysregulation. Skeletal Muscle. https://doi.org/10.1186/s13395-014-0022-6
2. Stamler JS, Meissner G (2001) Physiology of nitric oxide in skeletal muscle. Phys Rev 81:209–237
3. McConell GK, Rattigan S, Lee-Young RS, Wadley GD, Merry TL (2012) Skeletal muscle nitric oxide signaling and exercise: a focus on glucose metabolism. Am J Physiol Endocrino Metabol 303:E301–E307. https://doi.org/10.1152/ajpendo.00667.2011
4. Eu JP, Hare JM, Hess DT, Skaf M, Sun J, Cardenas-Navina I et al (2003) Concerted regulation of skeletal muscle contractility by oxygen tension and endogenous nitric oxide. Proc Natl Acad Sci U S A 100:15229–15234
5. Tengan CH, Rodrigues GS, Godinho RO (2012) Nitric oxide in skeletal muscle: role on mitochondrial biogenesis and function. Int J Mol Sci 13:17160–17184. https://doi.org/10.3390/ijms131217160
6. Cleeter MW, Cooper JM, Darleyusmar VM, Moncada S, Schapira AH (1994) Reversible inhibition of cytochrome-C-oxidase, the terminal enzyme of the mitochondrial respiratory-chain, by nitric-oxide—Implications for neurodegenerative diseases. FEBS Lett 345:50–54. https://doi.org/10.1016/0014-5793(94)00424-2

7. De Palma C, Morisi F, Cheli S, Pambianco S, Cappello V, Vezzoli M, Rovere-Querini P, Moggio M, Ripolone M, Francolini M, Sandri M, Clementi E (2012) Autophagy as a new therapeutic target in Duchenne muscular dystrophy. Cell Death Dis 3:e418. https://doi.org/10.1038/cddis.2012.159
8. Romanello V, Guadagnin E, Gomes L, Roder I, Sandri C, Petersen Y, Milan G, Masiero E, Del Piccolo P, Foretz M, Scorrano L, Rudolf R, Sandri M (2010) Mitochondrial fission and remodelling contributes to muscle atrophy. EMBO J 29:1774–1785. https://doi.org/10.1038/emboj.2010.60
9. Mammucari C, Milan G, Romanello V, Masiero E, Rudolf R, Del Piccolo P, Burden SJ, Di Lisi R, Sandri C, Zhao J, Goldberg AL, Schiaffino S, Sandri M (2007) FoxO3 controls autophagy in skeletal muscle in vivo. Cell Metab 6:458–471. https://doi.org/10.1016/j.cmet.2007.11.001
10. Sandri M, Sandri C, Gilbert A, Skurk C, Calabria E, Picard A, Walsh K, Schiaffino S, Lecker SH, Goldberg AL (2004) Foxo transcription factors induce the atrophy-related ubiquitin ligase atrogin-1 and cause skeletal muscle atrophy. Cell 117:399–412. https://doi.org/10.1016/S0092-8674(04)00400-3
11. Chao DS, Silvagno F, Bredt DS (1998) Muscular dystrophy in mdx mice despite lack of neuronal nitric oxide synthase. J Neurochem 71:784–789. https://doi.org/10.1046/j.1471-4159.1998.71020784.x
12. Crosbie RH, Straub V, Yun HY, Lee JC, Rafael JA, Chamberlain JS, Dawson VL, Dawson TM (1998) Campbell KP: mdx muscle pathology is independent of nNOS perturbation. Hum Mol Genet 7:823–829. https://doi.org/10.1093/hmg/7.5.823
13. Nuzzo JL, Pinto MD, Nosaka K, Steele J (2023) The eccentric: concentric strength ratio of human skeletal muscle in vivo: meta-analysis of the influences of sex, age, joint action, and velocity. Sports Med. https://doi.org/10.1007/s40279-023-01851-y
14. Hollander DB, Kraemer RR, Kilpatrick MW, Ramadan ZG, Reeves GV, Francois M et al (2007) Maximal eccentric and concentric strength discrepancies between young men and women for dynamic resistance exercise. J Strength Cond Res 21(1):34–40
15. Harden M, Bruce C, Wolf A, Hicks KM, Howatson G (2020) Exploring the practical knowledge of eccentric resistance training in high-performance strength and conditioning practitioners. Intl J Sports Sci Coaching 15(1):41–52
16. Brandenburg JP, Docherty D (2002) The effects of accentuated eccentric loading on strength, muscle hypertrophy, and neural adaptations in trained individuals. J Strength Cond Res 16(1):25–32
17. Coratella G, Milanese C, Schena F (2015) Unilateral eccentric resistance training: a direct comparison between isokinetic and dynamic constant external resistance modalities. Eur J Sport Sci 15(8):720–726
18. Doan BK, Newton RU, Marsit JL, Triplett-McBride NT, Koziris LP, Fry AC et al (2002) Effects of increased eccentric loading on bench press 1RM. J Strength Cond Res 16(1):9–13
19. English KL, Loehr JA, Lee SM, Smith SM (2014) Early-phase musculoskeletal adaptations to different levels of eccentric resistance after 8 weeks of lower body training. Eur J Appl Physiol 114(11):2263–2280
20. Montalvo S, Gruber LD, Gonzalez MP, Dietze-Hermosa MS, Dorgo S (2021) Effects of augmented eccentric load bench press training on one repetition maximum performance and electromyographic activity in trained powerlifters. J Strength Cond Res 35(6):1512–1519
21. Sarto F, Franchi MV, Rigon PA, Grigoletto D, Zoffoli L, Zanuso S et al (2020) Muscle activation during leg-press exercise with or without eccentric overload. Eur J Appl Physiol 120(7):1651–1656
22. Schroeder ET, Hawkins SA, Jaque SV (2004) Musculoskeletal adaptations to 16 weeks of eccentric progressive resistance training in young women. J Strength Cond Res 18(2):227–235
23. Sheppard JM, Young K (2010) Using additional eccentric loads to increase concentric performance in the bench throw. J Strength Cond Res 24(10):2853–2856
24. Tøien T, Pedersen Haglo H, Unhjem R, Hoff J, Wang E (2018) Maximal strength training: the impact of eccentric overload. J Neurophysiol 120(6):2868–2876

25. Wagle JP, Cunanan AJ, Carroll KM, Sams ML, Wetmore A, Bingham GE et al (2021) Accentuated eccentric loading and cluster set configurations in the back squat: a kinetic and kinematic analysis. J Strength Cond Res 35(2):420–427
26. McNeill C, Beaven CM, McMaster DT, Gill N (2020) Survey of eccentric-based strength and conditioning practices in sport. J Strength Cond Res 34(10):2769–2775
27. Durell DL, Pujol TJ, Barnes JT (2003) A survey of the scientific data and training methods utilized by collegiate strength and conditioning coaches. J Strength Cond Res 17(2):368–373
28. McCourt JL, Stearns-Reider KM, Mamsa H, Kannan P, Afsharinia MH, Shu C, Gibbs EM, Shin KM, Kurmangaliyev YZ, Schmitt LR, Hansen KC, Crosbie RH (2023) Multi-omics analysis of sarcospan overexpression in mdx skeletal muscle reveals compensatory remodeling of cytoskeleton-matrix interactions that promote mechanotransduction pathways. Skeletal Muscle. https://doi.org/10.1186/s13395-022-00311-x
29. Briguet A, Courdier-Fruh I, Foster M, Meier T, Magyar JP (2004) Histological parameters for the quantitative assessment of muscular dystrophy in the mdx-mouse. Neuromus Disord 14(10):675–682. [cited 2017 Mar 31]. Available from: http://www.ncbi.nlm.nih.gov/pubmed/15351425
30. Connolly AM, Keeling RM, Mehta S, Pestronk A, Sanes JR (2001) Three mouse models of muscular dystrophy: the natural history of strength and fatigue in dystrophin-, dystrophin/utrophin-, and laminin alpha2-deficient mice. Neuromus Disord 11(8):703–712
31. Tinsley J, Deconinck N, Fisher R, Kahn D, Phelps S, Gillis J-M et al (1998) Expression of full-length utrophin prevents muscular dystrophy in mdx mice. Nat Med 4(12):1441–1444. [cited 2018 Oct 10]. Available from: http://www.ncbi.nlm.nih.gov/pubmed/9846586
32. Lamar K-M, Bogdanovich S, Gardner BB, Gao QQ, Miller T, Earley JU et al (2016) Overexpression of latent TGFβ binding protein 4 in muscle ameliorates muscular dystrophy through myostatin and TGFβ. PLoS Genet 12(5):e1006019
33. Tjondrokoesoemo A, Schips T, Kanisicak O, Sargent MA, Molkentin JD (2016) Genetic overexpression of Serpina3n attenuates muscular dystrophy in mice. Hum Mol Genet 25(6):1192–1202
34. Heller KN, Montgomery CL, Shontz KM, Clark KR, Mendell JR, Rodino-Klapac LR (2015) Human alpha7 integrin gene (ITGA7) delivered by adeno-associated virus extends survival of severely affected dystrophin/utrophin-deficient mice. Hum Gene Ther 26(10):647–656
35. Mázala DAG, Pratt SJP, Chen D, Molkentin JD, Lovering RM, Chin ER (2015) SERCA1 overexpression minimizes skeletal muscle damage in dystrophic mouse models. Am J Physiol Cell Physiol 308(9):C699–C709
36. Baltgalvis KA, Jaeger MA, Fitzsimons DP, Thayer SA, Lowe DA, Ervasti JM (2011) Transgenic overexpression of γ-cytoplasmic actin protects against eccentric contraction-induced force loss in mdx mice. Skelet Muscle 1(1):32. [cited 2013 Jan 22]. Available from: http://www.pubmedcentral.nih.gov/articlerender.fcgi?artid=3214766&tool=pmcentrez&rendertype=abstract
37. Martin PT, Xu R, Rodino-Klapac LR, Oglesbay E, Camboni M, Montgomery CL et al (2009) Overexpression of Galgt2 in skeletal muscle prevents injury resulting from eccentric contractions in both mdx and wild-type mice. Am J Physiol Cell Physiol 296(3):C476–C488
38. Handschin C, Kobayashi YM, Chin S, Seale P, Campbell KP, Spiegelman BM (2007) PGC-1alpha regulates the neuromuscular junction program and ameliorates Duchenne muscular dystrophy. Genes Dev 21(7):770–783
39. Peter AK, Marshall JL, Crosbie RH (2008) Sarcospan reduces dystrophic pathology: stabilization of the utrophin-glycoprotein complex. J Cell Biol 183(3):419–427. [cited 2019 Nov 23]. Available from: http://www.ncbi.nlm.nih.gov/pubmed/18981229
40. Marshall JL, Oh J, Chou E, Lee JA, Holmberg J, Burkin DJ et al (2015) Sarcospan integration into laminin-binding adhesion complexes that ameliorate muscular dystrophy requires utrophin and α7 integrin. Hum Mol Genet 24(7):2011–2022. [cited 2022 Jun 16]. Available from: https://pubmed.ncbi.nlm.nih.gov/25504048/
41. Parvatiyar MS, Marshall JL, Nguyen RT, Jordan MC, Richardson VA, Roos KP et al (2015) Sarcospan regulates cardiac isoproterenol response and prevents Duchenne muscular dys-

trophy-associated cardiomyopathy. J Am Heart Assoc 4(12) [cited 2019 Nov 23]. Available from: http://www.ncbi.nlm.nih.gov/pubmed/26702077

42. Gibbs EM, Marshall JL, Ma E, Nguyen TM, Hong G, Lam JS et al (2016) High levels of sarcospan are well tolerated and act as a sarcolemmal stabilizer to address skeletal muscle and pulmonary dysfunction in DMD. Hum Mol Genet 25(24):5395–5406
43. Parvatiyar MS, Brownstein AJ, Kanashiro-Takeuchi RM, Collado JR, Dieseldorff Jones KM, Gopal J et al (2019) Stabilization of the cardiac sarcolemma by sarcospan rescues DMD-associated cardiomyopathy. JCI Insight 5. [cited 2019 May 7] Available from: http://www.ncbi.nlm.nih.gov/pubmed/31039133
44. Marshall JL, Holmberg J, Chou E, Ocampo AC, Oh J, Lee J et al (2012) Sarcospan-dependent Akt activation is required for utrophin expression and muscle regeneration. J Cell Biol 197(7):1009–1027. [cited 2019 Nov 23]. Available from: http://www.ncbi.nlm.nih.gov/pubmed/22734004
45. Mi H, Ebert D, Muruganujan A, Mills C, Albou LP, Mushayamaha T, et al. PANTHER version 16: a revised family classification, tree-based classification tool, enhancer regions and extensive API. Nucleic Acids Res 2021;49(D1):D394–403. [cited 2021 Dec 3]. Available from: https://academic.oup.com/nar/article/49/D1/D394/6027812 .
46. Deng H, Yang L, Wen P, Lei H, Blount P, Pan D (2020) Spectrin couples cell shape, cortical tension, and hippo signaling in retinal epithelial morphogenesis. J Cell Biol 219(4) [cited 2022 Jun 16]. Available from: https://pubmed.ncbi.nlm.nih.gov/32328630/
47. Cai X, Wang KC, Meng Z (2021) Mechanoregulation of YAP and TAZ in cellular homeostasis and disease progression. Front Cell Develop Biol 9:1333
48. Nghiem PP, Hoffman EP, Mittal P, Brown KJ, Schatzberg SJ, Ghimbovschi S et al (2013) Sparing of the dystrophin-deficient cranial sartorius muscle is associated with classical and novel hypertrophy pathways in GRMD dogs. Am J Pathol 183(5):1411–1424
49. Issahaku AR, Ibrahim MAA, Mukelabai N, Soliman MES (2023) Intermolecular and dynamic investigation of the mechanism of action of Reldesemtiv on fast skeletal muscle troponin complex toward the treatment of impaired muscle function. Protein J. https://doi.org/10.1007/s10930-023-10091-y
50. Russell AJ et al (2012) Activation of fast skeletal muscle troponin as a potential therapeutic approach for treating neuromuscular diseases. Nat Med 18(3):452–455. https://doi.org/10.1038/nm.2618
51. Clause KC et al (2012) Developing cardiac and skeletal muscle share fast-skeletal myosin heavy chain and cardiac troponin-I expression. PLoS One 7(7):e40725. https://doi.org/10.1371/journal.pone.0040725
52. Gomes AV, Potter JD, Szczesna-Cordary D (2002) The role of troponins in muscle contraction. IUBMB Life 54(6):323–333. https://doi.org/10.1080/15216540216037
53. Collibee SE et al (2018) Discovery of tirasemtiv, the first direct fast skeletal muscle troponin activator. ACS Med Chem Lett 9(4):354–358. https://doi.org/10.1021/acsmedchemlett.7b00546
54. Hwee DT et al (2014) Fast skeletal muscle troponin activator tirasemtiv increases muscle function and performance in the B6SJL-SOD1G93A ALS mouse model. PLoS One 9(5):e96921. https://doi.org/10.1371/journal.pone.0096921
55. Pettersen EF et al (2004) UCSF chimera–a visualization system for exploratory research and analysis. J Comput Chem 25:1605–1612. https://doi.org/10.1002/jcc.20084
56. Nikitin, S. (2014). Leap gradient algorithm. arXiv. https://doi.org/10.48550/arXiv.1405.5548.
57. Preobrazenski N, Islam H, Gurd BJ (2021) Molecular regulation of skeletal muscle mitochondrial biogenesis following blood flow-restricted aerobic exercise: a call to action. Eur J Appl Physiol. https://doi.org/10.1007/s00421-021-04669-6
58. Patterson SD, Hughes L, Warmington S et al (2019) Blood flow restriction exercise position stand: considerations of methodology, application, and safety. Front Physiol 10:1–15. https://doi.org/10.3389/fphys.2019.00533
59. Wernbom M, Augustsson J, Thomeé R (2006) Effects of vascular occlusion on muscular endurance in dynamic knee extension exercise at different submaximal loads. J Strength Cond Res 20:372–377. https://doi.org/10.1519/R-16884.1

60. Wernbom M, Järrebring R, Andreasson MA, Augustsson J (2009) Acute effects of blood flow restriction on muscle activity and endurance during fatiguing dynamic knee extensions at low load. J Strength Cond Res 23:2389–2395. https://doi.org/10.1519/JSC.0b013e3181bc1c2a
61. Hollander DB, Reeves GV, Clavier JD et al (2010) Partial occlusion during resistance exercise alters effort sense and pain. J Strength Cond Res 24:235–243. https://doi.org/10.1519/JSC.0b013e3181c7badf
62. Rossow LM, Fahs CA, Loenneke JP et al (2012) Cardiovascular and perceptual responses to blood-flow-restricted resistance exercise with differing restrictive cuffs. Clinl Physiol Funct Imag 32:331–337. https://doi.org/10.1111/j.1475-097X.2012.01131.x
63. Weatherholt A, Beekley M, Greer S et al (2013) Modified Kaatsu training: adaptations and subject perceptions. Med Sci Sports Exer 45:952–961. https://doi.org/10.1249/MSS.0b013e31827ddb1f
64. Centner C, Wiegel P, Gollhofer A, König D (2019) Effects of blood flow restriction training on muscular strength and hypertrophy in older individuals: a systematic review and meta-analysis. Sport Med 49:95–108. https://doi.org/10.1007/s40279-018-0994-1
65. Madarame H, Neya M, Ochi E et al (2008) Cross-transfer effects of resistance training with blood flow restriction. Med Sci Sports Exer 40:258–263. https://doi.org/10.1249/mss.0b013e31815c6d7e
66. Hughes L, Paton B, Rosenblatt B et al (2017) Blood flow restriction training in clinical musculoskeletal rehabilitation: a systematic review and meta-analysis. Br JSports Med 51:1003–1011. https://doi.org/10.1136/bjsports-2016-097071
67. Scott BR, Loenneke JP, Slattery KM, Dascombe BJ (2016) Blood flow restricted exercise for athletes: a review of available evidence. J Sci Med Sport 19:360–367. https://doi.org/10.1016/j.jsams.2015.04.014
68. Pignanelli C, Christiansen D, Burr JF (2021) Blood flow restriction training and the high-performance athlete: science to application. Journal of Applied Physiology. https://doi.org/10.1152/japplphysiol.00982.2020
69. Higgins JPT, Altman DG, Gotzsche PC et al (2011) The Cochrane Collaboration's tool for assessing risk of bias in randomised trials. BMJ 343:d5928–d5928. https://doi.org/10.1136/bmj.d5928
70. Jessee MB, Mattocks KT, Buckner SL et al (2018) Mechanisms of blood flow restriction: the new testament. Techniq Orthopaed 33:72–79. https://doi.org/10.1097/BTO.0000000000000252
71. Nishiyasu T, Tan N, Kondo N et al (1999) Near-infrared monitoring of tissue oxygenation during application of lower body pressure at rest and during dynamical exercise in humans. Acta Physiol Scand 166:123–130. https://doi.org/10.1046/j.1365-201x.1999.00548.x
72. Stucky F, Vesin JM, Kayser B, Uva B (2018) The effect of lower-body positive pressure on the cardiorespiratory response at rest and during submaximal running exercise. Front Physiol 9:1–11. https://doi.org/10.3389/fphys.2018.00034
73. Norrbom J, Sällstedt EK, Fischer H et al (2011) Alternative splice variant PGC-1α-b is strongly induced by exercise in human skeletal muscle. Am J Physiol Endocrinol Metabol 301:E1092–E1098. https://doi.org/10.1152/ajpendo.00119.2011
74. Christiansen D, Murphy RM, Bangsbo J et al (2018) Increased FXYD1 and PGC-1α mRNA after blood flow-restricted running is related to fibre type-specific AMPK signalling and oxidative stress in human muscle. Acta Physiol 223:e13045. https://doi.org/10.1111/apha.13045
75. Preobrazenski N, Islam H, Drouin PJ et al (2020) A novel gravity-induced blood flow restriction model augments ACC phosphorylation and PGC-1α mRNA in human skeletal muscle following aerobic exercise: a randomized crossover study. Appl Physiol Nutr Metab 45:641–649. https://doi.org/10.1139/apnm-2019-0641
76. Egan B, Carson BP, Garcia-Roves PM et al (2010) Exercise intensity-dependent regulation of peroxisome proliferator-activated receptor γ coactivator-1α mRNA abundance is associated with differential activation of upstream signalling kinases in human skeletal muscle. J Physiol 588:1779–1790. https://doi.org/10.1113/jphysiol.2010.188011

77. Gibala MJ, McGee SL, Garnham AP et al (2009) Brief intense interval exercise activates AMPK and p38 MAPK signaling and increases the expression of PGC-1α in human skeletal muscle. J Appl Physiol 106:929–934. https://doi.org/10.1152/japplphysiol.90880.2008
78. Bartlett JD, Hwa Joo C, Jeong T-S et al (2012) Matched work high-intensity interval and continuous running induce similar increases in PGC-1α mRNA, AMPK, p38, and p53 phosphorylation in human skeletal muscle. J Appl Physiol 112:1135–1143. https://doi.org/10.1152/japplphysiol.01040.2011
79. Norrbom J, Sundberg CJ, Ameln H et al (2004) PGC-1α mRNA expression is influenced by metabolic perturbation in exercising human skeletal muscle. J Appl Physiol 96:189–194. https://doi.org/10.1152/japplphysiol.00765.2003
80. Ferguson RA, Mitchell EA, Taylor CW et al (2021) Blood-flow-restricted exercise: strategies for enhancing muscle adaptation and performance in the endurance-trained athlete. Experim Physiol 53:EP089280. https://doi.org/10.1113/EP089280
81. Larsen S, Nielsen J, Hansen CN et al (2012) Biomarkers of mitochondrial content in skeletal muscle of healthy young human subjects. J Physiol 590:3349–3360. https://doi.org/10.1113/jphysiol.2012.230185
82. Kaijser L, Sundberg CJ, Eiken O et al (1990) Muscle oxidative capacity and work performance after training under local leg ischemia. J Appl Physiol 69:785–787. https://doi.org/10.1152/jappl.1990.69.2.785
83. Esbjörnsson M, Jansson E, Sundberg CJ et al (1993) Muscle fibre types and enzyme activities after training with local leg ischaemia in man. Acta Physiol Scand 148:233–241. https://doi.org/10.1111/j.1748-1716.1993.tb09554.x
84. Granata C, Jamnick NA, Bishop DJ (2018) Training-induced changes in mitochondrial content and respiratory function in human skeletal muscle. Sport Med 48:1809–1828. https://doi.org/10.1007/s40279-018-0936-y
85. Bennett H, Slattery F (2019) Effects of blood flow restriction training on aerobic capacity and performance. J Strength Cond Res 33:572–583. https://doi.org/10.1519/JSC.0000000000002963
86. Mitchell R, Mellows B, Sheard J, Antonioli M, Kretz O, Chambers D, Zeuner M-T, Tomkins JE, Denecke B, Musante L, Joch B, Debacq-Chainiaux F, Holthofer H, Ray S, Huber TB, Dengjel J, De Coppi P, Widera D, Patel K (2019) Secretome of adipose-derived mesenchymal stem cells promotes skeletal muscle regeneration through synergistic action of extracellular vesicle cargo and soluble proteins. Stem Cell Res Ther. https://doi.org/10.1186/s13287-019-1213-1
87. Mellows B, Mitchell R, Antonioli M, Kretz O, Chambers D, Zeuner M-T et al (2017) Protein and molecular characterization of a clinically compliant amniotic fluid stem cell-derived extracellular vesicle fraction capable of accelerating muscle regeneration through enhancement of angiogenesis. Stem Cells Develop 26(18):1316–1333
88. Lo Sicco C, Reverberi D, Balbi C, Ulivi V, Principi E, Pascucci L et al (2017) Mesenchymal stem cell-derived extracellular vesicles as mediators of anti-inflammatory effects: endorsement of macrophage polarization. Stem Cells Translat Med 6(3):1018–1028
89. Nakamura Y, Miyaki S, Ishitobi H, Matsuyama S, Nakasa T, Kamei N et al (2015) Mesenchymal-stem-cell-derived exosomes accelerate skeletal muscle regeneration. FEBS Lett 589(11):1257–1265
90. Garcia NA, Ontoria-Oviedo I, González-King H, Diez-Juan A, Sepúlveda P (2015) Glucose starvation in cardiomyocytes enhances exosome secretion and promotes angiogenesis in endothelial cells. PLoS One 10(9):e0138849
91. Jalabert, Audrey; Reininger, Laura; Berger, Emmanuelle; Coute, Yohann; Meugnier, Emmanuelle; Forterre, Alexis; Errazuriz-Cerda, Elizabeth, ; Geloen, Alain; Aouadi, Myriam; Bouzakri, Karim; Rieusset, Jennifer; Rome, SophieProfiling of ob/ob mice skeletal muscle exosome-like vesicles demonstrates combined action of miRNAs, proteins and lipids to modulate lipid homeostasis in recipient cells. Sci Rep (2021). doi: https://doi.org/10.1038/s41598-021-00983-3.

92. Rome S, Forterre A, Mizgier ML, Bouzakri K (2019) Skeletal muscle-released extracellular vesicles: state of the art. Front Physiol 10:929
93. Aswad H et al (2014) Exosomes participate in the alteration of muscle homeostasis during lipid-induced insulin resistance in mice. Diabetologia 57(10):2155–2164
94. Jalabert A et al (2016) Exosome-like vesicles released from lipid-induced insulin-resistant muscles modulate gene expression and proliferation of beta recipient cells in mice. Diabetologia 59(5):1049–1058
95. Rome S, Blandin A, Le Lay S (2021) Adipocyte-derived extracellular vesicles: state of the art. Int J Mol Sci 22:4
96. Casabona MG, Vandenbrouck Y, Attree I, Coute Y (2013) Proteomic characterization of Pseudomonas aeruginosa PAO1 inner membrane. Proteomics 13(16):2419–2423
97. Yu Y et al (2018) Adipocyte-derived exosomal MiR-27a induces insulin resistance in skeletal muscle through repression of PPARgamma. Theranostics 8(8):2171–2188
98. López-Bellón S, Rodríguez-López S, González-Reyes JA, Burón MI, de Cabo R, Villalba JM (2022) CYB5R3 overexpression preserves skeletal muscle mitochondria and autophagic signaling in aged transgenic mice. GeroScience. https://doi.org/10.1007/s11357-022-00574-8
99. Leduc-Gaudet JP et al (2021) Mitochondrial dynamics and mitophagy in skeletal muscle health and aging. Int J Mol Sci 22(15)
100. Martin-Montalvo A et al (2016) Cytochrome b5 reductase and the control of lipid metabolism and healthspan. NPI Aging Mechan Dis 2:16006
101. Camerino GM, Fonzino A, Conte E, De Bellis M, Mele A, Liantonio A, Tricarico D, Tarantino N, Dobrowolny G, Musarò A, Desaphy J-F, De Luca A, Pierno S (2019) Elucidating the contribution of skeletal muscle ion channels to amyotrophic lateral sclerosis in search of new therapeutic options. Sci Rep. https://doi.org/10.1038/s41598-019-39676-3
102. Kalmar B, Edet-Amana E, Greensmith L (2012) Treatment with a coinducer of the heat shock response delays muscle denervation in the SOD1-G93A mouse model of amyotrophic lateral sclerosis. Amyotrophic Lateral Sclerosis 13:378–392
103. Desaphy J-F et al (2013) Functional characterization of ClC-1 mutations from patients affected by recessive myotonia congenita presenting with different clinical phenotypes. Exp Neurol 248:530–540
104. Pierno S, De Luca A, Beck CL, George AL, Conte Camerino D (1999) Aging-associated down-regulation of ClC-1 expression in skeletal muscle: phenotypic-independent relation to the decrease of chloride conductance. FEBS Lett 449:12–16
105. Desaphy J-F, Carbonara R, Costanza T, Conte Camerino D (2014) Preclinical evaluation of marketed sodium channel blockers in a rat model of myotonia discloses promising antimyotonic drugs. Exp Neurol 255:96–102
106. Dobrowolny G et al (2018) Muscle expression of SOD1G93A triggers the dismantlement of neuromuscular junction via PKC-theta. Antioxid Redox Signaling 28:1105–1119
107. Reza MM et al (2017) Irisin is a pro-myogenic factor that induces skeletal muscle hypertrophy and rescues denervation-induced atrophy. Nat Commun 8:1104
108. Tricarico D et al (2006) Hybrid assemblies of ATP-sensitive K+ channels determine their muscle-type-dependent biophysical and pharmacological properties. Proc Natl Acad Sci U S A 103:1118–1123
109. Gambara G et al (2017) Gene expression profiling in slow-type calf soleus muscle of 30 days space-flown mice. PLoS One 12:e0169314
110. Ringner M (2008) What is principal component analysis? Nat Biotechnol 26:303–304
111. Tam BT, Pei XM, Yung BY, Yip SP, Chan LW, Wong CS, Siu PM (2015) Unacylated ghrelin restores insulin and autophagic signaling in skeletal muscle of diabetic mice. Pflügers Archiv Eur J Physiol. https://doi.org/10.1007/s00424-015-1721-5
112. DeFronzo RA, Tripathy D (2009) Skeletal muscle insulin resistance is the primary defect in type 2 diabetes. Diabetes Care 32(Suppl 2):S157–S163. https://doi.org/10.2337/dc09-S302
113. He C, Bassik MC, Moresi V, Sun K, Wei Y, Zou Z, An Z, Loh J, Fisher J, Sun Q, Korsmeyer S, Packer M, May HI, Hill JA, Virgin HW, Gilpin C, Xiao G, Bassel-Duby R, Scherer PE,

Levine B (2012) Exercise-induced BCL2-regulated autophagy is required for muscle glucose homeostasis. Nature 481(7382):511–515. https://doi.org/10.1038/nature10758
114. Lv P, Huang J, Yang J, Deng Y, Xu J, Zhang X, Li W, Zhang H, Yang Y (2014) Autophagy in muscle of glucose-infusion hyperglycemia rats and streptozotocin-induced hyperglycemia rats via selective activation of m-TOR or FoxO3. PLoS One 9(2):e87254. https://doi.org/10.1371/journal.pone.0087254
115. Yang L, Li P, Fu S, Calay ES, Hotamisligil GS (2010) Defective hepatic autophagy in obesity promotes ER stress and causes insulin resistance. Cell Metab 11:467–478
116. Kim KH, Jeong YT, Oh H, Kim SH, Cho JM, Kim YN, Kim SS, Kim Do H, Hur KY, Kim HK, Ko T, Han J, Kim HL, Kim J, Back SH, Komatsu M, Chen H, Chan DC, Konishi M, Itoh N, Choi CS, Lee MS (2013) Autophagy deficiency leads to protection from obesity and insulin resistance by inducing Fgf21 as a mitokine. Nat Med 19(1):83–92. https://doi.org/10.1038/nm.3014
117. Mann S, Abuelo A, Nydam DV, Leal Yepes FA, Overton TR, Wakshlag JJ (2016) Insulin signaling and skeletal muscle atrophy and autophagy in transition dairy cows either overfed energy or fed a controlled energy diet prepartum. J Compar Physiol B. https://doi.org/10.1007/s00360-016-0969-1
118. Glass DJ (2010) PI3 kinase regulation of skeletal muscle hypertrophy and atrophy. Curr Topics Microbiol Immunol 346:267–278. https://doi.org/10.1007/82_2010_78
119. Sandri M (2008) Signaling in muscle atrophy and hypertrophy. Physiology (Bethesda) 23:160–170. https://doi.org/10.1152/physiol.00041.2007
120. Kerestes M, Faigl V, Kulcsar M, Balogh O, Foldi J, Febel H, Chilliard Y, Huszenicza G (2009) Periparturient insulin secretion and whole-body insulin responsiveness in dairy cows showing various forms of ketone pattern with or without puerperal metritis. Domest Anim Endocrinol 37(4):250–261. https://doi.org/10.1016/j.domaniend.2009.07.003
121. Stipanuk MH (2009) Macroautophagy and its role in nutrient homeostasis. Nutr Rev 67(12):677–689. https://doi.org/10.1111/j.1753-4887.2009.00252.x
122. Cecconi F, Levine B (2008) The role of autophagy in mammalian development: cell makeover rather than cell death. Dev Cell 15(3):344–357. https://doi.org/10.1016/j.devcel.2008.08.012
123. Nisr RB, Shah DS, Ganley IG, Hundal HS (2019) Proinflammatory NFkB signalling promotes mitochondrial dysfunction in skeletal muscle in response to cellular fuel overloading. Cell Mol Life Sci. https://doi.org/10.1007/s00018-019-03148-8
124. Feve B, Bastard JP (2009) The role of interleukins in insulin resistance and type 2 diabetes mellitus. Nature Rev Endocrinol 5:305–311
125. Funaki M (2009) Saturated fatty acids and insulin resistance. J Med Investig 56:88–92
126. Kwon OS et al (2016) Intramyocellular ceramides and skeletal muscle mitochondrial respiration are partially regulated by toll-like receptor 4 during hindlimb unloading. Am J Physiol Regul Integr Compar Physiol 311:R879–R887
127. Remels AH et al (2010) TNF-alpha impairs regulation of muscle oxidative phenotype: implications for cachexia? FASEB J 24:5052–5062
128. Lipina C et al (2013) Mitochondrial substrate availability and its role in lipid-induced insulin resistance and proinflammatory signalling in skeletal muscle. Diabetes 62:3426–3436
129. Green CJ et al (2011) Counter-modulation of fatty acid-induced pro-inflammatory nuclear factor kappaB signalling in rat skeletal muscle cells by AMP-activated protein kinase. Biochemical J 435:463–474
130. Coll T et al (2008) Oleate reverses palmitate-induced insulin resistance and inflammation in skeletal muscle cells. J Biol Chem 283:11107–11116
131. Hajduch E et al (1998) Constitutive activation of protein kinase Bα (PKBα) by membrane targeting promotes glucose and system A amino acid transport, protein synthesis and GSK3 inactivation in L6 muscle cells. Diabetes 47:1006–1013
132. Sparks LM et al (2016) ANT1-mediated fatty acid-induced uncoupling as a target for improving myocellular insulin sensitivity. Diabetologia 59:1030–1039

133. Mathes S, Vanmunster M, Bloch W, Suhr F (2019) Evidence for skeletal muscle fiber type-specific expressions of mechanosensors. Cell Mol Life Sci. https://doi.org/10.1007/s00018-019-03026-3
134. Burkholder TJ (2007) Mechanotransduction in skeletal muscle. Front Biosci 12:174–191
135. Conceição MS, Chacon-Mikahil MPT, Telles GD et al (2016) Attenuated PGC-1α isoforms following endurance exercise with blood flow restriction. Med Sci Sports Exer 48:1699–1707. https://doi.org/10.1249/MSS.0000000000000970
136. Mitchell SJ et al (2016) Effects of sex, strain, and energy intake on hallmarks of aging in mice. Cell Metab 23(6):1093–1112
137. Mann S, Yepes FA, Overton TR, Wakshlag JJ, Lock AL, Ryan CM, Nydam DV (2015) Dry period plane of energy: effects on feed intake, energy balance, milk production, and composition in transition dairy cows. J Dairy Sci 98(5):3366–3382. https://doi.org/10.3168/jds.2014-9024

Skeletal Muscle Energy Metabolism & Contraction

6

Christopher Myers

6.1 Introduction by the Editor

Skeletal muscle contraction and energy metabolism are closely intertwined processes essential for maintaining physiological functions and supporting physical activity. Muscle contraction's biochemical basis involves converting chemical energy stored in adenosine triphosphate (ATP) into mechanical energy. This process is regulated by complex signaling pathways and is influenced by various factors, including diet, physical activity, and genetic predispositions.

Chapter 4 discussed the expression of the VDR in skeletal muscle and its essential relationship with Vitamin D in determining muscle mass and its functional capabilities. This chapter highlights recent studies demonstrating the significant role of vitamin D in skeletal muscle function, particularly concerning oxidative stress, energy metabolism, and maintaining an anabolic state. Furthermore, this line of research shows Vitamin D deficiency is associated with impaired mitochondrial function, increased oxidative stress, and a predisposition to muscle atrophy. Mechanistically, vitamin D influences muscle function through its receptor, VDR, and is implicated in various cellular processes, including mitochondrial oxidative phosphorylation and muscle cell proliferation and differentiation.

Skeletal muscle undergoes significant metabolic remodeling when exposed to environmental hypoxia, which has sparked considerable debate within the scientific community. This chapter highlights the review written by Horscroft and Murray that addresses the controversies surrounding hypoxia-induced changes in skeletal muscle energy metabolism. Their research suggests that exposure to low oxygen conditions leads to a selective decrease in the oxidation of fatty acids, while the uptake of glucose is either maintained or increased to support increased glycolytic activity.

C. Myers (✉)
3 Sports Science, 3 Sports Science LLC, Akron, OH, United States of America
e-mail: chris@3sportsscience.com

C. Myers (ed.), *Skeletal Muscle Physiology*,
https://doi.org/10.1007/978-3-031-47065-3_6

This adaptation might optimize ATP synthesis pathways under reduced oxidative metabolism, an essential survival mechanism in hypoxic environments. Despite these findings, the literature remains divided on whether environmental hypoxia prompts a switch from fatty acid oxidation to glucose oxidation within muscle mitochondria. The review also highlights that while oxidative processes in skeletal muscle are downregulated in response to hypoxia, glycolytic markers remain unchanged. Moreover, the review concludes that environmental hypoxia does not consistently alter the total capacity for skeletal muscle glycolysis. This comprehensive analysis underscores the complexity of skeletal muscle energy metabolism in response to environmental hypoxia, paving the way for a better understanding of physiological adaptations to low oxygen availability.

In human skeletal muscle, the buffering of ATP by phosphocreatine, through a reaction catalyzed by creatine kinase, allows for extended periods of activity. However, for activity to be sustained, phosphocreatine must be continuously regenerated. This chapter highlights the research by Barclay [1] that emphasizes that ATP breakdown rates can vary significantly, from 70 to 140 mM per minute during isometric contractions to 400 mM per minute during intense dynamic activities. The study underscores that the maximum rate of oxidative energy supply in untrained individuals is approximately 50 mM per minute, which is adequate to match an ATP breakdown rate of 100 mM per minute during contraction if the duty cycle is 0.5. Barclay also notes that during high-intensity exercises, the demand for ATP can surpass the supply from phosphocreatine regeneration and glycolysis, leading to a net decrease in phosphocreatine levels. The creatine kinase reaction is crucial for buffering ATP and signaling energy demand from ATP breakdown sites to mitochondria. The findings suggest that the maximum sustainable level of muscular activity is only a fraction of the potential rate of energy turnover in muscle, constrained by the peak rate of oxidative energy supply. Therefore, the capacity for sustained high-energy demand and muscle power output is limited, with endurance training and age-related or disease-related declines in oxidative and glycolytic energy supplies affecting this balance.

In recent research exploring the impacts of l-arginine and l-carnitine supplementation on metabolic health, researchers observed notable differences between prediabetic South Asian and Europid men. They found that l-arginine supplementation improved glucose metabolism and insulin sensitivity in Europid men but not South Asian men, suggesting a potential genetic or physiological variation in response to supplementation. Meanwhile, vegetarians, who generally have lower plasma carnitine levels due to dietary restrictions, showed an increase in both plasma and skeletal muscle carnitine content after l-carnitine supplementation. This increase, however, did not translate to improved muscle function or energy metabolism, indicating that while l-carnitine can replenish carnitine stores, it may not affect metabolic pathways in a manner that enhances physical performance or energy efficiency. These studies highlight the complex interaction between dietary supplements and metabolic health, suggesting that the benefits of l-arginine and l-carnitine supplementation can vary based on genetic background, existing health conditions, and diet. Further research is needed to understand these dynamics and tailor supplementation recommendations accordingly.

Furthermore, this chapter discusses research that identified that key molecules, AMP-activated protein kinase (AMPK) and peroxisome proliferator-activated receptor gamma coactivator 1-alpha (PGC-1α), as play crucial regulators in energy metabolism and mitochondrial function within skeletal muscle. AMPK functions as an energy sensor within cells, activating multiple pathways that ramp up ATP production whenever there's an energy shortfall. Researchers investigated the impact of prolonged fasting on the muscle energy metabolism of the raccoon dog (Nyctereutes procyonoides), a species known for its seasonal fat accumulation and extended winter fasts and used as a human assay analog. They focused on the expression of key metabolic regulators like AMP-activated protein kinase (AMPK), acetyl-CoA carboxylase (ACC), glucose transporter 4 (GLUT 4), and protein kinase B (Akt) in the hind limb muscles after 10 weeks of fasting. Despite the fasting, there were no significant changes in AMPK phosphorylation; however, total AMPK expression decreased, leading to a higher phosphorylation ratio in specific muscles.

Additionally, while GLUT 4, insulin receptor, and Akt expressions remained unchanged, Akt phosphorylation decreased in some muscles. These findings suggest that raccoon dogs possess a robust molecular system to maintain skeletal muscle function during long periods of fasting and inactivity. Furthermore, this animal model research may underscore the importance of molecular pathways in adapting to exercise and maintaining cellular energy homeostasis.

This chapter highlights the research demonstrating the intricate relationship between Vitamin D and skeletal muscle function. It underscores its vital role in regulating muscle health, with deficiencies linked to increased oxidative stress and muscle atrophy. Studies into skeletal muscle's metabolic remodeling under environmental hypoxia reveal adaptations that optimize ATP synthesis, emphasizing the body's resilience to low oxygen levels, and human skeletal muscle energy dynamics highlight the importance of the creatine kinase reaction in buffering ATP and signaling energy demand, illustrating the limitations of sustained muscle activity by oxidative energy supply. Investigations into the effects of l-arginine and l-carnitine supplementation on metabolic health reveal variability in response based on genetic and dietary factors, suggesting the need for personalized nutritional recommendations. Moreover, the study on the raccoon dog's adaptation to prolonged fasting sheds light on the molecular systems that maintain skeletal muscle function during periods of inactivity, offering insights into human muscle adaptation and the significance of maintaining cellular energy homeostasis.

6.2 Machine Generated Summaries

Disclaimer: The summaries in this chapter were generated from Springer Nature publications using extractive AI auto-summarization: An extraction-based summarizer aims to identify the most important sentences of a text using an algorithm and uses those original sentences to create the auto-summary (unlike generative AI). As the constituted sentences are machine selected, they may not fully reflect the body of the work, so we strongly advise that the original content is read and cited. The auto generated summaries were curated by the editor to meet Springer Nature publication standards. To cite this content, please refer to the original papers.

Machine generated keywords: energy metabolism, fatty acid, fatty, hsp, acid, energy, content, voltage, atp, vitamin, mitochondrial, stress, glucose, deficiency, muscle contraction.

Mechanisms of Vitamin D on Skeletal Muscle Function: Oxidative Stress, Energy Metabolism and Anabolic State [2]

This is a machine-generated summary of:

Dzik, Katarzyna Patrycja; Kaczor, Jan Jacek: Mechanisms of vitamin D on skeletal muscle function: oxidative stress, energy metabolism and anabolic state [2]

Published in: European Journal of Applied Physiology (2019)

Link to original: https://doi.org/10.1007/s00421-019-04104-x

If you want to cite the papers, please refer to the original.

For technical reasons we could not place the page where the original quote is coming from.

Abstract-Summary "This review provides a current perspective on the mechanism of vitamin D on skeletal muscle function with the emphasis on oxidative stress, muscle anabolic state and muscle energy metabolism."

"It focuses on several aspects related to cellular and molecular physiology such as VDR as the trigger point of vitamin D action, oxidative stress as a consequence of vitamin D deficiency."

"The interaction between vitamin D deficiency and mitochondrial function as well as skeletal muscle atrophy signalling pathways have been studied and clarified in the last years."

"Vitamin D deficiency is associated with oxidative stress in skeletal muscle that influences the mitochondrial function and affects the development of skeletal muscle atrophy."

"Vitamin D deficiency may contribute to the development of muscle atrophy."

"Based on the current knowledge we propose that vitamin D deficiency results from the loss of VDR function and it could be partly responsible for the development of neurodegenerative diseases in human beings."

Introduction

"The last decade brought a tremendous number of studies on vitamin D function in human body."

"Dr. Elmer McCollum and others officially termed this nutritional factor as vitamin D [3]."

"The aim of this review is to present the latest reports on skeletal muscle function and vitamin D status."

"The current review provides the evidence that deficiency of vitamin D through oxidative stress and disruption of mitochondrial function may affect the development of skeletal muscle atrophy."

Vitamin D deficiency

"The latest guidelines indicate the concentration lower than 20 ng/mL (50 nmol/L) as vitamin D deficiency."

"The true vitamin D deficiency should consider its bioavailability and, therefore, its binding to vitamin D binding protein (VDBP)."

"VDBP is the primary vitamin D carrier, binding 85–90% of circulating 25(OH)D and 1,25-dihydroxyvitamin D_3 and the remaining unbound 25(OH)D is considered bioavailable (either free or bound to albumin)."

"About 10–15% of total 25(OH)D is bound to albumin, in contrast to free 25(OH)D, which accounts for 1% of total circulating vitamin D [4]."

"There might be tremendous personal differences in bioavailable vitamin D for humans with a genetic mutation for VDBP or in VDBP-KD mouse models were shown to have lower 25(OH)D blood level [5, 6]."

Vitamin D physiology

"That 1α,25(OH)2 D_3 concentration is regulated by two vitamin D_3 regulating enzymes, CYP24A1 (cytochrome P450 family 24 subfamily A member 1), and CYP27B1 (25-hydroxyvitamin D-1-α-hydroxylase)."

"The study of Abboud and coworkers [7], which examined the concentration and time-dependent effects of calcitriol on the capacity of muscle cells to take up and release 25(OH) D_3, showed an evidence that skeletal muscle cells indeed contain a mobile pool of 25(OH) D_3 which accumulates from and returns to the extracellular environment."

"The authors postulated that if the capacity to hold 25(OH) D_3 out of the circulation in skeletal muscle is high when vitamin D status is falling in winter, 25(OH) D_3 might be protected from inactivating activity of CYP24A1 in the liver."

"It is important to emphasize that also in this hypothesis VDBP plays an important role in regulating bioavailability of vitamin D yet this time in skeletal muscle cells."

VDR in Musculoskeletal System

"Although, as mentioned before, VDR is predominantly expressed in fast twitch muscles, a study on human paraspinal, slow twice muscle shows that vitamin D deficiency induces its atrophy and decreases the concentration of intramyonucelar VDR and VDR gene expression level [8]."

"Latest study of Ryan and coworkers [9] demonstrated increased oxygen consumption rate (OCR) of skeletal muscle cells after treatment with1α,25$(OH)_2$ D_3, indicating vitamin D action in the regulation of mitochondrial oxygen consumption and dynamics."

"Hitherto studies postulated that the possible signalling pathway involved in muscle vitamin D function might require steroid receptor coactivator complex (Src), non-receptor tyrosine kinase, which has been shown to activate mitogen-activated protein kinases (MAPK) [10, 11] in various tissues."

"Recent study showed that VDR signalling enhanced by vitamin D treatment inhibited FOXO1 expression, nuclear translocation, and activity in C2C12 muscle cells."

Summary

"It is shown that vitamin D deficiency may induce paraspinal muscle atrophy and decreases the concentration of intramyonucelar VDR and gene expression of VDR."

"Vitamin D deficiency decreases oxygen consumption rate and induces disruption of mitochondrial function."

"It is very likely that vitamin D deficiency in the long run induces VDR ablation, ROS generation and in consequence deleterious effects on the mitochondrial function, which in turn leads to elevated muscle atrophy."

"We assume that vitamin D deficiency results from the loss of VDR function and it could be partly responsible for the development of neurodegenerative diseases in human beings."

Future Directions

"Given the important action of vitamin D on skeletal muscle tissue, a better understanding of the mechanisms involved in muscle atrophy is needed."

"There is a great need of a new insight into VDR expression and activation, biogenesis and the function of mitochondria as well as signalling pathways associated with progressive muscle atrophy in vitamin D deficiency."

"Even more, we suppose that supplementation with vitamin D to sufficient serum vitamin D level will: reduce ROS overproduction, increase VDR gene expression and protein content, improve mitochondrial function and inhibit the atrophy of muscle."

Impairments in Contractility and Cytoskeletal Organisation Cause Nuclear Defects in Nemaline Myopathy [12]

This is a machine-generated summary of:

Ross, Jacob A.; Levy, Yotam; Ripolone, Michela; Kolb, Justin S.; Turmaine, Mark; Holt, Mark; Lindqvist, Johan; Claeys, Kristl G.; Weis, Joachim; Monforte, Mauro; Tasca, Giorgio; Moggio, Maurizio; Figeac, Nicolas; Zammit, Peter S.; Jungbluth, Heinz; Fiorillo, Chiara; Vissing, John; Witting, Nanna; Granzier, Henk; Zanoteli, Edmar; Hardeman, Edna C.; Wallgren-Pettersson, Carina; Ochala, Julien: Impairments in contractility and cytoskeletal organisation cause nuclear defects in nemaline myopathy [12]

Published in: Acta Neuropathologica (2019)
Link to original: https://doi.org/10.1007/s00401-019-02034-8

If you want to cite the papers, please refer to the original.

For technical reasons we could not place the page where the original quote is coming from.

Abstract-Summary "We report novel pathological defects in skeletal muscle fibres of mouse models and patients with NM: irregular spacing and morphology of nuclei; disrupted nuclear envelope; altered chromatin arrangement; and disorganisation of the cortical cytoskeleton."

"We also establish the role of microtubule organisation in determining nuclear morphology, a phenomenon which is likely to contribute to nuclear alterations in this disease."

"Our results overlap with findings in diseases caused directly by mutations in nuclear envelope or cytoskeletal proteins."

"Given the important role of nuclear shape and envelope in regulating gene expression, and the cytoskeleton in maintaining muscle fibre integrity, our findings are likely to explain some of the hallmarks of NM, including contractile filament disarray, altered mechanical properties and broad transcriptional alterations."

Introduction

"The nuclei of skeletal muscle fibres are linked with various cytoskeletal components including non-sarcomeric/cytoplasmic actins, microtubules and intermediate filaments such as desmin."

"All three of these cytoskeletal networks have been implicated in the spacing and positioning of nuclei in models of skeletal muscle development: microtubules in the initial translocation/spacing of nuclei along the fibre [13–16], and actin and desmin in their movement to the fibre periphery [17]."

"The force-generating properties of NM muscle fibres are severely limited, we hypothesised that cytoskeletal components as well as nuclear function, positioning and integrity might be affected in this disease, and possibly contribute to pathology."

"Using single muscle fibres from mouse models and NM patients with mutations in ACTA1 or NEB, we found that myonuclei display a range of defects, including irregular spacing and morphology, abnormal nuclear envelope and altered chromatin distribution."

Materials and Methods

"Fibres were permeabilised in 0.1% triton-X/PBS for 10 min, washed 3× and blocked in 10% normal goat serum/PBS for 1 h. Fibres were then treated with primary antibodies in blocking solution overnight (β-tubulin) or for 3 h (lamin A, nesprin-1, pericentrin) at 4 °C."

"Analysis of nuclear number and spacing: Coordinates of myonuclei were identified in 3D within Z-stacks of muscle fibres."

"For statistical comparisons of nuclear organisation and nuclear shape in human subjects (column graphs), control data points were pooled, since no significant age-related differences were observed amongst healthy control subjects."

"For studies with animals, no significant differences were observed between animals of the same genotype, and as such, individual data points for both control and mutant/treated animals were pooled."

Results

"Our results indicate that severe defects in nuclear envelope and morphology can also occur in diseases caused by muscle contractile dysfunction."

"These results suggest that in NM, force impairment results in nuclear spacing and morphological alterations, and that enhancing force production restores these parameters."

"These results indicate that nuclear defects also occur in presumably non-genetic phenocopies of NM, and also provide some correlative evidence that a reduction in force-generating capacity of muscle might underlie these defects (although other aspects of myofibre pathology might also contribute)."

"We observe defects in nuclear morphology and spacing in NM patients, we sought to determine whether the microtubule network and/or its associated proteins were perturbed."

"These results indicate that at least in non-contracting isolated muscle fibres, microtubule disruption for this length of time has no major impact on nuclear positioning or the nuclear envelope."

Discussion

"We identify a range of nuclear and cytoskeletal defects in skeletal muscle fibres that occur as part of ACTA1 or NEB-related NM pathology."

"Given the severity of disease caused by mutations in genes encoding nuclear envelope proteins, it is likely that the array of nuclear defects that we describe in NM patients contributes to muscle dysfunction."

"Broad alterations in the transcriptional profile of skeletal muscle are observed in patients with NM (including in genes related to metabolism and calcium homeostasis), and this may partially result from these characteristic nuclear shape and envelope alterations and/or reorganised chromatin [18]."

"They might explain some of the features observed in NM, such as broad transcriptional changes and hindered muscle fibre growth (possibly related to alterations in nuclear envelope and chromatin organisation, which is likely to affect programmes of gene expression [19–21]), myofibrillar disarray (due to the roles of desmin and the nuclear envelope in sarcomere organisation [22, 23]) and altered fibre mechanical properties (due to disrupted cytoskeletal arrangements [24])."

Excitation Contraction Uncoupling by High Intracellular [Ca^{2+}] in Frog Skeletal Muscle: A Voltage Clamp Study [25]

This is a machine-generated summary of:

Olivera, J. Fernando; Pizarro, Gonzalo: Excitation contraction uncoupling by high intracellular [Ca^{2+}] in frog skeletal muscle: a voltage clamp study [25]

Published in: Journal of Muscle Research and Cell Motility (2016)

Link to original: https://doi.org/10.1007/s10974-016-9446-4

If you want to cite the papers, please refer to the original.

For technical reasons we could not place the page where the original quote is coming from.

Abstract-Summary "Raising the intracellular [Ca^{2+}] ([Ca^{2+}]$_i$) was previously found to produce uncoupling between the electrical depolarization of the transverse tubules and contraction in skinned muscle fibers."

"We study the effect of elevated [Ca^{2+}]$_i$ in voltage clamped cut fibers of frog skeletal muscle to establish how the charge movement, a measure of the activation of the dihydropyridine receptors (DHPR)-voltage sensors, and Ca^{2+} release, a consequence of the opening of the ryanodine receptor (RyR)-release channels, were affected. [Ca^{2+}]$_i$ was raised by various procedures (pharmacological release from the sarcoplasmic reticulum, application of high [Ca^{2+}]$_i$ intracellular solution, permeabilization of the plasma membrane by a Ca^{2+} ionophore) all of which produced impairment of excitation–contraction coupling."

"In approximately 40 % of the uncoupled cells we observed that the [Ca^{2+}]$_i$ transient continued to rise after the voltage clamp pulse was turned off."

"This loss of control by membrane voltage suggests that the uncoupled release channels might have another mechanism of activation, likely by Ca^{2+}."

Introduction

"The rise in intracellular Calcium concentration ($[Ca^{2+}]_i$) was found to produce EC uncoupling in skinned skeletal muscle fibers of toads and rats [26]."

"After the exposure to high $[Ca^{2+}]_i$ the cells ceased to contract in response to ionic depolarization of the TT membrane voltage but still responded fully upon the application of caffeine (2 mM) or low $[Mg^{2+}]_i$ (0.05 mM) to directly open RyR channels and mobilize Ca^{+2} from the SR."

"The uncoupling effect of the augmented $[Ca^{2+}]_i$ depended on the amplitude as well as on the duration of the treatment."

"Ca^{2+}-dependent activation of calpain was found to be a step in the mechanism of uncoupling [26–28]. $[Ca^{2+}]_i$ as low a 300 nM was able to activate the calpains involved in the uncoupling process [27]."

"The experiments in this work were designed to directly study the effect of increased $[Ca^{2+}]_i$ on charge movement and Ca^{2+} release flux in cut fibers of the frog under voltage clamp in a double Vaseline gap voltage clamp."

Methods

"Ethylene glycol-bis(b-aminoethyl ether)-N,N,N′,N′-tetraacetic acid (EGTA) was added to the internal solution at 0.5, 1 or 2 mM adding the necessary amount of $CaCl_2$ in order to yield a free $[Ca^{2+}]$ = 20 nM assuming an apparent K_D = 0.47 μM for EGTA, calculated with the program Maxchelator for pH = 7.0, temperature of 15 °C and 0.15 M ionic strength."

"In a series of experiments the Ca^{2+} ionophore A23187 was used from a 25 mM stock in DMSO and added to the external solution at a final concentration of 50 μM, with DMSO final dilution of 2/1000."

"Since EGTA was added to the internal solution at a rather high concentration (not less than 0.5 mM), its concentration or one of its rate constants were always fitted."

Results

"4-CmC applied alone (100 μM) increased $[Ca^{2+}]_i$ to a rather steady level during the application. $[Ca^{2+}]_i$ increased to values between 0.2 and 0.4 μM and was maintained for up to 20 min until the charge movement decayed significantly."

"We performed control experiments to study the effect of 100 μM of 4-CmC on charge movement preventing the increase in $[Ca^{2+}]_i$ by using a 50 mM EGTA intracellular solution to heavily buffer Ca^{2+}."

"In five fibers studied this way the charge per unit capacitance was reduced from 20 ± 1.8 to 6.9 ± 1.4 nC μF^{-1} (34 ± 6 % of reference), release from 14.6 ± 1.7 to 1.38 ± 0.3 mM s^{-1} (9.7 ± 2 % of reference) and the change in $[Ca^{2+}]_i$ due to drug treatment from 0.55 + 0.08 to 0.3 + 0.03 μM, (59 ± 11 % of the reference value)."

Discussion

"The experiments in which we increased $[Ca^{2+}]_i$ without mobilizing it from the SR also showed a more drastic reduction in release flux than in the charge per unit capacitance."

"If anything SR [Ca^{2+}] might have increased as a consequence of the increased [Ca^{2+}]$_i$, thus our measurements might underestimate the uncoupling between voltage sensors and release channels."

"In more than one half of the fibers that we studied the exposure to high [Ca^{2+}]$_i$ lead to a drastic increase in holding current that rendered the voltage clamp unreliable and the experiment was terminated."

"The loss of control occurred only in fibers that developed an important reduction in charge per unit capacitance and Ca^{2+} release after the exposure to high intracellular [Ca^{2+}]$_i$ by drug induced release or by means of an ionophore."

"This study shows the existence of EC uncoupling due to high [Ca^{2+}]$_i$ in cut fibers under voltage clamp."

The Effect of Heat Stress on Skeletal Muscle Contractile Properties [29]

This is a machine-generated summary of:

Locke, Marius; Celotti, Carlo: The effect of heat stress on skeletal muscle contractile properties [29]

Published in: Cell Stress and Chaperones (2013)

Link to original: https://doi.org/10.1007/s12192-013-0478-z

If you want to cite the papers, please refer to the original.

For technical reasons we could not place the page where the original quote is coming from.

Abstract-Summary "An elevated heat-shock protein (HSP) content protects cells and tissues, including skeletal muscles, from certain stressors."

"We determined if heat stress and the elevated HSP content that results is correlated with protection of contractile characteristics of isolated fast and slow skeletal muscles when contracting at elevated temperatures."

"To elevate muscle HSP content, one hindlimb of Sprague–Dawley rats (21–28 days old, 70–90 g) was subjected to a 15 min 42 °C heat-stress."

"HSP content was elevated in muscles from the heat-stressed limbs when compared with controls."

"For tetanic tension, both muscles also showed an increase in tension between 5 and 30 min when stimulated at 20 °C regardless of treatment but when stimulated at 42 °C no change was observed."

"No protective effect of an elevated HSP content was observed for either muscle."

"Although heat stress caused an elevation in HSP content, no protective effects were conferred to isolated contracting muscles."

Introduction

"An elevated HSP content might be beneficial to skeletal muscle during contractions at elevated temperatures."

"Very few studies have provided any evidence for an HSP-mediated protection to skeletal muscle during contraction."

"Studies using transgenic overexpression of Hsp72 have also suggested HSPs to be beneficial to contracting skeletal muscle [30]."

"In view of the fact that elevated muscle temperatures have detrimental effects on skeletal muscle contractile properties, coupled with the knowledge that HSPs provide a cross-tolerance effect to cardiac and skeletal muscle, it was of interest to determine if elevated skeletal muscle HSP content would aid skeletal muscles contracting at an elevated temperatures."

"The purpose of this study was to determine whether a prior heat stress and the resultant increase in HSP content altered the contractile characteristics of isolated fast (extensor digitorum longus) and slow (soleus) muscles when stimulated to contract at 42 °C."

Methods

"One hindlimb from each animal in the one-limb heat-stress group was subjected to a heat stress (42 °C for 15 min), while the contra-lateral (1LC; contra-lateral control) limb served as a control."

"Following the heat stress, animals were allowed to recover at room temperature for 24 h. Isolated muscles from the two groups were equally divided into a 20 °C (C20, n = 10; 1LC20, n = 10; 1LHS20, n = 10) group and a 42 °C (C42, n = 10; 1LC42, n = 10; 1LHS42, n = 10) group."

"The soleus and extensor digitorum longus (EDL) muscles were removed in random order from both [heat-stressed and non-heat-stressed (contra-lateral)] limbs of the one-limb heat-stressed groups, and from one limb from animals from the control group."

Results

"Following densitometric scanning and quantification of bands on the blots, a significant ($p < 0.05$) increase in Hsp72 content was observed in both the heat-stressed soleus and EDL when compared with contra-lateral control muscles."

"As time increased, the EDL muscle groups stimulated at 20 °C gradually increased P_0 such that, at 30 min, all 20 °C stimulated muscles were significantly ($p < 0.05$) greater from all 42 °C stimulated EDL muscles."

"A significant ($p < 0.05$) increase in EDL P_0 between 5 and 30 min was also detected for the all groups stimulated at 20 °C."

"Statistically significant ($p < 0.05$) differences in P_0 were observed between 5 and 30 min in all muscle groups stimulated at 20 °C, but not for any muscle group stimulated at 42 °C."

Discussion

"It was hypothesized that an increased HSP content following the heat stress would aid muscles during contraction at 42 °C."

"In support of this, most studies demonstrating striated muscle HSP-mediated protection have used models of cross-tolerance where exposure to "stressor A" is used to protect against "stressor B" and not against the original "stressor A." For example, an elevated HSP content from heat stress/shock appears to protect cardiac and skeletal muscle against an ischemic stress."

"The length of recovery (24 h) following heat stress and the temperatures selected for muscle stimulation may also have been important factors for the lack of protection observed."

"The temperature of 42 °C likely presents a considerable stress to the muscles, and therefore, the greater oxidative capacity of the soleus conferred a greater tolerance to the stressors than the EDL."

"A prior heat stress does not appear to benefit the contractile properties of the EDL and soleus muscles when contracting at temperatures of 20 °C or 42 °C."

Skeletal Muscle Energy Metabolism in Environmental Hypoxia: Climbing Towards Consensus [31]

This is a machine-generated summary of:

Horscroft, James A; Murray, Andrew J: Skeletal muscle energy metabolism in environmental hypoxia: climbing towards consensus [31]

Published in: Extreme Physiology & Medicine (2014)

Link to original: https://doi.org/10.1186/2046-7648-3-19

If you want to cite the papers, please refer to the original.

For technical reasons we could not place the page where the original quote is coming from.

Abstract-Summary "Skeletal muscle undergoes metabolic remodelling in response to environmental hypoxia, yet aspects of this process remain controversial."

"Broadly, environmental hypoxia has been suggested to induce: (i) a loss of mitochondrial density; (ii) a substrate switch away from fatty acids and towards other substrates such as glucose, amino acids and ketone bodies; and (iii) a shift from aerobic to anaerobic metabolism."

"To attempt to resolve some of the controversies, we performed a comprehensive review of the literature pertaining to hypoxia-induced changes in skeletal muscle energy metabolism."

"Environmental hypoxia does however induce a selective attenuation of fatty acid oxidation, whilst glucose uptake is maintained or increased, perhaps to support glycolysis in the face of a downregulation of oxidative metabolism, optimising the pathways of ATP synthesis for the hypoxic environment."

Review

"The aim of this review was to collate evidence pertaining to the remodelling of metabolic processes in mammalian skeletal muscle in vivo in response to environmental hypoxia, accounting for variations in degree and duration of hypoxic exposure."

"A single result for each of the four metabolic processes and mitochondrial density was inferred from each setting as follows: increase (where at least one biomarker of a process was significantly increased by hypoxia, and none decreased); decrease (where at least one biomarker of a process was significantly decreased by hypoxia, and none increased); unchanged (where at least one biomarker was measured and no biomarkers were significantly altered by hypoxia); and unclear (where at least one biomarker of a process was significantly increased and another significantly decreased)."

"We set out to understand the remodelling of metabolic processes in the mammalian skeletal muscle in vivo in response to environmental hypoxia, accounting for variations in degree and duration of hypoxic exposure."

"Whilst oxidative processes are selectively downregulated in the skeletal muscle following exposure to environmental hypoxia, in contrast to studies in cultured cells, glycolytic markers appear to remain largely unchanged."

Conclusions

"The literature suggests that skeletal muscle oxidative metabolism is lowered by exposure to environmental hypoxia, which may precede a loss in muscle mitochondrial density."

"The total capacity for skeletal muscle glycolysis is not consistently altered by environmental hypoxia."

"The literature is not clear on whether a hypoxia-induced substrate switch from fatty acid oxidation to glucose oxidation occurs within the mitochondria of skeletal muscle as it does in the hypoxic rat heart, for instance."

"Environmental hypoxia does however induce a selective attenuation of whole muscle fatty acid oxidation, whilst glucose uptake is maintained or increased, perhaps to support glycolytic flux in the face of a downregulation of oxidative metabolism, optimising the pathways of ATP synthesis for the hypoxic environment."

Voltage Sensing Mechanism in Skeletal Muscle Excitation-Contraction Coupling: Coming of Age or Midlife Crisis? [32]

This is a machine-generated summary of:

Hernández-Ochoa, Erick O.; Schneider, Martin F.: Voltage sensing mechanism in skeletal muscle excitation-contraction coupling: coming of age or midlife crisis? [32]

Published in: Skeletal Muscle (2018)

Link to original: https://doi.org/10.1186/s13395-018-0167-9

Abstract-Summary "Major advances in electrical and optical measurements, including muscle fiber voltage clamp to reveal membrane electrical properties, in conjunction with the development of electron microscopy to unveil structural details provided an elegant view of ECC in skeletal muscle during the last century."

"This surge of knowledge on structural and biophysical aspects of the skeletal muscle was followed by breakthroughs in biochemistry and molecular biology, which allowed for the isolation, purification, and DNA sequencing of the muscle fiber membrane calcium channel/transverse tubule (TT) membrane voltage sensor (Cav1.1) for ECC and of the muscle ryanodine receptor/sarcoplasmic reticulum Ca^{2+} release channel (RyR1), two essential players of ECC in skeletal muscle."

"In regard to the process of voltage sensing for controlling calcium release, numerous studies support the concept that the TT Cav1.1 channel is the voltage sensor for ECC, as well as also being a Ca^{2+} channel in the TT membrane."

Background

"In skeletal muscle, electrical impulses carried by the axons of motoneurons travel to the nerve endings at the muscle endplate (the muscle synapse), where these electrical signals are converted into chemical signals that produce depolarizing postsynaptic potentials at the neuromuscular junction sarcolemma of the muscle fiber [33, 34]."

"In all but a few "tonic" muscle fibers, these postsynaptic endplate potentials elicit a further depolarization of the muscle fiber, carried out by skeletal muscle voltage-gated sodium channels, initiating and propagating the muscle action potential [35–37]."

"The muscle action potential (AP) travels both longitudinally away from the fiber endplate along the muscle fiber surface sarcolemma and radially into the fiber via invaginations of the sarcolemma that form the transverse tubular (TT) system [38, 39]."

"The AP depolarization activates skeletal muscle voltage-gated calcium channels (Cav1.1; also known as dihydropyridine receptors, DHPR) [40]."

Introduction to Excitation-Contraction Coupling

"By the 1950s and 1960s, the processes that initiated and accompanied skeletal muscle contraction had been studied from several angles [41, 42]."

"These pioneers made important fundamental contributions to understanding ECC, including the following. (1) Hodgkin and Horowicz [43, 44] proposed that the event that normally induces muscle contraction is a change in membrane potential rather than the longitudinal spread of current along the fiber; they also showed that the development of tension was dependent on membrane potential and was described by a steep sigmoidal curve of tension as a function of membrane potential. (2) The experiments of Huxley and Taylor [39] showing activation at the Z disk in frog muscle fibers and of Huxley and Straub showing local activation at the A-I band junction in lizard muscle [45], together with the localization of the TT system at the Z disk in frog muscle [46] and at the A and I band junction in lizard muscle [47], indicated that the transverse tubules (TT) of the skeletal muscle fibers form the network which conducts the surface depolarization radially into a muscle fiber to initiate contraction. (3) Investigations started by Ringer [48] and continued by Heilbrunn [49], Kamada and Kinoshita [50], and others introduced the role of Ca^{2+} as key regulator of striated muscle activation."

"In keeping with the theme of "coming of age/midlife crisis" of the ECC voltage sensor, here we will first review the discovery and early functional studies of the ECC voltage sensor and its role and properties, largely carried out during the last quarter of the twentieth century."

"A detailed review of the cloning of the Cav1.1 and RyR, and identification of their skeletal muscle isoforms as the ECC voltage sensor and skeletal muscle SR Ca^{2+} release channel, respectively, is beyond the present scope and can be found elsewhere [51–54]."

Intramembrane Charge Movement and ECC

"Further below, from the original report, the charge movement kinetics and voltage dependence were generally in the range that would be appropriate for muscle contractile activation, so it was not unreasonable to identify these charge movements with the intramembrane movement of ECC voltage sensors in the TT membrane [55]."

"An immediate question that arose after the first detection of charge movement currents was whether the voltage sensor currents detected in muscle fibers were in fact the control system for depolarization-induced contractile activation."

"Two early studies addressed this question, using different pulse protocols to show that voltage sensor charge movement measurements can be used to closely predict the initiation of muscle contraction."

"These studies demonstrated a close correlation between the voltage sensor charge movement and just-detectable contractile activation of muscle fibers."

Molecular Components and Mechanisms in ECC

"Cav1.1 channels are principally expressed in the membrane of the TT system of adult skeletal muscle fibers and are members of a diverse family of voltage-dependent Ca^{2+} channels."

"The TT voltage sensor (CaV1.1) is believed to be the "ligand" that uniquely enables RyR1 opening in functioning skeletal muscle fibers."

"These types of experiments indicate that in functioning muscle fibers, the RyR1 SR Ca^{2+} release channels are essentially fully off when the voltage sensors are in the resting condition, but turn on strongly and rapidly during the AP or voltage clamp depolarization that activates the voltage sensors [56–58]."

"All the other ligands that modulate RyR1 channel activity in isolated membrane or protein preps may or may not also similarly modulate the RyR1 in a functioning fiber, but the TT voltage sensor serves as master regulator determining whether or not the channel can open at all."

Future Perspectives

"The Schneider and Chandler hypothesis that the excitatory signal passes from the TT to the SR membrane by way of charges moving in the TT membrane connecting with the junctional feet of the SR Ca^{2+} release channel initiated the path to answer this question."

"It is still unknown how this movement of charge, originating in Cav1.1, transfers a signal across to the RyR1 (feet) to trigger Ca^{2+} release."

"This question is particularly fascinating because of the as yet unknown molecular structure-function relationship between these components in two different membrane systems, the TT (Cav1.1 voltage sensors) and the SR (RyR1 Ca^{2+} release channels)."

Conclusions

"Some of the many remaining unanswered questions regarding ECC include the following: Are all four VSDs (I–IV) needed to activate the RyR1 Ca^{2+}channel, or is only a subset of charges involved?"

"Which VSDs contribute the voltage sensor element(s) for electromechanical coupling between the Cav1.1 and RyR1 Ca^{2+} release?"

"Which residues are moved during Cav1.1 channel activation?"

"Which residues are moved for RyR1 Ca^{2+} release channel activation?"

"What are the molecular determinants that mediate the electromechanical coupling between the VSD and RyR1 Ca^{2+} release?"

HSP72 Expression Is Specific to Skeletal Muscle Contraction Type [59]

This is a machine-generated summary of:

Bonello, John-Peter; Locke, Marius: HSP72 expression is specific to skeletal muscle contraction type [59]

Published in: Cell Stress and Chaperones (2019)

Link to original: https://doi.org/10.1007/s12192-019-00997-x

If you want to cite the papers, please refer to the original.

For technical reasons we could not place the page where the original quote is coming from.

Abstract-Summary "Exercise is capable of inducing the cellular stress response and increasing skeletal muscle heat shock protein (HSP) content."

"To determine the relationship between muscle contraction types, muscle damage, and HSP content, one tibialis anterior (TA) muscle from male Sprague-Dawley rats (n = 5/group) was electrically stimulated while actively lengthening (LC), shortening (SC), or remaining to stagnate (IC) for 15 repetitions (3 sets of five)."

"Twenty-four hours after stimulation, TA muscles were removed, processed, and assessed for damage and for HSP25 and HSP72 content."

"HSP72 content increased after all LCs conditions but not following ICs or SCs."

"HSP25 content remained unchanged following all contractions."

"Muscle damage was observed only after LCs and not after other contraction types."

"Muscle HSP72 content can be increased with as few as 5 maximal lengthening contractions and appears to be related to muscle damage."

Introduction

"These stressors are known to occur in skeletal muscle during exercise, it is perhaps not surprising that a number of studies [60–63], though not all [64, 65], have shown an increased muscle HSP content following exercise."

"Since all of the exercise-related cellular stressors ultimately originate from muscle contraction, it follows that different types of muscle contraction may differentially influence the CSR and thus HSP content."

"Using a rodent model, our previous work has demonstrated an increased muscle damage coupled with an elevated HSP content following 100 LCs but not following 100 SCs [66]."

"The purpose of this study was to examine a more physiologically relevant range (5–15 LCs) of muscle contractions, determining the minimum number of LCs required to elevate muscle HSP content while minimizing the effect of muscle damage."

Methods and Materials

"Following a similar methodology outlined by Pollock-Tahiri and Locke [67], transferred membranes were blocked in 5% BSA (5% BSA powder in Tris-buffered saline ((TBS) 500 mM NaCl, 20 mM Tris, pH 7.5) for 1 h. Blocked membranes underwent two 5-min wash cycles in Tween-Tris-buffer saline (TTBS; TBS plus 0.05% Tween 20) then incubated in a polyclonal antibody specific to HSP25 (ADI-SPA-801, Enzo, USA) and/or HSP72 (ADI-SPA-812-F, Enzo, USA) diluted 1:1000 in TTBS with 2% Blotto."

"Muscle pieces were placed in ethanol: Spurr's resin mixture (3:1) for 30 min, 1:1 for 1 h, and 1:3 for the last hour before being left to rotate overnight in 100% Spurr's resin."

"These data points from stimulated TA muscles were compared with contralateral (CL) TA values through a one-way ANOVA with Tukey's post hoc testing and Bartlett's variance testing."

Results

"The force-generating ability of each TA muscle was assessed pre- and post-tetanus for all contraction types as well as for the two additional LC groups."

"The observed pattern of force shown by the TA muscles subjected to LCs shows clear separations in force generation recovery/damage between contraction types as well as with the number of LCs."

"To determine the general morphology of TA muscles at 24 h after each contraction type, light microscopy was used to qualitatively assess muscle damage."

"Similar to light microscopy, the damage observed in the muscle fibers following LCs was infrequent."

"TA muscles subjected to 15 LCs showed a marked significant increase (17.7 ± 3.4 fold; $p < 0.01$) that differed from both SC and IC ($p < 0.001$) groups."

Discussion

"24 h following muscle contractions, 15 LCs resulted in a marked increase in TA muscle HSP72 content (but not HSP25 content), while 15 ICs or 15 SCs showed no increase in either HSP content."

"Since cells/tissues with an increased HSP content demonstrate cellular protection [68], it follows that exercises composed of specific contractions types may allow for increased muscle HSP72 content that may confer protection."

"While the exact reason for the lack of an increased HSP25 content remains unclear, a number of reasons such as exercise intensity and/or duration, the number of contractions, the extent of damage, fiber type composition, or even muscle species differences may account for differences observed between our model and others [69, 70]."

"This study demonstrated that as few as 5 LCs were capable of significantly increasing muscle HSP72 content with minimal observable muscle damage."

HIF Prolyl Hydroxylase Inhibition Protects Skeletal Muscle from Eccentric Contraction-Induced Injury [71]

This is a machine-generated summary of:

Billin, Andrew N.; Honeycutt, Samuel E.; McDougal, Alan V.; Kerr, Jaclyn P.; Chen, Zhe; Freudenberg, Johannes M.; Rajpal, Deepak K.; Luo, Guizhen; Kramer, Henning Fritz; Geske, Robert S.; Fang, Frank; Yao, Bert; Clark, Richard V.; Lepore, John; Cobitz, Alex; Miller, Ram; Nosaka, Kazunori; Hinken, Aaron C.; Russell, Alan J.: HIF prolyl hydroxylase inhibition protects skeletal muscle from eccentric contraction-induced injury [71].

Published in: Skeletal Muscle (2018).

Link to original: https://doi.org/10.1186/s13395-018-0179-5

If you want to cite the papers, please refer to the original.

For technical reasons we could not place the page where the original quote is coming from.

Abstract-Summary "Therapies that protect skeletal muscle or enhance repair would be valuable medical treatments."

"We tested the importance of HIF activation following skeletal muscle injury, in both a murine model and human volunteers, using prolyl hydroxylase inhibitors that stabilize and activate HIF."

"Using a mouse eccentric limb injury model, we characterized the protective effects of prolyl hydroxylase inhibitor, GSK1120360A. We then extended these studies to examine the impact of EPO modulation and infiltrating immune cell populations on muscle protection."

"We extended this study with an experimental medicine approach using eccentric arm exercise in untrained volunteers to measure the muscle-protective effects of a clinical prolyl hydroxylase inhibitor, daprodustat."

"Treatment of healthy human volunteers with high-dose daprodustat reduced accumulation of circulating damage markers following eccentric arm exercise, although we did not observe any diminution of functional deficits with compound treatment."

"The results of these experiments highlight a novel skeletal muscle protective effect of prolyl hydroxylase inhibition via HIF-mediated expression of iNOS in macrophages."

Introduction

"Both HIF1α and the HIF-target gene, VEGF, are elevated by strenuous exercise [72, 73], and VEGF stimulates muscle angiogenesis and myoblast proliferation [74]."

"Another HIF-target gene, erythropoietin (EPO), also stimulates myoblast proliferation and muscle recovery after injury."

"One of the primary mediators of HIF1α activity in muscle is local oxygen concentration, which can drop significantly during exercise or injury [75, 76]."

"Under normoxic conditions, oxygen-regulated HIF proteins (primarily HIF1α, HIF2α) are hydroxylated and targeted for proteasomal degradation by HIF-prolyl 4-hydroxylases (PHD1,2,3) which require molecular O_2 for activity."

"Under low oxygen tension, PHD activity is reduced, stabilizing HIF and allowing transcription of target genes [77]."

Results

"As EPO has been previously linked to improved muscle repair after injury, we tested whether the elevation of EPO by GSK360 was responsible for its protective effects [78]."

"We then treated mice daily with recombinant EPO, with or without an EPO neutralizing antibody, beginning 2 days prior to eccentric injury to elucidate the effects of GSK360-induced EPO production on muscle recovery."

"That EPO was not required for muscle protection with GSK360 indicated another mechanism through which PHD inhibition might affect muscle injury, possibly via HIF1α upregulation in muscle itself."

"Using myeloid HIF1α KO mice, we asked whether altering hypoxic signaling in immune cells would affect recovery from muscle injury."

"These data point to PHI mediating a protective effect in muscle via activation of HIF1α in macrophages and the subsequent upregulation of iNOS."

Discussion

"Higher doses (30 mg/kg) of GSK360 have been associated with HIF1α target gene activation in muscle, but our studies identified a more sensitive myeloid population that drives muscle protection in an iNOS-dependent manner without large scale transcriptional responses [79]."

"We presume that the modest inflammatory response and NO generation associated with eccentric contraction-induced muscle injury leads to a greater protective role for NO generated through iNOS."

"We demonstrate that the macrophage HIF1α compartment is essential for the protective effect of PHD inhibition, which is consistent with their presumed role in membrane repair after eccentric contraction-induced muscle injury [80]."

"The exact mechanism by which the hypoxic response generated by PHD inhibition in macrophages is driving the protection and recovery from the muscle injury is as yet undetermined, but it is likely to either be remodeling the inflammatory response or accelerating satellite cell activation [81]."

Conclusion

"We have uncovered a novel connection between HIF1α and iNOS that contributes to muscle protection after exercise-driven injury in mice."

"Evidence suggests that some elements of this protective effect may also be competent in healthy volunteers after eccentric injury, although there are clear differences in the magnitude of response between mouse and man."

"Our data implicates the HIF signaling axis as an important modulator of inflammatory activity during skeletal muscle repair and injury and further indicates that this mechanism may be of therapeutic importance in the treatment of acute muscle injury."

Materials and Methods

"After a 2-min rest, subjects were asked to perform five sets of six maximal eccentric contractions of the elbow flexors of the non-dominant arm to induce local muscle inflammation, pain and transient functional deficits."

"The primary objective of the study was to evaluate the protective effects of daprodustat on eccentric exercise-induced muscle injury by measuring MVC of the exercised arm up to 72 h after completion of eccentric exercise in subjects treated with daprodustat in comparison to subjects treated with placebo."

"Another secondary objective was to evaluate the protective effects of daprodustat on deficits in arm range of motion after eccentric exercise by measuring change in degree of motion and resting arm angle from post-exercise up to 72 h after completion of eccentric exercise in subjects treated with daprodustat in comparison to subjects treated with placebo."

Aging and Short-Term Calorie Restriction Differently Affect the Cardiac and Skeletal Muscle Expression of Genes Regulating Energy Substrate Utilization in Male Rats [82]

This is a machine-generated summary of:

Ławniczak, Aleksandra; Wrońska, Agata; Wierzbicki, Piotr; Kmieć, Zbigniew: Aging and short-term calorie restriction differently affect the cardiac and skeletal muscle expression of genes regulating energy substrate utilization in male rats [82]

Published in: Biogerontology (2022)

Link to original: https://doi.org/10.1007/s10522-022-09965-y

If you want to cite the papers, please refer to the original.
For technical reasons we could not place the page where the original quote is coming from.

Abstract-Summary "The study aim was to assess the effects of short-term calorie restriction (SCR) and refeeding on the expression of genes involved in the control of cardiac and skeletal muscle energy metabolism in old vs. young male rats."

"In SM, Cpt-I, Glut4, Pgc-1α, and Sirt3 mRNA levels were lower in old than young rats."
"In CM of only young rats SCR increased Cpt-I expression which remained elevated after refeeding."
"Upon SCR, the expression of Ppar-β/δ, Glut4, Pgc-1α, and Sirt3 in CM increased in young but not old rats, and refeeding re-established control levels."
"In SM of young rats SCR increased Ppar-β/δ and Pgc-1α, and decreased Sirt3 expression, whereas refeeding generally decreased these mRNA levels."
"In SM of old rats SCR decreased only Pgc-1α expression."
"SCR appears to increase the efficiency of glucose and fatty acid utilization in the cardiac muscle of young, but not old male rats."

Introduction

"Peroxisome proliferator-activated receptor-γ coactivator-1α (PGC-1α) is involved in glucose and fatty acid metabolism, fibre type switching in skeletal muscle, cardiac diseases, and other processes [83]."

"In both types of muscles, PGC-1α co-activates peroxisome proliferator-activated receptor beta/delta (PPAR-β/δ) to modulate the transcription of genes involved in fatty acid oxidation and oxidative phosphorylation, as well as glucose uptake [84–87]."

"PGC-1α and PPAR-β/δ co-activate the transcription of many genes, including the one encoding carnitine palmitoyltransferase- I (CPT-I), a crucial enzyme for oxidation of fatty acids as the primary energy source in muscles [88]."

"We focused on SCR's effects on cardiac and skeletal muscle expression levels of key genes involved in mitochondrial fatty acid β-oxidation (the rate-limiting enzyme Cpt-I and its transcriptional regulator Ppar-β/δ) and glucose uptake (Glut4 and its transcriptional regulator Pgc-1α)."

Materials and Methods

"The experiment was performed on young (4-month-old at the onset of the experiment) and old (24-month-old) male Wistar-Han rats maintained at Specific Pathogen Free conditions in the Academic Laboratory Animal Centre in Gdansk, Poland, at 22 °C under a 12 h/12 h light/dark cycle."

"Frozen tissues samples of the cardiac and skeletal muscles were homogenised (MagNALyser homogenizer, Roche Diagnostics, Indianapolis, IN, USA) and total

RNA was extracted from homogenates using the Total RNA kit (A&A Biotechnology, Gdynia, Poland)."

"Semi-quantification of SIRT3 protein levels was performed in duplicate for every young and old rats' group."

"CPT-I activity was measured in cardiac and skeletal muscles of control and SCR rats (5 animals per group)."

"The concentrations of leptin and insulin in blood sera were measured by rat-specific ELISA tests (Merck Millipore, St Charles, MO, USA)."

Results

"Refeeding of young rats increased the HDL-cholesterol to control levels, however, in old rats the concentration remained below control throughout 4 days of refeeding ($p < 0.01$)."

"The TG/HDL ratio decreased after SCR by 27% ($p < 0.05$) and 38% ($p < 0.001$) in young and old rats, respectively, and increased upon refeeding up to control levels."

"In old rats, SCR decreased Pgc-1α expression 2.5-fold, and refeeding for 4 days restored the control level."

"In cardiac muscle of young rats on SCR, the mRNA level of Pgc-1α strongly correlated with Glut4 expression ($r = 0.82$, $p = 0.00002$), however, not in the old SCR animals."

"SIRT3 protein expression in skeletal muscle of young rats was elevated after SCR and refeeding for 2 and 4 days by 45%, 75%, and 50%, respectively, as compared to control."

Discussion

"Our study compared transcriptional-level changes in response to SCR followed by ad libitum feeding for up to 4 days in cardiac and skeletal muscles of young and old male rats."

"In line with the elevated TG levels, we found that SCR increased gene expression and activity of CPT-I in the cardiac muscle of young but not old rats."

"We noted that SCR increased the expression of Ppar-β/δ (which activates the transcription of Cpt-I in the muscle) in cardiac and skeletal muscles of young but not old rats [89]."

"In the skeletal muscle of old rats in our study, SCR exacerbated the age-related decline in Pgc-1α expression, with parallel (though insignificant) reduction of Glut4 mRNA."

"Differently than in the cardiac muscle, the skeletal muscle expression of Sirt3 decreased after SCR of young rats, even though SIRT3 protein levels were elevated."

Mitochondrial Co-Chaperone Protein Tid1 Is Required for Energy Homeostasis during Skeletal Myogenesis [90]

This is a machine-generated summary of:

Cheng, Li-Hao; Hung, Kai-Feng; Lee, Te-Chang; Huang, Chih-Yang; Chiu, Wen-Ting; Lo, Jeng-Fan; Huang, Tung-Fu: Mitochondrial co-chaperone protein Tid1 is required for energy homeostasis during skeletal myogenesis [90]

Published in: Stem Cell Research & Therapy (2016)

Link to original: https://doi.org/10.1186/s13287-016-0443-8

If you want to cite the papers, please refer to the original.

For technical reasons we could not place the page where the original quote is coming from.

Abstract-Summary "Tid1 is a mitochondrial co-chaperone protein and its transcript is abundantly expressed in skeletal muscle tissues."

"In vitro induced differentiation assay of mouse myoblast C2C12 cells was applied to examine the physiological role of Tid1 during skeletal myogenesis."

"Transgenic mice with muscle specific (HSA-Cre) Tid1 deletion were established and examined to determine the physiological function of Tid1 during skeletal muscle development in vivo."

"Expression of Tid1 protein was upregulated in the differentiated C2C12 cells, and the HSA-Tid1$^{f/f}$ mice displayed muscular dystrophic phenotype."

"Consistent with the in vivo finding, we observed that downregulation of Tid1 not only reduced the ATP production but also abolished the differentiation ability of C2C12 cells by impairing the mitochondrial activity."

"Our results suggest that Tid1 deficiency reduces ATP production and abolishes mitochondrial activity, resulting in energy imbalance and promoting apoptosis of muscle cells during myogenesis."

Background

"Accumulating studies suggest that mitochondria are associated with the regulation of the skeletal muscle physiology [91, 92]."

"The PGC-1α is expressed in skeletal muscle and activates mitochondrial biogenesis and oxidative metabolism [93, 94]."

"It has been shown that Tid1 is required to maintain the steady-state distribution of mitochondrial membrane potential and is associated with the mtDNA [95]."

"The Tid1 transcripts are expressed in most of the adult tissues including heart, muscle, bone marrow, spleen, lymph node, and thymus [96]."

"The regulatory role of Tid1 in skeletal muscle homeostasis remains unclear."

"We demonstrated that Tid1 is required to maintain the physiological function in muscular development in vitro and in vivo."

"The significance of this study warrants further understanding of the function of Tid1 in muscular development."

"The transgenic mice generated in this study will be a valuable tool to elucidate the physiological function of Tid1 during normal muscular homeostasis."

Methods

"Protein extracts from C2C12 cells or mice tissues were prepared using RIPA buffer."

"The following procedures were performed according to the method described by Lo and others to generate Tid1$^{f/f}$ mice [97]."

"Deletion of the Tid1 gene in Tid1 floxed mice was achieved by crossing Tid1$^{f/f}$ mice with transgenic Cre mice."

"All genotyping of the Cre transgene and the Tid1-deficient mice was performed by polymerase chain reaction (PCR) using genomic DNA isolated from the tail tip."

"The mice were placed on a rotating rod running at different speeds and for different durations."

"For data collection, mice were pre-trained for 3 days and trained four times per day on three consecutive days."

"7-μm thick frozen sections from mice tissues were sliced in a freezing cryostat at −20 °C."

"An ATP measurement kit for cells (K791-100; BIOVISION, USA) was used to measure the ATP concentration according to the manufacturer's protocol."

Results

"We observed that the HSA-Tid1$^{f/f}$ mice died between days P8 and P10."

"Because the mice died between P8 and P10, examination of the muscular physiology of adult mice with Tid1 deficiency was prevented."

"We evaluated the muscular physiology of adult mice carrying Tid1 heterogeneity (HSA-Tid1$^{f/+}$)."

"These results suggest that deletion of the Tid1 gene causes dysfunction of muscle tissue of transgenic mice."

Discussion

"In order to further understand the role of Tid1, a mitochondrial co-chaperone, on mediating myogenesis, we determined the expression of Tid1 during the induced differentiation of myoblasts (C2C12 cells) in vitro."

"These results suggest that Tid1 deficiency impairs the mitochondrial activity, resulting in insufficient ATP production and cell apoptosis."

"The discrepancy between the expression of C-caspase 3 under the distinct downregulation of Tid1 and Caspase 3 during the in vitro induced differentiation may be caused by the distinct physiological effect after the gene-specific knockdown."

"These results suggest that downregulation of Tid1 would initially decrease ATP production along with increased cell death, followed by reduced mitochondrial biogenesis (reduction of PGC-1α)."

"These in vitro results imply that energy insufficiency of muscle cells mediated by Tid1 gene deletion (HAS-Tid1$^{f/f}$) is likely the reason for apoptotic cell death during the myogenesis and muscular dystrophy phenotype."

Conclusion

"Downregulation of Tid1 abolished the ability of C2C12 to differentiate via reduced ATP production, impaired mitochondrial activity, and induced apoptotic cell death."

"Tid1 deficiency reduces ATP production, impairs the mitochondrial activity of muscle cells during the myogenesis, and, consequently, causes muscle cell apoptosis."

"How Tid1 deficiency reduces ATP production and inhibits PGC-1α activity remains unclear."

"We will further investigate the molecular mechanisms mediated by Tid1 to regulate ATP production and PGC-1α activity during myogenesis."

Effect of L-Arginine on Energy Metabolism, Skeletal Muscle and Brown Adipose Tissue in South Asian and Europid Prediabetic Men: A Randomised Double-Blinded Crossover Study [98]

This is a machine-generated summary of:

Boon, Mariëtte R.; Hanssen, Mark J. W.; Brans, Boudewijn; Hülsman, Cindy J. M.; Hoeks, Joris; Nahon, Kimberly J.; Bakker, Charlotte; van Klinken, Jan B.; Havekes, Bas; Schaart, Gert; Jazet, Ingrid M.; Rensen, Patrick C. N.; van Marken Lichtenbelt, Wouter D.: Effect of l-arginine on energy metabolism, skeletal muscle and brown adipose tissue in South Asian and Europid prediabetic men: a randomised double-blinded crossover study [98]

Published in: Diabetologia (2018)

Link to original: https://doi.org/10.1007/s00125-018-4752-6

If you want to cite the papers, please refer to the original.

For technical reasons we could not place the page where the original quote is coming from.

Abstract-Summary "Individuals of South Asian origin are at increased risk of developing type 2 diabetes mellitus and associated comorbidities compared with Europids."

"We aimed to investigate the effects of 6 weeks of supplementation with l-arginine, a precursor of NO, on energy metabolism by BAT and skeletal muscle, as well as glucose metabolism in South Asian men compared with men of European descent."

"We included ten Dutch South Asian men (age 46.5 ± 2.8 years, BMI 30.1 ± 1.1 kg/m^2) and ten Dutch men of European descent, that were similar with respect to age and BMI, with prediabetes (fasting plasma glucose level 5.6–6.9 mmol/l or plasma glucose levels 2 h after an OGTT 7.8–11.1 mmol/l)."

"The primary study endpoint was the effect of l-arginine on BAT volume and activity."

"l-Arginine did not affect BMR, [^{18}F]FDG uptake by BAT or skeletal muscle respiration in either ethnicity."

"During OGTT, l-arginine lowered plasma glucose concentrations ($AUC_{0\text{–}2\,h}$ − 9%, $p < 0.01$), insulin excursion ($AUC_{0\text{–}2\,h}$ − 26%, $p < 0.05$) and peak insulin concentrations (−26%, $p < 0.05$) in Europid but not South Asian men."

"In skeletal muscle biopsies several respiration states were consistently lower in South Asian men compared with Europid men."

"l-Arginine supplementation does not affect BMR, [^{18}F]FDG uptake by BAT, or skeletal muscle mitochondrial respiration in Europid and South Asian overweight and prediabetic men."

"L-arginine improves glucose tolerance in Europids but not in South Asians."

"South Asian men have lower skeletal muscle oxidative capacity than men of European descent."

Introduction

"Poor skeletal muscle oxidative capacity may play a crucial role in the development of obesity-induced peripheral insulin resistance (and subsequently type 2 diabetes), so enhancing skeletal muscle mitochondrial function may improve whole-body glucose metabolism."

"Skeletal muscle and possibly also brown adipose tissue (BAT) are known to be involved in human energy metabolism."

"Enhancing BAT volume and activity could be a potential therapeutic approach to ameliorate type 2 diabetes risk, especially in South Asians."

"Increasing NO bioavailability might be a promising approach to enhance skeletal muscle mitochondrial function as well as BAT activity, thereby exerting positive effects on whole-body energy and substrate metabolism."

"The primary aim of the current study was to investigate the effects of 6 weeks of l-arginine supplementation on energy metabolism by BAT in overweight prediabetic Dutch South Asian and Dutch Europid men."

Methods

"Prediabetes was defined as either fasting plasma glucose levels between 5.6 and 6.9 mmol/l or plasma glucose levels 2 h after an OGTT between 7.8 and 11.1 mmol/l [99]."

"Participants ingested either l-arginine (9 g/day l-arginine divided over three servings; Argimax; Hankintatukku, Karkkila, Finland) or visually identical placebo tablets (Hankintatukku) for 6 weeks in a randomised double-blind crossover design, with a 4 week washout period in between."

"At least an hour after the muscle biopsy, an OGTT was performed and body composition was determined by means of dual x-ray absorptiometry (Discovery A; Hologic, Bedford, MA, USA)."

"Mixed model analysis with treatment, ethnicity and baseline glucose levels as fixed effects and subject-specific intercepts as random effects was used to assess the effect of l-arginine on glucose and insulin excursions after adjustment for baseline glucose levels."

Results

"South Asians were shorter than Europids (1.76 ± 0.02 vs 1.81 ± 0.02 m), although the difference did not reach statistical significance ($p = 0.06$)."

"When analysing both ethnic groups together, l-arginine did not change BAT activity and volume (data not shown)."

"Radiodensity of this supraclavicular BAT depot (expressed in Hounsfield units) was not significantly different between ethnicities or after l-arginine treatment (data not shown)."

"Although differences were not statistically significant, Europids had higher peak glucose levels ($p = 0.10$) and peak glucose time ($p = 0.10$) after receiving placebo, compared with South Asians, but had a lower Matsuda index ($p = 0.004$), the latter pointing to enhanced insulin sensitivity."

"Since skeletal muscle mitochondrial function is closely associated with peripheral insulin sensitivity [100], we next assessed the effect of l-arginine on skeletal muscle respiratory capacity in the South Asian and Europid men."

Discussion

"We show that l-arginine, the precursor of NO, does not affect energy expenditure, cold-induced BAT activity or skeletal muscle mitochondrial respiration in Europid and South Asian men with prediabetes."

"Our finding that l-arginine treatment improves glucose metabolism in overweight and obese prediabetic Europid men is in line with previous studies."

"An interesting result in the current study is the lack of effect of l-arginine on glucose metabolism in men of South Asian ethnicity."

"A recent study has shown that endothelial cells isolated from South Asian men display lower expression levels of endothelial NOS, one of the enzymes that converts l-arginine into NO [101], compared with cells isolated from a matched control group of European origin [102]."

"We show that 6 weeks of l-arginine supplementation does not affect BMR, [^{18}F]FDG uptake by BAT or skeletal muscle mitochondrial respiration in Europid and South Asian overweight and obese prediabetic men."

"L-arginine improves glucose tolerance in Europid men but not in South Asian men."

Effect of L-Carnitine Supplementation on the Body Carnitine Pool, Skeletal Muscle Energy Metabolism and Physical Performance in Male Vegetarians [103]

This is a machine-generated summary of:

Novakova, Katerina; Kummer, Oliver; Bouitbir, Jamal; Stoffel, Sonja D.; Hoerler-Koerner, Ulrike; Bodmer, Michael; Roberts, Paul; Urwyler, Albert; Ehrsam, Rolf; Krähenbühl, Stephan: Effect of l-carnitine supplementation on the body carnitine pool, skeletal muscle energy metabolism and physical performance in male vegetarians [103]

Published in: European Journal of Nutrition (2015)

Link to original: https://doi.org/10.1007/s00394-015-0838-9

If you want to cite the papers, please refer to the original.

For technical reasons we could not place the page where the original quote is coming from.

Abstract-Summary "Vegetarians ingest less carnitine and carnitine precursors and have lower plasma carnitine concentrations than omnivores."

"Principle aims of the current study were to assess the plasma and skeletal muscle carnitine content and physical performance of male vegetarians and matched omnivores under basal conditions and after l-carnitine supplementation."

"Before carnitine supplementation, vegetarians had a 10 % lower plasma carnitine concentration, but maintained skeletal muscle carnitine stores compared to omnivores."

"Sub-maximal exercise (75 % VO_2max for 1 h) revealed no significant differences between vegetarians and omnivores (respiratory exchange ratio, blood lactate and muscle metabolites)."

"Supplementation with l-carnitine significantly increased the total plasma carnitine concentration (24 % in omnivores, 31 % in vegetarians) and the muscle carnitine content in vegetarians (13 %)."

"Vegetarians have lower plasma carnitine concentrations, but maintained muscle carnitine stores compared to omnivores."

"Oral l-carnitine supplementation normalizes the plasma carnitine stores and slightly increases the skeletal muscle carnitine content in vegetarians, but without affecting muscle function and energy metabolism."

Introduction

"One nutrient that is almost totally devoid in a vegetarian diet is carnitine [104]."

"The consequences of limited carnitine ingestion upon the health of vegetarians have received relatively little attention to date [105, 106]."

"Long-term vegetarians have lower plasma carnitine concentrations than omnivores and a markedly reduced urinary carnitine excretion [107]."

"These findings have been confirmed and also a slight reduction in the skeletal muscle carnitine content of vegetarians has been reported [108]."

"Considering the existing literature about carnitine homeostasis in vegetarians and the important role of carnitine in energy metabolism, we decided to study the effect of long-term oral treatment with carnitine on the body carnitine pool, fuel metabolism and physical performance of vegetarians and omnivores."

"We hypothesized that treatment with carnitine would increase plasma and possibly also skeletal muscle carnitine concentrations in vegetarians and would thereby improve skeletal muscle energy metabolism and physical performance."

Materials and Methods

"One week later, the participants underwent a second session of sub-maximal exercise testing at 75 % of VO_2max which was followed by blood collection and muscle biopsy as described above."

"Primary outcome measure of the study was the effect of carnitine supplementation on the plasma and muscle total carnitine concentration, VO_2max, and on the muscle content of markers of energy metabolism (phosphocreatine, lactate and glycogen) before and 1 h after exercise at 75 % VO_2max in vegetarians."

"Subjects were asked to report to the laboratory, following an overnight fast, and performed 60 min of sub-maximal (two-legged) cycling exercise at a cadence of 70–85 rpm against a pre-determined resistance, which equated to 75 % of their VO_2max determined before."

"Blood samples were obtained from the subject's non-dominant arm before and after the 24-hour urine collection and before and after the sub-maximal exercise by venipuncture into heparinized."

Results

"Carnitine supplementation had no significant effect on the muscle glycogen content in vegetarians [difference post- to pre-carnitine supplementation −6 (95 % CI −42–30) and 0 (95 % CI −30–30) mmol × kg^{-1} dry weight pre- and post-exercise, respectively] or omnivores [difference post- to pre-carnitine supplementation −27 (95 % CI −121–67) and 27 (95 % CI −2–56) mmol × kg^{-1} dry weight pre- and post-exercise, respectively]."

"Carnitine supplementation had no significant effect on the muscle lactate content in vegetarians [difference post- to pre-carnitine supplementation −0.4 [95 % CI −1.2–0.4) and −0.2 (95 % CI −5.6–5.2) mmol × kg^{-1} dry weight pre- and post-exercise, respectively] or omnivores [difference post- to pre-carnitine supplementation −0.7 (95 % CI −2.7–1.3) and −0.6 (95 % CI −9.3–8.1) mmol × kg^{-1} dry weight pre- and post-exercise, respectively]."

Discussion

"To the current study, Stephens and others [109] found an approximately 20% decrease in the carnitine skeletal muscle content in vegetarians compared to omnivores."

"In the current study, vegetarians treated with carnitine showed not only the expected increase in the carnitine plasma concentration, but also an approximately 13 % increase in the skeletal muscle carnitine content."

"In the recent study of Stephens and others [108], it was not possible, however, to acutely increase the skeletal muscle carnitine content in vegetarians with the combination of high carnitine plasma concentrations and insulin."

"Although we did not determine the PDC activity, the fact that carnitine administration had no significant effect on the skeletal muscle (and blood) lactate concentrations and the skeletal muscle glycogen content in vegetarians compared to omnivores indicates the observed increase in the skeletal muscle carnitine content in vegetarians was without effect on skeletal muscle carbohydrate metabolism."

Whole-Body Insulin Resistance and Energy Expenditure Indices, Serum Lipids, and Skeletal Muscle Metabolome in a State of Lipoprotein Lipase Overexpression [110]

This is a machine-generated summary of:

Nishida, Yuichiro; Nishijima, Kazutoshi; Yamada, Yosuke; Tanaka, Hiroaki; Matsumoto, Akiko; Fan, Jianglin; Uda, Yoichi; Tomatsu, Hajime; Yamamoto, Hiroyuki; Kami, Kenjiro; Kitajima, Shuji; Tanaka, Keitaro: Whole-body insulin resistance and energy expenditure indices, serum lipids, and skeletal muscle metabolome in a state of lipoprotein lipase overexpression [110]

Published in: Metabolomics (2021)

Link to original: https://doi.org/10.1007/s11306-021-01777-4

If you want to cite the papers, please refer to the original.

For technical reasons we could not place the page where the original quote is coming from.

Abstract-Summary "The main purpose of the current study was to examine the effects of the overexpression of LPL on the skeletal muscle metabolomic profiles to test our hypothesis that the mitochondrial oxidative metabolism would be activated in the skeletal muscle of LPL transgenic rabbits and that the higher mitochondrial oxidative metabolism activity would confer better phenotypic metabolic outcomes."

"Under a HFD, insulin resistance index was measured using the intravenous glucose tolerance test, and total energy expenditure (TEE) was measured by doubly-labeled water in control and LPL transgenic rabbits (n = 12, each group)."

"A metabolite set enrichment analysis (MSEA) with muscle metabolites and a false discovery rate $q < 0.2$ was performed to identify significantly different metabolic pathways between the 2 groups."

"The triglycerides and free fatty acid levels and insulin resistance index were lower, whereas the TEE was higher in the LPL transgenic rabbits than in the control rabbits."

"The MSEA revealed that the TCA cycle and proteinogenic amino acid metabolism pathways were significantly different between the 2 groups ($P < 0.05$)."

"In the MSEA, all four selected metabolites for the TCA cycle (2-oxoglutaric acid, citric acid, malic acid, fumaric acid), as well as eight selected metabolites for proteinogenic amino acid metabolism (asparagine, proline, methionine, phenylalanine, histidine, arginine, leucine, isoleucine) were consistently increased in the transgenic rabbits compared with control rabbits, suggesting that these two metabolic pathways were activated in the transgenic rabbits."

"The current results suggest that the overexpression of LPL may lead to increased activities of TCA cycle and proteinogenic amino acid metabolism pathways in the skeletal muscle, and these enhancements may play an important role in the biological mechanisms underlying the anti-obesity/anti-diabetes features of LPL overexpression."

Introduction

"We therefore investigated LPL transgenic rabbits in the current study, because they resemble physically active humans (who are assumed to be resistant to obesity and insulin resistance) with regard to their healthy metabolism [111–113]."

"No studies have yet examined the effect of LPL overexpression on the skeletal muscle metabolome in LPL transgenic rabbits that show anti-obesity and anti-diabetes characteristics."

"We hypothesized that the overexpression of LPL would lead to the activation of mitochondrial oxidative metabolism in the skeletal muscle, and such activation would be associated with better metabolic phenotype outcomes, such as better values for the percent fat, serum lipids, whole-body insulin resistance, and energy expenditure."

"To test this hypothesis, we performed a metabolomics analysis (CE-TOFMS) of skeletal muscle, along with the intravenous glucose tolerance test to examine insulin resistance index, and the doubly-labeled water (DLW) method to assess percent fat and total energy expenditure in the LPL transgenic and control rabbits under a HFD."

Materials and Methods

"After adding 1500 μL of 50% (v/v) acetonitrile solution containing internal standards (20 μmol/L l-methionine sulfone and 5 μmol/L D-camphor-10-sulfonic acid for cation and anion analyses, respectively) to each frozen muscle sample, they were homogenized (4500 rpm, 4 °C, 60 s, 10 times) using a Micro Smash MS-100R (TOMY Seiko Co., Ltd., Tokyo, Japan)."

"To assess the whole-body insulin resistance, all rabbits (n = 12, each group) were fasted for 16 h and an intravenous glucose tolerance test was performed as previously described [114] in the 14th week of HFD administration (at 38 weeks old)."

"The TEE was measured by DLW method during the period of standard chow diet (24 weeks old) and at the 15th week of HFD (39 weeks old) in the control and LPL transgenic rabbits (n = 12, each group), as previously described [115]."

Results

"Serum triglycerides were markedly lower in the LPL transgenic rabbits than in the control rabbits, while the levels of total cholesterol, LDL-cholesterol, and HDL-cholesterol were not markedly different between the two groups."

"The LPL transgenic rabbits had lower serum concentrations of T_3 than the control rabbits, while the T_4 levels were similar between the two groups."

"These amino acids were also consistently increased in the transgenic rabbits, suggesting that the proteogenomic amino acid metabolism was upregulated in the muscle of LPL transgenic rabbits."

"None of the eight proteogenomic amino acids were related to the percent fat, serum free fatty acid, or TEE values."

Discussion

"The correlation analyses revealed that higher levels of citric acid (selected for the TCA cycle in the MSEA) and methionine (selected for proteogenomic amino acid metabolism in the MSEA) in skeletal muscle were both related to a better whole-body insulin resistance."

"Activation of the TCA cycle (based on the increased levels of selected four metabolites) observed in the muscle of LPL transgenic rabbits might be due to the activation of PPAR-α, although we did not assess the PPAR-α level in the current study."

"We are unable to explain how LPL overexpression is linked to increased levels of muscle metabolites involved in proteogenomic amino acid metabolism and can only speculate that these amino acids are also involved in the TCA cycle in the muscle of LPL transgenic rabbits."

"The elevated levels of asparagine and phenylalanine (that were selected for proteogenomic amino acid metabolism) observed in the LPL transgenic rabbits resemble the adaptation of skeletal muscle to exercise training to enhance exercise capacity in humans."

Maintenance of Skeletal Muscle Energy Homeostasis During Prolonged Wintertime Fasting in the Raccoon Dog (Nyctereutes Procyonoides) [116]

This is a machine-generated summary of:

Kinnunen, Sanni; Mänttäri, Satu; Herzig, Karl-Heinz; Nieminen, Petteri; Mustonen, Anne-Mari; Saarela, Seppo: Maintenance of skeletal muscle energy homeostasis during prolonged wintertime fasting in the raccoon dog (Nyctereutes procyonoides) [116]

Published in: Journal of Comparative Physiology B (2015)

Link to original: https://doi.org/10.1007/s00360-015-0893-9

If you want to cite the papers, please refer to the original.

For technical reasons we could not place the page where the original quote is coming from.

Abstract-Summary "The raccoon dog (Nyctereutes procyonoides) is a canid species with autumnal fattening and prolonged wintertime fasting."

"During acute fasting, AMPK promotes fatty acid oxidation and enhances glucose uptake."

"We evaluated the effects of prolonged fasting on muscle energy metabolism in farm-bred raccoon dogs."

"Total and phosphorylated AMPK and acetyl-CoA carboxylase (ACC), glucose transporter 4 (GLUT 4), insulin receptor and protein kinase B (Akt) protein expressions of hind limb muscles were determined by Western blot after 10 weeks of fasting."

"Fasting had no effects on AMPK phosphorylation, but total AMPK expression decreased in m. rectus femoris, m. tibialis anterior and m. extensor digitorum longus resulting in a higher phosphorylation ratio."

"Fasting did not influence GLUT 4, insulin receptor or Akt expression, but Akt phosphorylation was lower in m. flexor digitorum superficialis and m. extensor digitorum longus."

"The studied muscles were resistant to prolonged fasting indicating that raccoon dogs have an effective molecular regulatory system for preserving skeletal muscle function during wintertime immobility and fasting."

Introduction

"Skeletal muscle is highly adaptive in the utilization of energy sources and able to switch between glucose and fatty acids depending on the energy status of the body [117, 118]."

"AMPK acts as an initial sensor for fasting-induced adaptations in skeletal muscle and enables an efficient shift between different energy substrates [119]."

"The preferred energy-producing pathway in skeletal muscle is fatty acid oxidation, whereas glucose uptake and usage are down-regulated."

"AMPK is known to respond to acute energy deprivation by enhancing both glucose uptake and free fatty acid (FFA) oxidation [120]."

"We also measured plasma hormones involved in the regulation of energy homeostasis (insulin, leptin and ghrelin) and to evaluate potential structural changes in skeletal muscle, we determined the fiber-type composition in selected muscles of the raccoon dog."

Materials and Methods

"All animals received water/ice ad libitum and the control group was fasted overnight before sampling."

"The left hind limb was removed whole and stored in dry ice for later muscle sample preparation."

"The plasma insulin and leptin concentrations were measured with the Human Insulin Specific RIA kit and Multi-Species Leptin RIA kit (Millipore, St. Charles, MO, USA)."

"Samples were obtained to measure the muscle glycogen concentrations and for histological studies."

"Samples containing equal amounts of protein (4.7 μg) were electrophoretically separated using 4–12 % separating gel (Amersham ECL™ Gel, GE Healthcare Bio-Science, Uppsala, Sweden) at 160 V for 75 min."

"Samples from the same muscle were always run on the same gel."

"The secondary antibody (blotting grade affinity purified, AP-conjugated, rabbit anti-goat, Bio-Rad, Hercules, CA, USA) incubation time was 2 h at room temperature."

Results

"The body mass of the fasted raccoon dogs was 33 % lower after 9 weeks of food deprivation; the control group lost 19 % of their body weight."

"The average body mass of the fasted raccoon dogs was significantly lower ($P < 0.05$) compared to the control animals."

"The corresponding fiber types are I, IIA and IIX, respectively."

"MHC IIb (type IIB) was absent in all studied muscles."

Discussion

"In the raccoon dogs, prolonged fasting did not have significant effects on the AMPK phosphorylation in the skeletal muscle."

"Parallel to the decreased AMPK levels, total ACC expression was reduced in the same muscles and the fasted raccoon dogs had higher pACC/ACC ratio in EDL."

"As leptin is known to increase AMPK activity in skeletal muscle leading to increased fatty acid oxidation [121], the high leptin levels could participate in sustaining muscle energy homeostasis throughout the prolonged fast, as observed in the elevated pAMPK/AMPK ratio in TA and EDL."

"Our results show that, in the muscles of raccoon dog, protein expressions of GLUT 4 and IR were unaltered even after prolonged fasting with decreased plasma insulin concentrations."

"Our results indicate that prolonged fasting does not significantly affect muscle AMPK phosphorylation in the raccoon dog."

Energy Demand and Supply in Human Skeletal Muscle [1]

This is a machine-generated summary of:

Barclay, C. J.: Energy demand and supply in human skeletal muscle [1]

Published in: Journal of Muscle Research and Cell Motility (2017)

Link to original: https://doi.org/10.1007/s10974-017-9467-7

If you want to cite the papers, please refer to the original.

For technical reasons we could not place the page where the original quote is coming from.

Abstract-Summary "Buffering of ATP by phosphocreatine, a reaction catalysed by creatine kinase, extends the duration of activity possible but sustained activity depends on continual regeneration of PCr."

"The rate of ATP breakdown ranges from 70 to 140 mM min^{-1} during isometric contractions of various intensity to as much as 400 mM min^{-1} during intense, dynamic activity."

"The maximum rate of oxidative energy supply in untrained people is ~50 mM min^{-1} which, if the contraction duty cycle is 0.5 as is often the case in cyclic activity, is sufficient to match an ATP breakdown rate during contraction of 100 mM min^{-1}."

"During brief, intense activity the rate of ATP turnover can exceed the rates of PCr regeneration by combined oxidative and glycolytic energy supply, resulting in a net decrease in PCr concentration."

"The creatine kinase reaction plays an important role not only in buffering ATP but also in communicating energy demand from sites of ATP breakdown to the mitochondria."

Introduction

"When the rate of energy supply is sufficient to match the energy demand (i.e. when the muscle is in an energetic steady-state), the amount of PCr regenerated over a full cycle of extension/flexion cycle is equal to the amount of PCr, and thus ATP, broken down during the contraction (extension) phase alone [122]."

"In muscle, the ATP formed in reaction 8 contributes to the regeneration of PCr (reaction 4) so that the net reaction for glycolytic energy supply is: During high intensity exercise, energy demand can greatly exceed the capacity of oxidative energy supply (see above) and the difference is accounted for by a net decrease in PCr and glycolytic regeneration of PCr, neither of which require O_2."

Conclusions

"It is apparent that the highest possible rates of energy demand greatly exceed the capacities of oxidative and glycolytic energy supply and can only be met by breakdown of PCr."

"The limited capacity of that process to supply energy restricts how long high rates of energy demand, and by inference high muscle power output, can be sustained."

"The level of muscular activity that can be maintained for prolonged periods is a small fraction of the possible rate of energy turnover by muscle and is limited by the maximum rate of oxidative energy supply."

"Endurance training can extend the maximum rate of oxidative energy supply whereas advanced age and various systemic pathologies (e.g. pulmonary and cardiac diseases) can restrict energy supply by both oxidative and glycolytic processes [123]."

Bibliography

1. Barclay CJ (2017) Energy demand and supply in human skeletal muscle. J Mus Res Cell Motil. https://doi.org/10.1007/s10974-017-9467-7
2. Dzik KP, Kaczor JJ (2019) Mechanisms of vitamin D on skeletal muscle function: oxidative stress, energy metabolism and anabolic state. Eur J Appl Physiol. https://doi.org/10.1007/s00421-019-04104-x
3. McCollum EF, Simmonds N, Becker JE, Shipley PG (1922) Studies on experimental rickets: and experimental demonstration of the existence of a vitamin which promotes calcium deposition. J Biol Chem 53:293–312
4. Bikle DD, Gee E, Halloran B, Kowalski MA, Ryzen E, Haddad JG (1986) Assessment of the free fraction of 25-hydroxyvitamin D in serum and its regulation by albumin and the vitamin D-binding protein. J Clin Endocrinol Metabol 63(4):954–959
5. Fu L, Yun F, Oczak M, Wong BY, Vieth R, Cole DE (2009) Common genetic variants of the vitamin D binding protein (DBP) predict differences in response of serum 25-hydroxyvitamin D [25(OH)D] to vitamin D supplementation. Clin Biochem 42(10–11):1174–1177
6. Jones KS, Assar S, Harnpanich D, Bouillon R, Lambrechts D, Prentice A et al (2014) 25(OH) D2 half-life is shorter than 25(OH)D3 half-life and is influenced by DBP concentration and genotype. J Clin Endocrinol Metab 99(9):3373–3381
7. Abboud M, Puglisi DA, Davies BN, Rybchyn M, Whitehead NP, Brock KE et al (2013) Evidence for a specific uptake and retention mechanism for 25-hydroxyvitamin D (25OHD) in skeletal muscle cells. Endocrinology 154(9):3022–3030
8. Bang WS, Lee DH, Kim KT, Cho DC, Sung JK, Han IB, Kim DH, Kwon BK, Kim CH, Park KS, Park MK, Seo SY, Seo YJ (2018) Relationships between vitamin D and paraspinal muscle: human data and experimental rat model analysis. Spine J 18:1053–1061. https://doi.org/10.1016/j.spinee.2018.01.007
9. Ryan ZC, Craig TA, Folmes CD, Wang X, Lanza IR, Schaible NS et al (2016) 1alpha,25-dihydroxyvitamin D3 regulates mitochondrial oxygen consumption and dynamics in human skeletal muscle cells. J Biol Chem 291(3):1514–1528

10. Li Z, Hosoi Y, Cai K, Tanno Y, Matsumoto Y, Enomoto A et al (2006) Src tyrosine kinase inhibitor PP2 suppresses ERK1/2 activation and epidermal growth factor receptor transactivation by X-irradiation. Biochem Biophys Res Commun 341(2):363–368
11. Thobe BM, Frink M, Choudhry MA, Schwacha MG, Bland KI, Chaudry IH (2006) Src family kinases regulate p38 MAPK-mediated IL-6 production in Kupffer cells following hypoxia. Am J Physiol. Cell Physiol 291(3):C476–C482
12. Ross JA, Levy Y, Ripolone M, Kolb JS, Turmaine M, Holt M, Lindqvist J, Claeys KG, Weis J, Monforte M, Tasca G, Moggio M, Figeac N, Zammit PS, Jungbluth H, Fiorillo C, Vissing J, Witting N, Granzier H, Zanoteli E, Hardeman EC, Wallgren-Pettersson C, Ochala J (2019) Impairments in contractility and cytoskeletal organisation cause nuclear defects in nemaline myopathy. Acta Neuropathologica. https://doi.org/10.1007/s00401-019-02034-8
13. Falcone S, Roman W, Hnia K, Gache V, Didier N, Lainé J et al (2014) N-WASP is required for Amphiphysin-2/BIN1-dependent nuclear positioning and triad organization in skeletal muscle and is involved in the pathophysiology of centronuclear myopathy. EMBO Mol Med 6:1455–1475. https://doi.org/10.15252/emmm.201404436
14. Folker ES, Schulman VK, Baylies MK (2014) Translocating myonuclei have distinct leading and lagging edges that require Kinesin and Dynein. Development 141:355–366. https://doi.org/10.1242/dev.095612
15. Gimpel P, Lee YL, Sobota RM, Calvi A, Koullourou V, Patel R et al (2017) Nesprin-1α-dependent microtubule nucleation from the nuclear envelope via Akap450 is necessary for nuclear positioning in muscle cells. Curr Biol 27:2999–3009.e9. https://doi.org/10.1016/j.cub.2017.08.031
16. Metzger T, Gache V, Xu M, Cadot B, Folker ES, Richardson BE et al (2012) MAP and kinesin-dependent nuclear positioning is required for skeletal muscle function. Nature 484:120–124. https://doi.org/10.1038/nature10914
17. Roman W, Martins JP, Carvalho FA, Voituriez R, Abella JVG, Santos NC et al (2017) Myofibril contraction and cross-linking drive nuclear movement to the periphery of skeletal muscle. Nat Cell Biol. https://doi.org/10.1038/ncb3605
18. Sanoudou D, Haslett JN, Kho AT, Guo S, Gazda HT, Greenberg SA et al (2003) Expression profiling reveals altered satellite cell numbers and glycolytic enzyme transcription in nemaline myopathy muscle. Proc Natl Acad Sci U S A 100:4666–4671. https://doi.org/10.1073/pnas.0330960100
19. Hetzer MW (2010) The nuclear envelope. Online 1411:241–254. https://doi.org/10.1101/cshperspect.a000539
20. Meinke P, Schirmer EC (2015) LINC'ing form and function at the nuclear envelope. FEBS Lett 589:2514–2521. https://doi.org/10.1016/j.febslet.2015.06.011
21. Webster M, Witkin KL, Cohen-Fix O (2009) Sizing up the nucleus: nuclear shape, size and nuclear-envelope assembly. J Cell Sci 122:1477–1486. https://doi.org/10.1242/jcs.037333
22. Auld AL, Folker ES (2016) Nucleus-dependent sarcomere assembly is mediated by the LINC complex. Mol Biol Cell 27:2351–2359. https://doi.org/10.1091/mbc.E16-01-0021
23. Capetanaki Y, Milner DJ, Weitzer G (1997) Desmin in muscle formation and maintenance: knockouts and consequences. Cell Struct Funct 22:103–116. https://doi.org/10.1247/csf.22.103
24. Kerr JP, Robison P, Shi G, Bogush AI, Kempema AM, Hexum JK et al (2015) Detyrosinated microtubules modulate mechanotransduction in heart and skeletal muscle. Nat Commun 6:8526. https://doi.org/10.1038/ncomms9526
25. Olivera JF, Pizarro G (2016) Excitation contraction uncoupling by high intracellular [Ca^{2+}] in frog skeletal muscle: a voltage clamp study. J Mus Res Cell Motil. https://doi.org/10.1007/s10974-016-9446-4
26. Lamb GD, Junankar PR, Stephenson DG (1995) Raised intracellular [Ca2+] abolishes excitation-contraction coupling in skeletal muscle fibres of rat and toad. J Physiol 489:349–362
27. Murphy RM, Verburg E, Lamb GD (2006) Ca2 + activation of diffusible and bound pools of μ-calpain in rat skeletal muscle. J Physiol 576:595–612

28. Verburg E, Murphy RM, Richard I, Lamb GD (2009) Involvement of calpains in Ca2+-induced disruption of excitation–contraction coupling in mammalian skeletal muscle fibers. Am J Physiol Cell Physiol 296:C1115–C1122
29. Locke M, Celotti C (2013) The effect of heat stress on skeletal muscle contractile properties. Cell Stress Chaperones. https://doi.org/10.1007/s12192-013-0478-z
30. Liu CC, Lin CH, Lin CY, Lee CC, Lin MT, Wen HC (2013) Transgenic overexpression of heat shock protein 72 in mouse muscle protects against exhaustive exercise-induced skeletal muscle damage. J Formos Med Assoc 112:24–30
31. Horscroft JA, Murray AJ (2014) Skeletal muscle energy metabolism in environmental hypoxia: climbing towards consensus. Extreme Physiol Med. https://doi.org/10.1186/2046-7648-3-19
32. Hernández-Ochoa EO, Schneider MF (2018) Voltage sensing mechanism in skeletal muscle excitation-contraction coupling: coming of age or midlife crisis? Skeletal Muscle. https://doi.org/10.1186/s13395-018-0167-9
33. Hubbard JI (1973) Microphysiology of vertebrate neuromuscular transmission. Physiol Rev. 53(3):674–723
34. Katz B (1996) Neural transmitter release: from quantal secretion to exocytosis and beyond. The Fenn Lecture. J Neurocytol 25(12):677–686
35. Adrian RH, Marshall MW (1977) Sodium currents in mammalian muscle. J Physiol. 268(1):223–250
36. Adrian RH, Peachey LD (1973) Reconstruction of the action potential of frog sartorius muscle. J Physiol. 235(1):103–131
37. Gage PW, Eisenberg RS (1969) Action potentials, afterpotentials, and excitation-contraction coupling in frog sartorius fibers without transverse tubules. J Gen Physiol. 53(3):298–310
38. Adrian RH, Costantin LL, Peachey LD (1969) Radial spread of contraction in frog muscle fibres. J Physiol. 204(1):231–257
39. Huxley AF, Taylor RE (1958) Local activation of striated muscle fibres. J Physiol. 144(3):426–441
40. Rios E, Brum G (1987) Involvement of dihydropyridine receptors in excitation-contraction coupling in skeletal muscle. Nature. 325(6106):717–720
41. Sandow A (1965) Excitation-contraction coupling in skeletal muscle. Pharmacol Rev. 17(3):265–320
42. Dulhunty AF (2006) Excitation-contraction coupling from the 1950s into the new millennium. Clin Exp Pharmacol Physiol. 33(9):763–772
43. Hodgkin AL, Horowicz P (1960) The effect of sudden changes in ionic concentrations on the membrane potential of single muscle fibres. J Physiol. 153:370–385
44. Hodgkin AL, Horowicz P (1960) Potassium contractures in single muscle fibres. J Physiol. 153:386–403
45. Huxley AF, Straub RW (1958) Local activation and interfibrillar structures in striated muscle. J Physiol. 143:40P
46. Porter KR, Palade GE (1957) Studies on the endoplasmic reticulum. III. Its form and distribution in striated muscle cells. J Biophys Biochem Cytol. 3(2):269–300
47. Robertson JD (1956) Some features of the ultrastructure of reptilian skeletal muscle. J Biophys Biochem Cytol. 2(4):369–380
48. Ringer S (1887) Regarding the action of lime, potassium and sodium salts on skeletal muscle. J Physiol. 8:20–24
49. Heilbrunn LV (1940) The action of calcium on muscle protoplasm. Physiol Zool. 13:88–94
50. Kamada T, Kinoshita H (1943) Disturbances initiated from the naked surface of muscle protoplasm. Japan J Zool 10:469–493
51. Catterall WA (2000) Structure and regulation of voltage-gated Ca2+ channels. Annu Rev Cell Dev Biol. 16:521–555
52. Catterall WA, Wisedchaisri G, Zheng N (2017) The chemical basis for electrical signaling. Nat Chem Biol. 13(5):455–463

53. Lanner JT, Georgiou DK, Joshi AD, Hamilton SL (2010) Ryanodine receptors: structure, expression, molecular details, and function in calcium release. Cold Spring Harb Perspect Biol. 2(11):a003996
54. Meissner G (2017) The structural basis of ryanodine receptor ion channel function. J Gen Physiol. 149(12):1065–1089
55. Schneider MF, Chandler WK (1973) Voltage dependent charge movement of skeletal muscle: a possible step in excitation-contraction coupling. Nature. 242(5395):244–246
56. Schneider MF (1999) Ca2+ sparks in frog skeletal muscle: generation by one, some, or many SR Ca2+ release channels? J Gen Physiol. 113(3):365–372
57. Rios E, Brum G (2002) Ca2+ release flux underlying Ca2+ transients and Ca2+ sparks in skeletal muscle. Front Biosci. 7:d1195–d1211
58. Csernoch L, Zhou J, Stern MD, Brum G, Rios E (2004) The elementary events of Ca2+ release elicited by membrane depolarization in mammalian muscle. J Physiol. 557(Pt 1):43–58
59. Bonello, John-Peter; Locke, Marius HSP72 expression is specific to skeletal muscle contraction type. Cell Stress Chaperon (2019). doi: https://doi.org/10.1007/s12192-019-00997-x
60. Locke M, Noble EG, Atkinson BG (1991) Inducible isoform HSP70 is constitutively expressed in a muscle fiber type specific pattern. Am J Phys 261(5):774–779. https://doi.org/10.1152/ajpcell.1991.261.5.C774
61. Thompson HS, Scordilis SP, Clarkson PM, Lohrer WA (2001) A single bout of eccentric exercise increases HSP72 and HSC/HSP70 in human skeletal muscle. Acta Physiol Scand 171(2):187–193. https://doi.org/10.1046/j.1365-201x.2001.00795.x
62. Milne KJ, Noble EG (2002) Exercise-induced elevation of HSP70 is intensity dependent. J Appl Physiol 93(2):561–568. https://doi.org/10.1152/japplphysiol.00528.2001
63. Thompson HS, Magnard EB, Morales ER, Scordilis SP (2003) Exercise-induced HSP27, HSP70 and MAPK responses in human skeletal muscle. Acta Physiol Scand 178(1):61–72. https://doi.org/10.1046/j.1365-201X.2003.01112.x
64. Puntschart A, Vogt M, Widmer HR, Hoppeler H, Billeter R (1996) Hsp70 expression in human skeletal muscle after exercise. Acta Physiol Scand 157(4):411–417. https://doi.org/10.1046/j.1365-201X.1996.512270000.x
65. Kim KB, Kim MH, Lee DJ (2004) The effect of exercise in cool, control and hot environments on cardioprotective HSP70 induction. J Physiol Anthropol Appl Hum Sci 23(6):225–230
66. Holwerda A, Locke M (2014) HSP25 and HSP72 content in rat skeletal muscle following controlled shortening and lengthening contractions. Appl Physiol Nutr Metab 39(12):1380–1387. https://doi.org/10.1139/apnm-2014-0118
67. Pollock-Tahiri E, Locke M (2017) The cellular stress response of rat skeletal muscle following lengthening contractions. Appl Physiol Nutr Metab 42(7):708–715. https://doi.org/10.1139/apnm-2016-0556
68. Morimoto R (1998) Regulation of the heat shock transcriptional response: cross talk between a family of heat shock factors, molecular chaperones, and negative regulators. Genes Dev 12(24):3788–3796
69. Paulsen G, Lauritzen F, Bayer ML, Kalhovde JM, Ugelstad I, Owe SG, Hallén J, Bergersen LH (1985) Raastad T (2009) Subcellular movement and expression of HSP27, alphaB-crystallin, and HSP70 after two bouts of eccentric exercise in humans. J Appl Physiol 107(2):570–582. https://doi.org/10.1152/japplphysiol.00209.2009
70. Thompson HS, Clarkson PM, Scordilis SP (2002) The repeated bout effect and heat shock proteins: intramuscular HSP27 and HSP70 expression following two bouts of eccentric exercise in humans. Acta Physiol Scand 174(1):47–56. https://doi.org/10.1046/j.1365-201x.2002.00922.x
71. Billin AN, Honeycutt SE, McDougal AV, Kerr JP, Chen Z, Freudenberg JM, Rajpal DK, Luo G, Kramer HF, Geske RS, Fang F, Yao B, Clark RV, Lepore J, Cobitz A, Miller R, Nosaka K, Hinken AC, Russell AJ (2018) HIF prolyl hydroxylase inhibition protects skeletal muscle from eccentric contraction-induced injury. Skelet Mus. https://doi.org/10.1186/s13395-018-0179-5

72. Tang K, Breen EC, Wagner H, Brutsaert TD, Gassmann M, Wagner PD (2004) HIF and VEGF relationships in response to hypoxia and sciatic nerve stimulation in rat gastrocnemius. Respir Physiol Neurobiol. 144(1):71–80
73. Gustafsson T, Puntschart A, Kaijser L, Jansson E, Sundberg CJ (1999) Exercise-induced expression of angiogenesis-related transcription and growth factors in human skeletal muscle. Am J Phys. 276(2 Pt 2):H679–H685
74. Rissanen TT, Vajanto I, Hiltunen MO, Rutanen J, Kettunen MI, Niemi M et al (2002) Expression of vascular endothelial growth factor and vascular endothelial growth factor receptor-2 (KDR/Flk-1) in ischemic skeletal muscle and its regeneration. Am J Pathol. 160(4):1393–1403
75. Richardson RS, Noyszewski EA, Kendrick KF, Leigh JS, Wagner PD (1995) Myoglobin O2 desaturation during exercise. Evidence of limited O2 transport. J Clin Invest. 96(4):1916–1926
76. Scheerer N, Dehne N, Stockmann C, Swoboda S, Baba HA, Neugebauer A et al (2013) Myeloid hypoxia-inducible factor-1alpha is essential for skeletal muscle regeneration in mice. J Immunol. 191(1):407–414
77. Schofield CJ, Ratcliffe PJ (2004) Oxygen sensing by HIF hydroxylases. Nat Rev Mol Cell Biol. 5(5):343–354
78. Jia Y, Suzuki N, Yamamoto M, Gassmann M, Noguchi CT (2012) Endogenous erythropoietin signaling facilitates skeletal muscle repair and recovery following pharmacologically induced damage. FASEB J. 26(7):2847–2858
79. Bao W, Qin P, Needle S, Erickson-Miller CL, Duffy KJ, Ariazi JL et al (2010) Chronic inhibition of hypoxia-inducible factor prolyl 4-hydroxylase improves ventricular performance, remodeling, and vascularity after myocardial infarction in the rat. J Cardiovasc Pharmacol. 56(2):147–155
80. Tidball JG, Wehling-Henricks M (2007) Macrophages promote muscle membrane repair and muscle fibre growth and regeneration during modified muscle loading in mice in vivo. J Physiol. 578(Pt 1):327–336
81. Ogle ME, Segar CE, Sridhar S, Botchwey EA (2016) Monocytes and macrophages in tissue repair: implications for immunoregenerative biomaterial design. Exp Biol Med (Maywood). 241(10):1084–1097
82. Ławniczak A, Wrońska A, Wierzbicki P, Kmieć Z (2022) Aging and short-term calorie restriction differently affect the cardiac and skeletal muscle expression of genes regulating energy substrate utilization in male rats. Biogerontology. https://doi.org/10.1007/s10522-022-09965-y
83. Cheng CF, Ku HC, Lin H (2018) PGC-1alpha as a pivotal factor in lipid and metabolic regulation. Int J Mol Sci 19:3447. https://doi.org/10.3390/ijms19113447
84. Arany Z, He H, Lin J, Hoyer K, Handschin C, Toka O, Ahmad F, Matsui T, Chin S, Wu PH, Rybkin II, Shelton JM, Manieri M, Cinti S, Schoen FJ, Bassel-Duby R, Rosenzweig A, Ingwall JS, Spiegelman BM (2005) Transcriptional coactivator PGC-1 alpha controls the energy state and contractile function of cardiac muscle. Cell Metab 1:259–271. https://doi.org/10.1016/j.cmet.2005.03.002
85. Kleiner S, Nguyen-Tran V, Baré O, Huang X, Spiegelman B, Wu Z (2009) PPAR{delta} agonism activates fatty acid oxidation via PGC-1{alpha} but does not increase mitochondrial gene expression and function. J Biol Chem 284:18624–18633. https://doi.org/10.1074/jbc.M109.008797
86. Park JS, Holloszy JO, Kim K, Koh JH (2020) Exercise training-induced PPARbeta increases PGC-1alpha protein stability and improves insulin-induced glucose uptake in rodent muscles. Nutrients 12:652. https://doi.org/10.3390/nu12030652
87. Wang P, Liu J, Li Y, Wu S, Luo J, Yang H, Subbiah R, Chatham J, Zhelyabovska O, Yang Q (2010) Peroxisome proliferator-activated receptor delta is an essential transcriptional regulator for mitochondrial protection and biogenesis in adult heart. Circ Res 106:911–919. https://doi.org/10.1161/CIRCRESAHA.109.206185

88. Liu HY, Zheng G, Zhu H, Woldegiorgis G (2007) Hormonal and nutritional regulation of muscle carnitine palmitoyltransferase I gene expression in vivo. Arch Biochem Biophys 465:437–442. https://doi.org/10.1016/j.abb.2007.06.026
89. Dressel U, Allen TL, Pippal JB, Rohde PR, Lau P, Muscat GE (2003) The peroxisome proliferator-activated receptor beta/delta agonist, GW501516, regulates the expression of genes involved in lipid catabolism and energy uncoupling in skeletal muscle cells. Mol Endocrinol 17:2477–2493. https://doi.org/10.1210/me.2003-0151
90. Cheng L-H, Hung K-F, Lee T-C, Huang C-Y, Chiu W-T, Lo J-F, Huang T-F (2016) Mitochondrial co-chaperone protein Tid1 is required for energy homeostasis during skeletal myogenesis. Stem Cell Res Ther. https://doi.org/10.1186/s13287-016-0443-8
91. Moyes CD, Mathieu-Costello OA, Tsuchiya N, Filburn C, Hansford RG (1997) Mitochondrial biogenesis during cellular differentiation. Am J Physiol. 272:C1345–C1351
92. Toledo FG, Watkins S, Kelley DE (2006) Changes induced by physical activity and weight loss in the morphology of intermyofibrillar mitochondria in obese men and women. J Clin Endocrinol Metab. 91:3224–3227. https://doi.org/10.1210/jc.2006-0002
93. Puigserver P, Wu Z, Park CW, Graves R, Wright M, Spiegelman BM (1998) A cold-inducible coactivator of nuclear receptors linked to adaptive thermogenesis. Cell. 92:829–839
94. Wu Z, Puigserver P, Andersson U et al (1999) Mechanisms controlling mitochondrial biogenesis and respiration through the thermogenic coactivator PGC-1. Cell 98:115–124. https://doi.org/10.1016/S0092-8674(00)80611-X
95. Ng AC, Baird SD, Screaton RA (2014) Essential role of TID1 in maintaining mitochondrial membrane potential homogeneity and mitochondrial DNA integrity. Mol Cell Biol. 34:1427–1437. https://doi.org/10.1128/MCB.01021-13
96. Trentin GA, Yin X, Tahir S, Lhotak S, Farhang-Fallah J, Li Y et al (2001) A mouse homologue of the Drosophila tumor suppressor l(2)tid gene defines a novel Ras GTPase-activating protein (RasGAP)-binding protein. J Biol Chem. 276:13087–13095. https://doi.org/10.1074/jbc.M009267200
97. Lo JF, Hayashi M, Woo-Kim S, Tian B, Huang JF, Fearns C et al (2004) Tid1, a cochaperone of the heat shock 70 protein and the mammalian counterpart of the Drosophila tumor suppressor l(2)tid, is critical for early embryonic development and cell survival. Mol Cell Biol. 24:2226–2236
98. Boon MR, Hanssen MJW, Brans B, Hülsman CJM, Hoeks J, Nahon KJ, Bakker C, van Klinken JB, Havekes B, Schaart G, Jazet IM, Rensen PCN, van Marken Lichtenbelt WD (2018) Effect of l-arginine on energy metabolism, skeletal muscle and brown adipose tissue in South Asian and Europid prediabetic men: a randomised double-blinded crossover study. Diabetologia. https://doi.org/10.1007/s00125-018-4752-6
99. American Diabetes Association (2014) Standards of medical care in diabetes--2014. Diabetes Care 37(Suppl 1):S14–S80. https://doi.org/10.2337/dc14-S014
100. Schrauwen-Hinderling VB, Kooi ME, Hesselink MK et al (2007) Impaired in vivo mitochondrial function but similar intramyocellular lipid content in patients with type 2 diabetes mellitus and BMI-matched control subjects. Diabetologia 50(1):113–120. https://doi.org/10.1007/s00125-006-0475-1
101. Alderton WK, Cooper CE, Knowles RG (2001) Nitric oxide synthases: structure, function and inhibition. Biochem J. 357:593–615. https://doi.org/10.1042/0264-6021:3570593
102. Cubbon RM, Yuldasheva NY, Viswambharan H et al (2014) Restoring Akt1 activity in outgrowth endothelial cells from South Asian men rescues vascular reparative potential. Stem Cells 32(10):2714–2723. https://doi.org/10.1002/stem.1766
103. Novakova K, Kummer O, Bouitbir J, Stoffel SD, Hoerler-Koerner U, Bodmer M, Roberts P, Urwyler A, Ehrsam R, Krähenbühl S (2015) Effect of l-carnitine supplementation on the body carnitine pool, skeletal muscle energy metabolism and physical performance in male vegetarians. Eur J Nutrit. https://doi.org/10.1007/s00394-015-0838-9
104. Pormsila W, Krahenbuhl S, Hauser PC (2010) Determination of carnitine in food and food supplements by capillary electrophoresis with contactless conductivity detection. Electrophoresis 31:2186–2191. https://doi.org/10.1002/elps.200900692

105. Etzioni A, Levy J, Nitzan M, Erde P, Benderly A (1984) Systemic carnitine deficiency exacerbated by a strict vegetarian diet. Arch Dis Child 59:177–179
106. Rebouche CJ, Lombard KA, Chenard CA (1993) Renal adaptation to dietary carnitine in humans. Am J Clin Nutr 58:660–665
107. Lombard KA, Olson AL, Nelson SE, Rebouche CJ (1989) Carnitine status of lactoovovegetarians and strict vegetarian adults and children. Am J Clin Nutr 50:301–306
108. Stephens FB, Marimuthu K, Cheng Y, Patel N, Constantin D, Simpson EJ, Greenhaff PL (2011) Vegetarians have a reduced skeletal muscle carnitine transport capacity. Am J Clin Nutr 94:938–944. https://doi.org/10.3945/ajcn.111.012047
109. Stephens FB, Wall BT, Marimuthu K, Shannon CE, Constantin-Teodosiu D, Macdonald IA, Greenhaff PL (2013) Skeletal muscle carnitine loading increases energy expenditure, modulates fuel metabolism gene networks and prevents body fat accumulation in humans. J Physiol 591:4655–4666. https://doi.org/10.1113/jphysiol.2013.255364
110. Nishida Y, Nishijima K, Yamada Y, Tanaka H, Matsumoto A, Fan J, Uda Y, Tomatsu H, Yamamoto H, Kami K, Kitajima S, Tanaka K (2021) Whole-body insulin resistance and energy expenditure indices, serum lipids, and skeletal muscle metabolome in a state of lipoprotein lipase overexpression. Metabolomics. https://doi.org/10.1007/s11306-021-01777-4
111. Nikkila EA, Taskinen MR, Rehunen S, Harkonen M (1978) Lipoprotein lipase activity in adipose tissue and skeletal muscle of runners: Relation to serum lipoproteins. Metabolism 27(11):1661–1667. https://doi.org/10.1016/0026-0495(78)90288-3
112. Plaisance EP, Grandjean PW, Mahurin AJ (2009) Independent and combined effects of aerobic exercise and pharmacological strategies on serum triglyceride concentrations: A qualitative review. Phys Sportsmedicine 37(1):11–19. https://doi.org/10.3810/psm.2009.04.1678
113. Kersten S (2014) Physiological regulation of lipoprotein lipase. Biochimica et Biophysica Acta 1841(7):919–933. https://doi.org/10.1016/j.bbalip.2014.03.013
114. Kitajima S, Morimoto M, Liu E, Koike T, Higaki Y, Taura Y et al (2004) Overexpression of lipoprotein lipase improves insulin resistance induced by a high-fat diet in transgenic rabbits. Diabetologia 47(7):1202–1209. https://doi.org/10.1007/s00125-004-1429-0
115. Yamada Y, Blanc S, Nishida Y, Nishijima K, Ebine N, Shriver T et al (2013) Validity of doubly labeled water in obese subjects: questioning the validity of any technique requires an indisputable accuracy of the reference method. Am J Physiol Endocrinol Metab 305(9):E1178–E1180. https://doi.org/10.1152/ajpendo.00355.2013
116. Kinnunen S, Mänttäri S, Herzig K-H, Nieminen P, Mustonen A-M, Saarela S (2015) Maintenance of skeletal muscle energy homeostasis during prolonged wintertime fasting in the raccoon dog (Nyctereutes procyonoides). J Compar Physiol B. https://doi.org/10.1007/s00360-015-0893-9
117. Muoio DM, Dohm GL, Fiedorek FT Jr, Tapscott EB, Coleman RA (1997) Leptin directly alters lipid partitioning in skeletal muscle. Diabetes 46:1360–1363
118. Kelley DE, Goodpaster BH (2001) Skeletal muscle triglyceride: an aspect of regional adiposity and insulin resistance. Diabetes Care 24:933–941
119. Cantó C, Jiang LQ, Deshmukh AS, Mataki C, Coste A, Lagouge M, Zierath JR, Auwerx J (2010) Interdependence of AMPK and SIRT1 for metabolic adaptation to fasting and exercise in skeletal muscle. Cell Metab 11:213–219
120. Hardie DG, Scott JW, Pan DA, Hudson ER (2003) Management of cellular energy by the AMP-activated protein kinase system. FEBS Lett 546:113–120
121. Minokoshi Y, Kim Y-B, Peroni OD, Fryer LGD, Müller C, Carling D, Kahn BB (2002) Leptin stimulates fatty-acid oxidation by activating AMP-activated protein kinase. Nature 415:339–343
122. Paul RJ (1983) Physical and biochemical energy balance during an isometric tetanus and steady state recovery in frog sartorius at 0 °C. J Gen Physiol 81:337–354
123. Lanza IR, Befroy DE, Kent-Braun JA (2005) Age-related changes in ATP-producing pathways in human skeletal muscle in vivo. J Appl Physiol 99:1736–1744
124. Catterall WA (1991) Excitation-contraction coupling in vertebrate skeletal muscle: a tale of two calcium channels. Cell. 64(5):871–874

125. Siivonen L (1972) The Raccoon dog. In: Siivonen L (ed) Mammals in Finland. Otava, Keuruu, pp 140–148. [in Finnish]
126. Boska M (1994) ATP production rates as a function of force level in the human gastrocnemius/soleus using 31P MRS. Magn Reson Med 32:1–10
127. Lewis DB, Barclay CJ (2014) Efficiency and cross-bridge work output of skeletal muscle is decreased at low levels of activation. Pflügers Archiv 466:599–609

Skeletal Muscle Formation, Regeneration, and Recovery from Injury

7

Christopher Myers

7.1 Introduction by the Editor

In this chapter, we embark on a journey through the remarkable world of skeletal muscle formation, regeneration, and recovery from injury. Skeletal muscles possess a remarkable ability to regenerate and repair themselves after injury. This chapter delves into the cellular and molecular mechanisms that underpin this ability, focusing on the role of muscle stem cells in the regeneration and repair of skeletal muscles.

The formation of skeletal muscle, known as myogenesis, is a highly orchestrated process involving myoblasts' differentiation and fusion to form multinucleated myofibers. Central to muscle regeneration are muscle stem cells, often called satellite cells. These cells remain quiescent under normal conditions but become activated upon muscle injury. This chapter looks at research that establishes that the Klotho hormone has a pivotal role in activating muscle stem cells in regenerating damaged myofibers and re-establishing muscle function. The mechanisms that regulate the activation, proliferation, and differentiation of muscle stem cells.

The muscle repair and recovery process are not limited to the mere replacement of damaged cells; it is an intricate dance of various cellular processes. This chapter investigates how intracompartmental pressure and mitochondrial networking affect skeletal muscle regeneration. Also, this chapter looks at the importance of angiogenesis.

This chapter illuminates the awe-inspiring capacity of skeletal muscles to regenerate and recover from injury. Through the lens of muscle stem cells, myofiber formation, and the orchestrated interplay of various cellular components, this chapter offers a holistic understanding of the muscle's regenerative journey from injury to

C. Myers (✉)
3 Sports Science, 3 Sports Science LLC, Akron, OH, United States of America
e-mail: chris@3sportsscience.com

C. Myers (ed.), *Skeletal Muscle Physiology*,
https://doi.org/10.1007/978-3-031-47065-3_7

recovery. Doing so lays the foundation for developing therapeutic strategies that harness the innate powers of muscle regeneration and repair.

The elucidation of skeletal muscle mechanics and regeneration continues to be a significant field of study within muscle physiology. This chapter covers many different lines of research investigating skeletal muscle regeneration and recovery after injury. One crucial line of research is the critical role of satellite cells and the intriguing properties of titin. Titin, identified as a pivotal element for muscle stability, especially during active states and eccentric contractions, opens new avenues for understanding muscle elasticity and force production. Concurrently, advancements in therapeutic approaches, such as low-level laser therapy (LLLT) and the application of neuregulin-1β (NRG-1β), demonstrate promising results in enhancing muscle repair and intrafusal muscle fiber formation, respectively. These innovative treatments improve collagen organization and inflammation response and highlight the potential for targeted muscle regeneration therapies. Moreover, exploring satellite cells' functionality, particularly their role in muscle regeneration and the impact of aging and signaling pathways on their efficacy, provides a more profound comprehension of muscle repair mechanisms. This chapter, therefore, aims to bridge the gap between fundamental skeletal muscle physiology research and practical therapeutic applications, offering insights into potential strategies for optimizing muscle regeneration.

The role of titin, introduced and discussed in Chap. 1, emerges as a central piece in this puzzle. Titin, known to increase its stiffness during active muscle states, could provide the necessary stability across the entire length spectrum of muscle fibers, including during eccentric contractions beyond actin-myosin overlap. However, much more needs to be understood about this "third filament." This chapter highlights the review and opinion of Dr. Walter Herzon, who conducts fundamental skeletal muscle physiology research. His hypotheses on titin interaction on force product on sarcomeres and individual myofibrils prompt researchers to reevaluate muscle elasticity and its contribution to force production, particularly in non-traditional contraction states.

This chapter highlights studies that evaluated the effects of low-level laser therapy (LLLT) on the tibialis anterior muscle of elderly rats post-cryoinjury, observing improved collagen organization and reduced inflammation when treated with infrared LLLT. The study concluded that LLLT positively influences muscle repair by enhancing collagen distribution and promoting blood vessel and immature fiber formation. This study supports the current use of LLLT in recovery methodologies seen in elite human performance programs. Additionally, research on the impact of neuregulin-1β (NRG-1β) in a neuromuscular coculture of dorsal root ganglion explants and skeletal muscle cells revealed that NRG-1β promotes neurite outgrowth and the formation of intrafusal muscle fibers by modulating the phenotype of tyrosine kinase receptor C (TrkC) phenotypic DRG neurons. This data suggests NRG-1β's potential in therapeutic applications for muscle fiber regeneration post-injury. Both studies collectively offer insights into therapies that could ameliorate muscle repair processes in elderly rats and highlight NRG-1β's role in enhancing muscle fiber formation through specific neuronal interactions, which could eventually produce new therapies for humans.

Satellite cells are an aspect of skeletal muscle remodeling. These skeletal muscle stem cells play an indispensable role in their regeneration, showcasing diverse gene expressions and phenotypic traits. The regeneration process hinges on the interplay of signaling pathways, cytoskeletal rearrangements, the critical role of miRNAs, and the contributions from non-satellite cells like immune cells and fibro-adipogenic progenitor cells. Animal model research highlighted in this chapter demonstrates that satellite cells express the Pax7 transcription factor, crucial for their postnatal maintenance and muscle regeneration, with their functionality waning due to factors like aging, which alters extracellular matrix composition and cell interactions. This impaired functionality is exacerbated by upregulated Wnt3a signaling in aged muscles, hindering satellite cell function. Satellite cell heterogeneity is evident from genetic analysis, with a notable distinction between Pax7+/YFP− stem cells and more committed Pax7+/YFP+ progenitor cells, pointing to diverse roles in muscle regeneration. Asymmetric divisions of satellite cells, governed by Notch signaling, emphasize the nuanced control of muscle regeneration, underpinning the complexity of satellite cell behavior. The regenerative journey involves crucial steps from the recruitment of inflammatory cells to the differentiation of satellite cells into myogenic progenitor cells, all orchestrated within a dynamic stem cell niche. The role of miRNAs in regulating satellite cell function illustrates the sophisticated post-transcriptional regulation underlying muscle regeneration. Non-myogenic cells, including immune cells and fibro-adipogenic progenitors, also contribute significantly to the regenerative process, showcasing a complex ecosystem of interactions vital for muscle healing. This summary underscores the multifaceted roles of satellite cells and associated factors in skeletal muscle regeneration, highlighting the intricate balance and interactions necessary for effective muscle repair and regeneration.

This chapter illuminates the awe-inspiring capacity of skeletal muscles to regenerate and recover from injury. Through the lens of muscle stem cells, myofiber formation, and the orchestrated interplay of various cellular components, this chapter offers a holistic understanding of the muscle's regenerative journey from injury to recovery. Doing so lays the foundation for developing therapeutic strategies that harness the innate powers of muscle regeneration and repair.

7.2 Machine Generated Summaries

Disclaimer: The summaries in this chapter were generated from Springer Nature publications using extractive AI auto-summarization: An extraction-based summarizer aims to identify the most important sentences of a text using an algorithm and uses those original sentences to create the auto-summary (unlike generative AI). As the constituted sentences are machine selected, they may not fully reflect the body of the work, so we strongly advise that the original content is read and cited. The auto generated summaries were curated by the editor to meet Springer Nature publication standards. To cite this content, please refer to the original papers.

Machine generated keywords: regeneration, injury, stem, muscle regeneration, stem cell, muscle stem, cell, regenerate, muscle injury, repair, regeneration skeletal, formation, muscle repair, recovery, myofiber

Klotho Expression Is a Prerequisite for Proper Muscle Stem Cell Function and Regeneration of Skeletal Muscle [1]

This is a machine-generated summary of:

Ahrens, Hellen E.; Huettemeister, Judith; Schmidt, Manuel; Kaether, Christoph; von Maltzahn, Julia: Klotho expression is a prerequisite for proper muscle stem cell function and regeneration of skeletal muscle [1]

Published in: Skeletal Muscle (2018)

Link to original: https://doi.org/10.1186/s13395-018-0166-x

If you want to cite the papers, please refer to the original.

For technical reasons we could not place the page where the original quote is coming from.

Abstract-Summary "We investigate the functional role of the anti-aging hormone Klotho for muscle stem cell function after cardiotoxin-induced injury of skeletal muscle using a klotho hypomorphic mouse line, which is characterized by a premature aging phenotype."

"We perform floating single myofiber cultures with their adjacent muscle stem cells to investigate the interplay between canonical Wnt signaling and Klotho function."

"We demonstrate that addition of recombinant Klotho protein inhibits aberrant excessive Wnt signaling in aged muscle stem cells thereby restoring their functionality."

"The anti-aging hormone Klotho counteracts aberrant canonical Wnt signaling in muscle stem cells and might be one of the naturally occurring inhibitors of canonical Wnt signaling in skeletal muscle."

Background

"Diseases such as muscular dystrophy and other conditions such as natural aging are major factors influencing muscle stem cell function [2–4]."

"Muscle stem cell number and functionality decline with age caused by intrinsic and extrinsic changes [4]."

"It is known that systemic factors and changes in the muscle stem cell niche can influence regeneration of skeletal muscle in the aged [5]."

"We investigated the effect of Klotho on regeneration of the skeletal muscle, muscle stem cell function, and the interplay between canonical Wnt signaling and sKlotho in muscle stem cells."

"We show that klotho hypomorphic mice display disturbed muscle stem cell function as well as reduced regenerative capacity."

"We identify sKlotho as one of the modulators of muscle stem cell function and thereby regeneration of skeletal muscle, potentially by inhibiting aberrant canonical Wnt signaling, e.g., in the context of aging."

Methods

"Immunoblot analyses were performed using the following antibodies: MuRF1/2/3 (rabbit, 1:1000, ab172479, AbCam), Atrogin/Fbxo32 (rabbit, 1:1000, ab168372, AbCam), non-phospho (active) β-Catenin (Ser33/37/Thr41) (D13A1) (rabbit, 1:1000, 8814S, Cell Signaling), phospho (inactive) β-Catenin (Ser33/37/Thr41) (rabbit, 1:1000, 9561S, Cell Signaling), Klotho (goat, 1:2000, AF1819, R&D Systems), Tubulin (mouse, 1:500, T9026, Sigma-Aldrich), GAPDH (G-9) (mouse, 1:200, sc-365062, Santa Cruz), peroxidase goat anti-rabbit (1:1000, P0448, Dako), peroxidase rabbit anti-goat (1:1000, P0449, Dako), and peroxidase goat anti-mouse (1:1500, P0447, Dako), including Pierce™ ECL Western blotting substrate (32106, Thermo Scientific) and Immobilon™ Western chemiluminescent HRP substrate (WBKLS0500, Millipore)."

"After 48 h, cells were fixed with 2% PFA for 5 min followed by subsequent immunofluorescence staining."

"Single myofibers were isolated from EDL muscle by digestion in DMEM with 0.2% collagenase (from Clostridium hystolyticum, Sigma) and cultured in DMEM supplemented with 20% FBS and 1% chick embryo extract (US biological) as described previously [6]."

Results

"We first compared the number, activation, and differentiation potential of muscle stem cells of ΔKlotho and control mice."

"Addition of sKlotho results in an increase in the number of clusters per fiber of ΔKlotho mice suggesting that sKlotho and not mKlotho is important for muscle stem cell function."

"Expression of klotho mRNA is reduced in myofibers with adjacent muscle stem cells from old mice directly after isolation and after 72 h of culture when the number of muscle stem cells on the myofibers is increased."

"sKlotho might be an effective natural inhibitor of canonical Wnt signaling in muscle stem cells in the young and loss of klotho expression in the aged might be one of the causes for increased canonical Wnt signaling and thereby impeded muscle stem cell functionality in the aged."

Discussion

"Canonical Wnt signaling plays an important role in muscle stem cell function [7]."

"Differentiation of muscle stem cells is controlled by canonical Wnt signaling, especially by signaling through the ligand Wnt3a."

"Aged skeletal muscle has been shown to display increased canonical Wnt signaling resulting in a conversion of muscle stem cells from a myogenic to a fibrogenic lineage [8]."

"We could demonstrate that inhibition of aberrant canonical Wnt signaling in the aged by Dkk1 had very similar effects compared to addition of sKlotho further supporting the notion that sKlotho is a naturally occurring inhibitor of canonical Wnt signaling in muscle stem cells."

"It is likely that sKlotho not only inhibits canonical Wnt signaling but also TFG-β1 signaling in aged skeletal muscle thereby maintaining muscle stem cell function."

Conclusions

"We conclude that reduction of sKlotho is an important contributor to loss of muscle stem cell function in the aged, most likely through loss of inhibition of canonical Wnt signaling."

"Replenishing the levels of sKlotho is a promising therapeutic approach for stimulation of muscle regeneration in the aged."

The Association Between Intracompartmental Pressure and Skeletal Muscle Recovery After Tibial Diaphyseal Fractures: An Ambispective Cohort Study [9]

This is a machine-generated summary of:

Tian, Shengjie; Chang, Shimin; Lu, Yaogang; Zhu, Jianhua; Kong, Xuqiang: The association between intracompartmental pressure and skeletal muscle recovery after tibial diaphyseal fractures: an ambispective cohort study [9]

Published in: Journal of Orthopaedics and Traumatology (2021)

Link to original: https://doi.org/10.1186/s10195-021-00579-7

If you want to cite the papers, please refer to the original.

For technical reasons we could not place the page where the original quote is coming from.

Abstract-Summary "Not only is this increased ICP the manifestation of skeletal muscle injury, but it induces further deterioration of the injury."

"The aim of this study was to assess the association between short-term ICP elevation and long-term skeletal muscle recovery after severe limb trauma."

"In this single-center ambispective cohort study, we retrospectively screened and recruited a cohort of tibial diaphyseal fracture patients with integrated ICP data during the early post-traumatic period, and performed a prospective observational study to evaluate their skeletal muscle recovery through long-term follow-up and MR imaging after the removal of the implants."

"We analyzed the association between ICP elevation and skeletal muscle recovery using statistical methods."

"Statistically significant associations were observed between the ICP parameters and the MR imaging parameters when simple linear regression analysis was performed."

"Short-term ICP elevation was associated with long-term skeletal muscle recovery following tibial diaphyseal fracture, especially for ICP data that integrated time factors."

Introduction

"In a tibial diaphyseal fracture, bleeding, edema, and exudation are confined within those compartments, resulting in increased intracompartmental pressure (ICP) [10, 11]."

"MR imaging is a powerful tool for evaluating and quantifying skeletal muscle volume and internal structures [12, 13]."

"Metal internal fixators such as intramedullary nails can interfere with MR imaging, which limits its application for evaluating the recovery of skeletal muscle in tibial diaphyseal fracture patients with intramedullary nails [14, 15]."

"This meant that there was an opportunity for us to use MR imaging to assess the recovery of skeletal muscle following tibial diaphyseal fracture."

"We conducted a prospective observational study to evaluate their skeletal muscle recovery through long-term follow-up and MR imaging after the removal of the implants."

"The purpose of the current study was to assess the association between short-term ICP elevation and long-term skeletal muscle recovery after severe limb trauma."

Materials and Methods

"The present study was approved by the Ethics Committee of Shanghai University of Medicine & Health Sciences Affiliated Zhoupu Hospital (2017-C-003), and the patients were fully informed about the purpose and procedure of this study before they underwent ICP monitoring, intramedullary nail removal, and MR imaging."

"The inclusion criteria were as follows: (1) tibial diaphyseal fracture as defined by AO/OTA classification code 42; (2) fixation of the tibial diaphyseal fracture with intramedullary nails; (3) complete ICP data for the anterior compartment after the trauma, with the maximum ICP above 30 mmHg and the minimum ΔP (ΔP = diastolic blood pressure minus ICP) below 50 mmHg; (4) aged over 18 at the time of injury; and (5) the patient was willing to have the intramedullary nail removed after fracture healing."

"To evaluate the association of ICP elevation with the recovery of skeletal muscle, simple linear regression analyses of various ICP data and MR imaging data were performed."

Results

"In this ambispective cohort study, 51 patients met the criteria to be recruited, and a total of 46 patients underwent intramedullary nail removal and MR imaging after 11–30 (19.23 ± 5.01) months of follow-up."

"Two cases presented delayed union after intramedullary nail fixation; in those cases, the distal interlocking screws were removed and union was achieved after dynamization."

"The removal of the intramedullary nail was completely successful in all patients; there were no complications such intramedullary nails or screws remaining in the body, vascular or nerve injuries, hematoma, or infection."

"The 46 patients underwent MR imaging 6–11 (8.48 ± 1.53) weeks after the removal of the internal fixator."

Discussion

"We conducted an ambispective observational cohort study based on MR imaging after the removal of implants to investigate the relationship between short-term ICP elevation and long-term skeletal muscle recovery after severe limb trauma."

"Linear regression analysis reconfirmed the importance of the ICP elevation duration, as indicated by the evidence that the maximum ICP and minimum ΔP (which do not take time into account) were less strongly associated with skeletal muscle recovery, while the accumulated ΔP (which takes time into account) produced the highest determination coefficients for CSAR and T2SIR, explaining 62.1% and 59.1% of the variance in CSAR and T2SIR, respectively."

"This could make the reliability of the estimated specificity of using ICP elevation to predict skeletal muscle recovery questionable in patients with high accumulated ICP and ΔP."

"Further investigations that focus on larger cohorts and use more accurate and comprehensive approaches to investigate skeletal muscle recovery are needed to explore the underlying mechanism for the impact of elevated ICP on skeletal muscle recovery as well as the functional implications for limbs."

Conclusions

"This ambispective cohort study demonstrated that short-term ICP elevation was associated with long-term skeletal muscle recovery, and that the accumulated ΔP, which accounts for both the ICP data and time, was the ICP parameter most closely associated with the recovery of skeletal muscle."

"Firm conclusions such as the accumulated ΔP threshold for skeletal muscle recovery could not be drawn from this study due to its size, but the study can provide ideas and thoughts for future research."

Mitochondrial Network Remodeling: An Important Feature of Myogenesis and Skeletal Muscle Regeneration [16]

This is a machine-generated summary of:

Rahman, Fasih Ahmad; Quadrilatero, Joe: Mitochondrial network remodeling: an important feature of myogenesis and skeletal muscle regeneration [16]

Published in: Cellular and Molecular Life Sciences (2021)

Link to original: https://doi.org/10.1007/s00018-021-03807-9

If you want to cite the papers, please refer to the original.

For technical reasons we could not place the page where the original quote is coming from.

Abstract-Summary "In skeletal muscle, the robust remodeling of the mitochondrial network is observed, particularly after injury where large portions of the tissue/cell structures are damaged."

"The significance of mitochondrial remodeling in regulating skeletal muscle regeneration has been widely studied, with alterations in mitochondrial remodeling processes leading to incomplete regeneration and impaired skeletal muscle function."

"Important questions related to mitochondrial remodeling and skeletal muscle regeneration still remain unanswered and require further investigation."

"This review will discuss the known molecular mechanisms of mitochondrial network remodeling, as well as integrate these mechanisms and discuss their relevance in myogenesis and regenerating skeletal muscle."

Introduction

"Mitochondrial turnover is also important in initiating the differentiation program of stem cells to reconstruct damaged tissues."

"Skeletal muscle tissue sustains damage due to daily wear and tear, and from numerous disease states, with its repair being intimately dependent upon the turnover of mitochondria within resident skeletal muscle stem cells (satellite cells) [17, 18]."

"The study of mitochondrial dynamics and turnover in regulating skeletal muscle stem cell function and skeletal muscle tissue regeneration is still in its infancy."

"Recent studies have shown the importance of these processes in maintaining skeletal muscle stem cell and skeletal muscle tissue function."

"This review will shed light on the molecular mechanisms of mitochondrial dynamics/turnover, their regulation through two major classes of sensors, and their known relevance in skeletal muscle stem cell differentiation and skeletal muscle tissue regeneration."

Mitochondrial Biogenesis

"NRFs are key components in regulating mitochondrial biogenesis, with NRF1 and nuclear factor erythroid 2 like 2 (NFE2L2 or NRF2) being the major targets of PPARGC1A. The significance of these nuclear factors has been established in developmental studies where the knockout of either of these factors results in embryonic lethality [19, 20]."

"Compared to NRFs, the physiological function of ESRRs in mitochondrial biogenesis is less studied in its relationship with PPARGC1A. PPARGC1A can target, bind, and activate ESRRA to regulate the expression of genes involved in OXPHOS, fatty acid oxidation, mitochondrial membrane transport, and mitochondrial DNA replication [21, 22]."

"PPARGC1B has been shown to exhibit similar properties to PPARGC1A and was thought to be an important regulator of mitochondrial biogenesis [23, 24]."

"Similar to PPARGC1A, PPARGC1B is able to partner with NRF1 and ESRRA to enhance the expression of mitochondrial transcription factors and trigger robust mitochondrial biogenesis [25–27]."

Mitochondrial Fission and Fusion

"The basis of mitochondrial fission is the constriction and scission of the mitochondrial membrane, resulting in two mitochondria [28, 29]."

"DNM1L migration to the mitochondria is accompanied by its interaction with adaptor proteins found on the outer mitochondrial membrane (OMM)."

"It is well understood that DNM1L mediates fission through its interaction with adaptor proteins found on the OMM, whereas the mechanism of IMM fission has yet to be fully determined."

"DNM1L, OPA1, MFN1/2 and their related proteins are major mediators of mitochondrial dynamics; however, other mediators have also been shown to play an important role in this process."

"The majority of mitochondrial fission and fusion events occur at ER-mito contact sites [30]."

"A subpopulation of fission proteins (DNM1L, MFF, and FIS1) have also been identified on the ER and were shown to be involved in priming mitochondria for division."

Mitochondrial Degradation

"The most studied mechanism of mitophagy is orchestrated by two proteins: the Ser/Thr kinase, PTEN-induced kinase 1 (PINK1), and the E3-ubitquitin ligase parkin (PRKN) [31, 32]."

"The addition of a phosphorylated ubiquitin plays an essential role in activating PRKN and inducing mitophagy even in the absence of PINK1-mediated PRKN phosphorylation [33, 34]."

"Although the PINK1/PRKN pathway of mitophagy is the most studied, other mechanisms exist and the contribution of these PRKN-independent mechanisms may be more important in certain types of cells or tissues."

"BNIP3 has been shown to anchor and inhibit the breakdown of OMM bound PINK1, thus increasing full-length PINK1 accumulation and concomitant PINK1/PRKN-mediated mitophagy [35]."

"Upon proteasomal rupture of the OMM, the LIR motif of PHB2 interacts with MAP1LC3 to induce mitophagy that is independent of PRKN [36]."

"PHB2 has also been shown to participate in PINK1/PRKN-mediated mitophagy."

Interplay Between Mitochondrial Biogenesis, Fission/Fusion, and Mitophagy

"AMPK-mediated phosphorylation of PPARGC1A activates the co-transcription factor, which upregulates genes responsible for ATP production and mitochondrial biogenesis, as discussed earlier."

"Activated AMPK triggers the phosphorylation of adaptor protein MFF that enhances the recruitment of DNM1L to the OMM, thus, promoting mitochondrial fission [37]."

"The energy sensors AMPK and SIRTs play similar roles with respect to mitochondrial turnover such that they promote biogenesis and mitophagy to clear away damaged mitochondria and repopulate the cell with healthy mitochondria."

"In non-cancer cells, Ca^{2+} influx results in MAPK1/3 activation and subsequent mitochondrial fission through the phosphorylation of Ser616 on DNM1L [38]."

"It is possible that under cellular stress the MAPK1/3-mediated DNM1L activation and MFN1 inhibition favours mitochondrial fission and may be an important preceding step in cell death signaling."

"Mitochondrial depolarization can recruit calcium-activated CAMK1 to the OMM to promote PINK1/PRKN-mediated mitophagy [39]."

Mitochondrial Interplay During Skeletal Muscle Regeneration

"Myoblast-specific knockout of DNM1L does not alter overall satellite cell content or affect the regeneration program following cardiotoxin (CTX)-induced skeletal muscle damage [40], suggesting that DNM1L does not impair satellite cell function during regeneration [40]."

"This may better elucidate the function of the initial fission/mitophagy processes in satellite cells during skeletal muscle regeneration."

"Similar to C2C12 cell culture experiments, MYOD1 and MYOG are elevated at approximately 3 days post-injury in skeletal muscle tissue and are also synchronous with upregulation of mitochondrial biogenesis genes, including Ppargc1b, Pprc1, Nrf1, Nfe2l2, Esrra, and Tfam [17, 41]."

"Investigation into OPA1 during skeletal muscle regeneration may provide novel insight into its role in maintaining mitochondrial morphology during the different phases of the regeneration process."

"It is unknown if mitochondrial biogenesis markers continue to increase as MRFs decline during the second half of the regeneration process (i.e., after satellite cell differentiation and formation of de novo myotube/myofibres), or whether mitochondrial respiration is altered in regenerating myofibres."

Concluding Remarks

"In the case of skeletal muscle stem cells, the precise mechanism and role of these processes has yet to be fully elucidated, and greater emphasis is needed in this regard."

"The present data highlights the importance of mitochondrial dynamic and turnover processes in facilitating skeletal muscle regeneration."

"As this field continues to emerge, it may provide valuable information for the development of inhibitor/activator molecules to enhance skeletal muscle regeneration, particularly in diseased states through the targeting of these various mitochondrial mechanisms."

GDF11 Treatment Attenuates the Recovery of Skeletal Muscle Function After Injury in Older Rats [42]

This is a machine-generated summary of:

Zhou, Yu; Sharma, Neel; Dukes, David; Myzithras, Maria B.; Gupta, Priyanka; Khalil, Ashraf; Kahn, Julius; Ahlberg, Jennifer S.; Hayes, David B.; Franti, Michael; Criswell, Tracy: GDF11 Treatment Attenuates the Recovery of Skeletal Muscle Function After Injury in Older Rats [42]

Published in: The AAPS Journal (2016)

Link to original: https://doi.org/10.1208/s12248-016-0024-x

If you want to cite the papers, please refer to the original.

For technical reasons we could not place the page where the original quote is coming from.

Abstract-Summary "Loss of skeletal muscle mass and function results in loss of mobility for elderly patients."

"Growth differentiation factor 11 (GDF11) has been proposed as a "youthful" circulating factor that can restore cardiac, neural, and skeletal muscle functions in aging animals."

"We used a complex rat model of skeletal muscle injury that physiologically mimics injuries seen in patients; to investigate the ability of GDF11 and to enhance skeletal muscle regeneration after injury in older rats."

"GDF11 impaired the recovery of skeletal muscle function in older rats after injury."

Introduction

"Growth differentiation factor-11 (GDF11) is a TGFβ family member that is highly homologous to another TGFβ family member, myostatin (GDF8), a known inhibitor of skeletal muscle hypertrophy."

"Further reports from high impact journals have suggested that this supposed age-dependent decrease in circulating GDF11, but not myostatin, can be rescued via exogenous administration of GDF11 and reverse age-related defects in the heart, brain, and skeletal muscle [43, 44]."

"They demonstrated that circulating GDF11 increased with age and inhibited skeletal muscle regeneration [45] Data from Rodgers and others demonstrated that the circulating molar concentration of myostatin is 500× higher than that of GDF11 [46]."

"This model provides a unique opportunity to test the effect of GDF11 treatment on the recovery of skeletal muscle function in older animals in a physiologically relevant model of injury."

"We provide data showing that GDF11 treatment resulted in increased fibrosis in injured muscle and decreased recovery of function after CS injury."

Materials and Methods

"GDF11 was reconstituted with 30 mM acetate, pH 5.0 buffer, and total protein concentration was confirmed using a Pierce BCA Protein assay kit (Thermo Scientific, Waltham, MA)."

"Circulating levels of GDF11, in both vehicle and GDF11-treated animals, were determined by MSD assays of flash frozen serum samples as previously described [47]."

"Contractile function of the left anterior crural muscles was assessed in vivo 7 and 14 days after injury by measuring maximal isometric torque as a function of stimulation frequency after direct stimulation of the peroneal nerve and compared to baseline function measured prior to injury as previously described [48]."

"Quantitative PCR (qPCR) was performed in 20-mL reactions in 96-well plates using cDNA samples generated from 12.5 ng of total RNA as previously described [49]."

Results

"Adult male Lewis rats were injected IP with 0.1 mg/kg/day GDF11 or vehicle alone for 28 days prior to CS injury using a protocol similar to what was described by Sinha and others, in the original Science paper demonstrating GDF11 rescue of age-dependent dysfunction in skeletal muscle [50]."

"The TA and EDL muscles are both in the anterior muscle compartment and tend to be similarly affected by the CS injury, whereas the Sol muscle is from a posterior muscle compartment that is not significantly damaged in our model and thus serves as a control."

"We have previously characterized and compared the timing of these events in young and old rats after CS injury and demonstrated that young rats efficiently recover muscle mass and function, while older animals have a delayed or deficient recovery [49]."

"At day 7 after injury, vehicle- and GDF11-treated muscles showed 10% maximum isometric torque as compared to baseline function measurements taken prior to injury (data not shown)."

Discussion

"With the known background of GDF11 and myostatin, it was surprising to find several recent high profile studies that have put forth GDF11 as a circulating factor that has the ability to "rejuvenate" old tissue."

"By Sinha and others, GDF11 was shown to exert its effect on the aged skeletal muscle satellite cell population by reducing DNA damage and caspase activation, by increasing myogenic differentiation, and by increasing the integration capacity of transplanted satellite cells [50]."

"This group then goes on to demonstrate inhibition of myoblast fusion in vitro and muscle regeneration in vivo after cardiotoxin injury in GDF11-treated mice, (41) and, in direct contrast to the Sinha and others study, they demonstrated that circulating GDF11 increased, rather than decreased, in old mice."

"Due to the controversy over the ability of GDF11 to influence skeletal muscle regeneration, we examined the effect of GDF11 treatment on skeletal muscle regeneration in older rats after CS injury."

Angiogenesis Precedes Myogenesis During Regeneration Following Biopsy Injury of Skeletal Muscle [51]

This is a machine-generated summary of:

Jacobsen, Nicole L.; Morton, Aaron B.; Segal, Steven S.: Angiogenesis precedes myogenesis during regeneration following biopsy injury of skeletal muscle [51]

Published in: Skeletal Muscle (2023)

Link to original: https://doi.org/10.1186/s13395-023-00313-3

If you want to cite the papers, please refer to the original.

For technical reasons we could not place the page where the original quote is coming from.

Abstract-Summary "Myofibers and microvessels regenerate from satellite cells and from surviving microvessel fragments, respectively, to restore intact muscle."

"In these models, efferocytosis removes cellular debris while basal laminae persist to provide guidance during myofiber and microvessel regeneration."

"It is unknown whether the spatiotemporal coupling between myofiber and microvascular regeneration persists when muscle tissue is completely removed and local guidance cues are lost."

"To test whether complete removal of skeletal muscle tissue affects the spatiotemporal relationship between myogenesis and angiogenesis during regeneration, subthreshold volumetric muscle loss was created with a biopsy punch (diameter, 2 mm) through the center of the gluteus maximus (GM) in adult mice."

"Confocal imaging and histological analyses of whole-mount GM preparations and tissue cross-sections assessed the growth of microvessels and myofibers into the wound."

"Regenerating microvessels advanced from the edges of the wound into the matrix by 7 dpi."

"Regenerated myofibers and microvessels were disorganized compared to the uninjured muscle."

"Following punch biopsy of adult skeletal muscle, regenerating microvessels span the wound and become perfused with blood prior to myofiber regeneration."

"The loss of residual guidance cues with complete tissue removal disrupts the spatiotemporal correspondence between microvascular and myofiber regeneration."

"We conclude that angiogenesis precedes myogenesis during regeneration following subthreshold volumetric muscle loss."

Background

"Injury to skeletal muscle damages myofibers, triggers an inflammatory response, and fragments capillary networks [52–54]."

"In established models of muscle injury and regeneration [52, 55–58], basal laminae persist following efferocytosis, which provides cell type-specific guidance to capillaries and myofibers during regeneration [59–63]."

"Whether microvascular regeneration and myofiber regeneration following injury may proceed sequentially has not been resolved."

"As an alternative approach to test the interdependence of microvascular and myofiber regeneration following acute injury, a punch biopsy (diameter, 2 mm) was performed through the mouse gluteus maximus (GM) muscle."

"This injury is below the critical size threshold of VML that prevents the regeneration of intact muscle [64]."

Methods

"The GM was dissected (as described below) at 1 dpi, and images of EBD staining were acquired with a 4 × objective on an E800 microscope coupled to a DS-Fi3 camera using Elements software (Nikon; Tokyo, Japan)."

"Exposed tissue on the mouse was covered with plastic film (Glad Press n' Seal) to prevent dehydration, and the preparation was transferred to the stage of a Nikon E600FN microscope, where the GM was equilibrated for 30 min while continuously superfused with bbPSS at 3 mL/min."

"The GM was removed from a Cdh5-mTmG mouse and placed in a custom imaging chamber with the ventral surface facing the objective to optimize the resolution of the microvasculature."

"For tissue cross-sections, trimmed GM samples were transferred to a cryomold containing OCT compound with 2–0 black silk suture (length, 2 mm) positioned adjacent to the injury for a visual reference."

Results

"Creating a circular hole through the center of the GM resulted in minimal collateral damage to the surrounding tissue as evidenced by nominal uptake of EBD into the ends of myofibers bordering the wound at 1 dpi (Supp."

"Intravital microscopy revealed that only those microvessels having the largest diameters were continuously perfused; most segments contained stationary red blood cells or plasma alone at 14 dpi (Supp.)"

"Regenerating myofibers were smaller in width (i.e., cross-sectional area) than in surrounding (uninjured) tissue."

"To confirm that revascularization occurred prior to myofiber regeneration following the biopsy, GM cross-sections (thickness, 10 μm) were immunolabeled for CD31 to identify ECs, myosin heavy chain (MyHC) for contractile myofibers, laminin to detect basal laminae, and DAPI for nuclei."

"The decrease from 14 to 21 dpi reflects the growth of myofibers into the wound, which localized with nascent microvessels as regeneration advanced."

Discussion

"Key findings are that (1) a provisional matrix infiltrated with $CD45^+$ immune cells and PDGFRα^+ FAPs spanned the void within 1 dpi; (2) nascent microvessels began extending into the matrix from the edges of the wound by 7 dpi, developed into branching microvascular networks that spanned the wound and were perfused with blood by 14 dpi; and (3) myofibers began regenerating into the wound from severed ends by 14 dpi and spanned the wound by 21 dpi."

"Previous studies of muscle injury have shown that regeneration of myofibers and microvessels are concomitant processes mediated by multifaceted crosstalk between SCs, ECs, immune cells, and FAPs in a constantly changing microenvironment [54, 56, 57, 65–70]."

"Upon injury, FAPs enter the cell cycle to proliferate, localize to the site of tissue damage, and work in concert with ECs to promote myofiber regeneration [65, 71, 72]."

Defective Angiogenesis in CXCL12 Mutant Mice Impairs Skeletal Muscle Regeneration [73]

This is a machine-generated summary of:

Hardy, David; Fefeu, Mylène; Besnard, Aurore; Briand, David; Gasse, Paméla; Arenzana-Seisdedos, Fernando; Rocheteau, Pierre; Chrétien, Fabrice: Defective angiogenesis in CXCL12 mutant mice impairs skeletal muscle regeneration [73]

Published in: Skeletal Muscle (2019)

Link to original: https://doi.org/10.1186/s13395-019-0210-5

If you want to cite the papers, please refer to the original.

For technical reasons we could not place the page where the original quote is coming from.

Abstract-Summary "During muscle regeneration, the chemokine CXCL12 (SDF-1) and the synthesis of some specific heparan sulfates (HS) have been shown to be critical."

"To investigate the role of the CXCL12/HS proteoglycan interactions in the pathophysiology of muscle regeneration, we performed two models of muscle injuries (notexin and freeze injury) in mutant CXCL12$^{Gagtm/Gagtm}$ mice, where the CXCL12 gene having been selectively mutated in critical binding sites of CXCL12 to interact with HS."

"Histological, cytometric, functional transcriptomic, and ultrastructure analysis focusing on the satellite cell behavior and the vessels were conducted on muscles before and after injuries."

"We showed that despite normal histology of the resting muscle and normal muscle stem cell behavior in the mutant mice, endothelial cells displayed an increase in the angiogenic response in resting muscle despite the downregulated transcriptomic changes induced by the CXCL12 mutation."

"The regenerative capacity of the CXCL12-mutated mice was only delayed after a notexin injury, but a severe damage by freeze injury revealed a persistent defect in the muscle regeneration of CXCL12 mutant mice associated with vascular defect and fibroadipose deposition with persistent immune cell infiltration."

"The present study shows that CXCL12 is crucial for proper muscle regeneration."

Background

"Skeletal muscle owns its remarkable capacity of regeneration to satellite cells (SCs), the main stem cells of this tissue [74]."

"Recent studies have also described the potential involvement of the CXCL12-CXCR4 pathway in the muscle repair process."

"An increase of CXCL12 was shown to enhance regeneration of injured skeletal muscles by inducing stem cell mobilization and by increasing the migration of myoblasts [75]."

"The CXCL12-CXCR4 pathway was shown to be upregulated in response to skeletal muscle damage and a CXCR4 antagonist induced a delay in muscle regeneration [75]."

"The administration of CXCL12 could accelerate the skeletal muscle repair process [75]."

"We therefore investigated the role of the CXCL12/HS proteoglycan interactions in the pathophysiology of muscle regeneration, focusing on the SC behavior and in vascular abnormalities."

"The disruption of the CXCL12/GAG interactions may transform the muscle into a non-regenerative tissue."

Materials and Methods

"The sections were then routinely stained with hematoxylin-eosin, Sirius Red, or Oil Red O. To preserve GFP fluorescence for the 3D study of blood vessel organization, whole TA muscles were fixed in 10% neutral-buffered formalin for 2 h, then cryopreserved in 40% sucrose overnight at 4 °C before freezing in OCT (Tissue-Tek®, Sakura® Finetek, CA, USA)."

"To assess the residual capacity of endothelial cells to respond to a normal or modified CXCL12 signaling gradient, attractor cells (SCs), which have better angiogenic properties than others as previously described [67], from Tg:Pax7nGFP WT and KI mice were FACS isolated and culture expanded in 22 μm filtered 1:1 DMEM Glutamax (Gibco): MCDB (Sigma-Aldrich; M6770) containing 20% fetal bovine serum (FBS), plated on Matrigel (BD Biosciences; #354234)."

"Non-parametric rank tests were separately performed per cell type to test for a mutation effect, using the Generally Applicable Gene-Set/Pathway Analysis in gage package."

Results

"To confirm the role of the CXCL12/HS disruption on EC behavior, we performed an in vivo angiogenesis assay using Matrigel plugs containing either KI-Pax7 or WT-Pax7 SCs grafted to either WT or KI recipient mice."

"To test this hypothesis, we studied the muscle regeneration in WT and KI mice."

"To investigate whether the CXCL12 mutation could have an effect during muscle regeneration, we began our investigation with a notexin (NTX) injection injury model that is less toxic for ECs and SCs [52]."

"This specific timing of the muscle regeneration was not observed in the KI mice."

"The muscle fibers were also affected in the KI mice at 1 month post-FI."

"To further study the impact of the CXCL12 mutation on the vascular network during muscle regeneration, we also performed a FI in WT-Flk1 and KI-Flk1 mice."

Discussion

"Although this in vitro phenotype has not been studied in exogenous CXCL12-free conditions, it is indicative of a lack of cumulative alterations in the KI SCs during muscle damage."

"Contrary to the appearance of a normal phenotype of the KI SCs, the KI mice exhibited minor vascular abnormalities in the non-challenged muscle, including an increase in the number of sprouting vessels."

"Although the potent interaction between the SCs and the ECs during in vivo angiogenesis assay has already been shown [66, 67, 76], the specific activation and response of the KI vessels only to the KI SCs, which are capable of producing and releasing mutated CXCL12, as we showed, or other CXCL12-regulated factors, remains to be explored."

"To study the importance of the CXCL12/HS interaction during muscle regeneration, we chose two injury models differing in severity and tissue layer location."

Conclusion

"The present study shows that CXCL12 is crucial for proper muscle regeneration."

"Whether comparable skews towards fibrosis are observed in otherwise regenerating models such as hepatectomy following CXCL12/HS interaction disruption should also warrant further studies."

"Understanding whether unbound CXCL12 alone triggers a profibrotic activity, HS-bound CXCL12 an antifibrotic activity, or whether an indirect mechanism is at play such as the displacement from HS disrupting the pattern of bound molecules could offer new insight in the occurrence of pathological scarring."

Adult Stem Cells at Work: Regenerating Skeletal Muscle [77]

This is a machine-generated summary of:

Schmidt, Manuel; Schüler, Svenja C.; Hüttner, Sören S.; von Eyss, Björn; von Maltzahn, Julia: Adult stem cells at work: regenerating skeletal muscle [77]

Published in: Cellular and Molecular Life Sciences (2019)

Link to original: https://doi.org/10.1007/s00018-019-03093-6

If you want to cite the papers, please refer to the original.

For technical reasons we could not place the page where the original quote is coming from.

Abstract-Summary "Satellite cells, the stem cells of skeletal muscle, are indispensable for skeletal muscle regeneration."

"We discuss the properties of satellite cells, including heterogeneity regarding gene expression and/or their phenotypic traits and the contribution of satellite cells to skeletal muscle regeneration."

"We also summarize the process of regeneration with a specific emphasis on signaling pathways, cytoskeletal rearrangements, the importance of miRNAs, and the contribution of non-satellite cells such as immune cells, fibro-adipogenic progenitor cells, and PW1-positive/Pax7-negative interstitial cells."

Introduction

"The paired box transcription factor Pax7 is expressed in all satellite cells, and is required for their postnatal maintenance and regeneration of skeletal muscle [78–80]."

"Ablation of satellite cells under homeostatic conditions in the adult does not seem to lead to muscle atrophy or myopathies without induced injury, suggesting that their main function in the adult is the regeneration of skeletal muscle [81, 82]."

"Although skeletal muscle can fully regenerate multiple times in healthy adult mice and men, the functionality of satellite cells declines in the course of various degenerative diseases or states such as aging resulting in impaired regeneration of skeletal muscle [3, 83–87]."

"Such impaired functionality results from an altered extracellular matrix composition, misbalanced cell–cell interactions with residing cells of the skeletal muscle or changes of systemic factors such as loss of the anti-aging hormone Klotho [5, 88–90]."

"Wnt3a signaling is upregulated in aged skeletal muscle, thereby impairing satellite cell function, resulting in disturbed regeneration [91]."

Identification and Characterization of Satellite Cells

"Satellite cells have also been termed muscle stem cells (MuSCs), probably due to the fact that the term satellite cell also refers to specific glial cells in the brain."

"These characteristics demonstrate that satellite cells are bonafide muscle stem cells."

"In adult skeletal muscle, all satellite cells express the paired box transcription factor Pax7, being essential for satellite cell function [78–80, 92], while subsets also express Pax3 [93] or myogenic regulatory factor 5 (Myf5) [94]."

"Of the genetic markers listed above, Pax7 is the canonical biomarker for satellite cells, since it is expressed in quiescent and activated satellite cells in most model species including mice, men, zebrafish, and chicken [55]."

Adult Satellite Cells: A Heterogeneous Cell Population

"For investigating satellite cell heterogeneity is based on genetic analysis using a Myf5-Cre reporter mouse line (R26R-YFP/Myf5-Cre)."

"Approximately 10% of satellite cells have never expressed myf5, as demonstrated by the absence of the reporter YFP (yellow fluorescent protein) in Pax7-positive cells [94]."

"Upregulation of myf5 and induction of the YFP reporter gene were demonstrated in YFP^- satellite daughter cells, further suggesting that $Pax7^+/YFP^-$ cells are a rare stem cell sub-population of satellite cells, while $Pax7^+/YFP^+$ cells are more committed progenitor cells [94]."

"Following activation of satellite cells—for instance after injury—both populations of satellite cells (YFP^+ and YFP^-) are proliferating and undergo planar cell divisions, thereby generating two identical daughter cells (identical in terms of YFP expression)."

Asymmetric Divisions of Satellite Cells

"Asymmetric satellite cell divisions are controlled by various signaling pathways, including Notch signaling [95]."

"Of interest is that Notch signaling seems to be responsible for asymmetric divisions of YFP$^-$ satellite cells when using the R26R-YFP/Myf5-cre reporter mouse line, since the Notch effectors Notch3 and Delta1 are asymmetrically distributed in the daughter cells [94], the YFP$^-$ satellite cells are expressing the Notch ligand Delta1, while the YFP$^+$ satellite cells express the Notch receptor Notch3 [94]."

"The importance of Notch signaling for asymmetric satellite cell divisions is further supported by the fact that the Notch antagonist Numb is also asymmetrically distributed in different mouse models used to discriminate the different satellite cell subpopulations [96–98]."

"Besides asymmetric distribution of signaling molecules and receptors, DNA strands are also asymmetrically segregated during satellite cell division."

"Control of cell polarity plays an important role in the regulation of asymmetric satellite cell divisions."

Regeneration of Skeletal Muscle

"The first inflammatory cells to be recruited to the damaged muscle are the neutrophils."

"Upon injury, quiescent satellite cells enter the cell cycle and begin to express MyoD, migrate to the site of injury, and either fuse with damaged myofibers or become myogenic progenitor cells."

"A highly orchestrated interplay between the stem cell niche and the satellite cell and other supporting cells is essential for proper regeneration of skeletal muscle."

"Examples for changes in the satellite cell niche and changes in interactions with the satellite cell niche during aging include the loss of Fibronectin, altered β1-Integrin activity, or reduced levels of the anti-aging hormone Klotho, all of which result in impaired regeneration of skeletal muscle [5, 88, 89, 99]."

"A brief overview of how signaling pathways affect satellite cell function and regeneration of skeletal muscle is given."

Wnt Signaling During Regeneration of Skeletal Muscle

"Wnt signaling drives development of skeletal muscle and is one of the key signaling pathways involved in regeneration of skeletal muscle. [7]."

"Multiple Wnt ligands are expressed during regeneration of skeletal muscle in a temporally controlled manner."

"A recent publication describes the importance of R-spondin, a known modulator of canonical Wnt signaling for proper differentiation of myogenic progenitor cells during regeneration, further emphasizing the importance of proper Wnt signaling for regeneration of skeletal muscle [100–102]."

"The importance of balanced canonical Wnt signaling for regeneration of skeletal muscle is further highlighted by a recent study from Rudolf and others, who demonstrated that disruption or activation of β-catenin in adult satellite cells impairs the regeneration process [103]."

Notch Signaling in Satellite Cells

"The Notch ligands are sequestered to the surface of the myofibers, thereby controlling satellite cell proliferation and differentiation [91, 95, 97, 98, 104]."

"The interplay between Notch and the transmembrane receptor Syndecan3 expressed in satellite cells controls the maintenance of the satellite cell pool and myofiber size after regeneration [105]."

"The importance of proper Notch signaling for satellite cell maintenance is further emphasized by two studies, demonstrating that expression of the downstream effector Recombination Signal Binding Protein for Immunoglobulin Kappa J Region (RBPJ) in satellite cells is a prerequisite for maintaining satellite cell quiescence [104, 106]."

"Cross talk of Notch signaling with the vascular niche is also important for regulating satellite cell quiescence—reminiscent of the role of cell adhesion molecules such as Cadherins, which control the transition from quiescence to activation through interaction with the myofiber niche [107, 108]."

Regulation of Satellite Cell Function by miRNAs

"Posttranscriptional regulation by miRNAs has been associated with maintenance of quiescence, activation, and differentiation of satellite cells [109, 110]."

"About 351 miRNAs are differentially expressed when comparing quiescent and activated satellite cells, underscoring their importance for activation of satellite cells [111]."

"It has been shown that myf5 mRNA is targeted by miR-31, which is highly expressed in quiescent satellite cells."

"Only a few miRNAs have been identified, that are downregulated during differentiation, amongst them miR-125b targeting Insulin-like growth factor II (IGF-II) in myoblasts and miR-221 and miR-222 controlling cell cycle exit [112, 113]."

"miRNAs control several signaling pathways important for satellite cell functionality."

"The impact of miRNAs on cell fate decisions is further reflected by the function of miR-133 through the transcriptional regulator Prdm16 (PR domain containing 16), to regulate cell fate choices between the closely related muscle progenitors and brown adipocytes."

Changes in the Cytoskeleton of Myogenic Cells During Regeneration

"In resting skeletal muscle, quiescent SCs express high levels of α7- and β1-Integrin, which interact with the extracellular matrix (ECM) and regulate satellite cell fate [114, 115]."

"The large protein superfamily of Integrins is essential for migration, assembly of the ECM and is mostly associated with focal adhesion sites [116–118]."

"The importance of proper signaling via Integrins can also be appreciated by the fact that β1-Integrin (Itgb1) signaling is required for activation of satellite cells via the fibroblast growth factor 2 (FGF2) interaction in aged and dystrophic mice [89]."

"Besides Integrins, work on the Dystroglycan complex revealed important functions of Dystrophin and Dystroglycan in activated satellite cells and asymmetric satellite stem cell divisions in concert with Par1b [2]."

"FAK has an essential role in myoblast fusion and differentiation through interaction with β1-Integrin in Integrin clusters associated with Vinculin and the actin-binding protein Talin [119–121]."

Non-myogenic Cells Involved in Regeneration of Skeletal Muscle

"Those can be divided into two groups, those having myogenic potential such as myoendothelial cells, Pax7-negative Pw1+ interstitial cells (PICs) ($PW1^+/Pax7^-$ interstitial cells) and $Twist2^+$ cells, contributing directly to the generation of new muscle fibers and those indirectly supporting regeneration."

"Several types of immune cells have been shown to be important for proper regeneration of skeletal muscle, e.g., eosinophils, neutrophils, and M1 and M2 macrophages, with the eosinophils being the major source of IL-4 [55, 122]."

"The other non-myogenic cells important for regeneration of skeletal muscle are the FAPs, which are located in the interstitium and can be identified by expression of PDGFRα (platelet-derived growth factor receptor-α)."

"Not only do they contribute to muscle fiber formation during regeneration of skeletal muscle, but also secrete factors such as FGF-2 and IGF-1, known to promote satellite cell functionality [123]."

Concluding Remarks and Future Perspectives

"Satellite cells are the main drivers of skeletal muscle regeneration."

"The finely tuned balance between the states of quiescence, activation, and differentiation is a prerequisite for proper regeneration."

"Alongside cell intrinsic signaling, interactions with other cell types and the extracellular matrix play an important role in controlling these processes."

"The biggest challenge will be to gain a comprehensive understanding of how those processes interact and how they are altered in age and disease."

Macrophage Depletion Impairs Skeletal Muscle Regeneration: the Roles of Pro-fibrotic Factors, Inflammation, and Oxidative Stress [124]

This is a machine-generated summary of:

Xiao, Weihua; Liu, Yu; Chen, Peijie: Macrophage Depletion Impairs Skeletal Muscle Regeneration: the Roles of Pro-fibrotic Factors, Inflammation, and Oxidative Stress [124]

Published in: Inflammation (2016)

Link to original: https://doi.org/10.1007/s10753-016-0438-8

If you want to cite the papers, please refer to the original.

For technical reasons we could not place the page where the original quote is coming from.

Abstract-Summary "Macrophages play complex roles in the regeneration of skeletal muscle."

"The roles of macrophages, especially the mechanisms involved, in the regeneration of muscle contusion are still not fully understood."

"We hypothesize that the depletion of macrophages impairs skeletal muscle regeneration and that pro-fibrotic factors, inflammation, and oxidative stress may be involved in the process."

"To test these hypotheses, we constructed a muscle contusion injury and a macrophage depletion model and followed it up with morphological and gene expression analyses."

"The data showed that fibrotic scars were formed in the muscle of contusion injury, and they deteriorated in the mice of macrophage depletion."

"These results suggest that macrophage depletion impairs skeletal muscle regeneration and that pro-fibrotic factors, inflammation, and oxidative stress may play important roles in the process."

Introduction

"The process of healing injured skeletal muscle in animal models is made up of three distinct phases: degeneration and inflammation, regeneration, and fibrosis [125, 126]."

"Regeneration, which is the next phase, normally takes place 5 to 10 days post-injury and, it comprises phagocytosis of the damaged tissue and regeneration of the injured muscle [58]."

"In the final phase, fibrosis (scar tissue formation), which seems to be the final result of the muscle repair process, hinders full muscle regeneration."

"Several studies have shown that macrophages play complex roles in an injured skeletal muscle and may be involved in all the three phases of regeneration, as described above [126, 127]."

"We hypothesize that depletion of macrophages impairs skeletal muscle regeneration and that pro-fibrotic factors, inflammation, and oxidative stress may be involved in the process."

Methods

"The mice were randomly divided into two groups, muscle contusion group (group S) and macrophage depleted group (group T)."

"At different time points (12 h, 1 day, 3 days, 5 days, 7 days, and 14 days) post-injury, the mice were killed by cervical dislocation while under anesthesia, and gastrocnemius muscles were harvested (eight mice per time point)."

"The GMs from the mice treated by clodronate-containing or control liposomes were surgically removed on days 1 and 3 after injury for evaluation (six mice per group)."

"At the time point of 14 days post-injury, the right gastrocnemius muscles were collected and embedded in paraffin (six mice per group)."

"To measure the area of fibrotic tissue in the injury sites, Masson's trichrome staining (total collagen staining) was performed (six mice per group)."

Results

"The data showed that CD68 mRNA increased significantly on days 1 and 3 post-injury."

"Interestingly, CD68 mRNA in injured muscles from the mice treated with CL-liposomes was significantly higher than that of the contusion group at days 5, 7, and 14 post-injury (1.81-fold, $p = 0$."

"The data showed that TGF-beta 1 mRNA increased significantly at 1, 3, and 7 days post-injury."

"As compared to the contusion group, macrophage depletion significantly inhibited TGF-beta 1 mRNA levels at 1 day post-injury."

"Macrophage depletion significantly increased the expression of myostatin at 1 and 3 days post-injury."

"A high expression of TNF-alpha and IL-1beta was observed in the mice of the macrophage depletion group at the 14 days post-injury."

"The data showed that gp91phox mRNA levels increased significantly at the time point of 3 days (11.44-fold, $p = 0.000$), 5 days (3.51-fold, $p = 0.020$) and 7 days (3.75-fold, $p = 0.045$) post-injury."

Discussion

"Due to the potent roles of pro-fibrotic factors in the fibrosis of tissues [70], we speculated that the pro-fibrotic factors may be involved in the fibrosis of injured skeletal muscles after macrophage depletion."

"On the 14 days post-injury, TGF-beta 1 in the muscle of the mice from macrophage depletion was still significantly higher than T_0 and S_{14d}."

"These data suggest that TGF-beta and myostatin may be involved in the fibrosis of injured skeletal muscles induced by macrophage depletion."

"These data suggest that inflammatory cytokines and chemokines may be involved in the fibrosis of skeletal muscles induced by macrophage depletion."

"It suggests that ROS mediated by NADPH oxidase may be involved in the fibrosis of skeletal muscles induced by macrophage depletion."

"These data suggest that pro-fibrotic factors produced by M1 macrophages, oxidative stress-related enzymes from M2 macrophages, and inflammatory cytokines from M1 and M2 macrophages may be involved in the fibrosis of skeletal muscles post-injury."

Conclusions

"Macrophage depletion induced a smaller size of regenerating myofiber as well as a more fibrotic scar formation after muscle contusion."

"Macrophage depletion could delay the expression of pro-fibrotic factors, inflammatory cytokines, chemokines, and oxidative stress-related enzymes."

"Most of them were still significantly higher in the later stage of regeneration."

"These results suggest that macrophage depletion impairs skeletal muscle regeneration and that pro-fibrotic factors, inflammation, and oxidative stress may be involved in the process."

Hydrogels for Skeletal Muscle Regeneration [128]

This is a machine-generated summary of:

Fischer, Kristin M.; Scott, Tracy E.; Browe, Daniel P.; McGaughey, Tyler A.; Wood, Caroline; Wolyniak, Michael J.; Freeman, Joseph W.: Hydrogels for Skeletal Muscle Regeneration [128]

Published in: Regenerative Engineering and Translational Medicine (2020)

Link to original: https://doi.org/10.1007/s40883-019-00146-x

If you want to cite the papers, please refer to the original.

For technical reasons we could not place the page where the original quote is coming from.

Abstract-Summary "Hydrogels are described as soft crosslinked polymeric networks with high water content that simulates the body's natural aqueous environment."

"Natural, semi-synthetic, synthetic, and composite hydrogels for skeletal muscle regeneration are discussed."

"This category was followed by a discussion of composite hydrogels, defined in this review as at least one synthetic and one natural polymer combined to form a hydrogel, and these are the next most favored materials."

"Synthetic polymer hydrogels came in third with semi-synthetic polymers, chemically modified natural polymers, being the least common."

"While many of the hydrogels show promise for skeletal muscle regeneration, continued investigation is needed in order to regenerate a functional muscle tissue replacement."

"Skeletal muscle tissue engineering focuses on regenerating large amounts of skeletal muscle tissue lost due to tumor removal, traumatic injuries, and/or disease."

"Neither natural repair processes by the body nor current medical interventions are able to completely restore function after volumetric muscle loss."

"This review paper summarizes the research from 2013 to early 2018 using hydrogels, a soft material with a high water content, as a tool to regenerate muscle."

"The review is categorized into hydrogels made from natural materials, semi-synthetic materials, synthetic materials, and composite materials (at least one natural and one synthetic material combined)."

Introduction

"Satellite cells only account for about 1–5% of muscle cells, and if not enough are not able to accumulate at the injury site, scar tissue may form."

"Other potential treatment options include cell transplantation, gene therapy, and growth factors just to name a few; however, none of these treatments result in functional restoration of the muscle tissue [129, 130]."

"The chosen biomaterial should be biocompatible, biodegradable, and have correct chemical and mechanical properties to encourage cellular growth and fusion to form parallel, aligned muscle fibers to better mimic native architecture and have contractile strength consistent with that of native skeletal muscle [130, 131]."

"Skeletal muscle TE is a relatively new field with the overall goal to replace the damaged muscle with a functional scaffold that is engineered to regenerate new tissue with properties similar to the native one."

Natural Hydrogels

"KOS hydrogels had significantly more multinucleated human muscle cells compared with Matrigel™, collagen, and uncoated tissue culture polystyrene (TCPS) while KTN hydrogels were only significantly greater compared with uncoated TCPS after 4 days."

"Baker and others and Passipieri and others built on this work by combining KTN hydrogels with combinations of IGF-1, bFGF, and muscle precursor cells (MPCs) prior to induced VML injury in either mice or rat TA muscles."

"Yi and others combined alginate with gelatin and heparin (Alg-G-H) to form a hydrogel that was then coated with muscle ECM to increase cell adherence."

"Human skeletal muscle progenitor cells (hSMPCs) were cultured on these hydrogels and assessed for myogenic protein expression and myotube formation."

"Ding and others combined chitosan, β-glycerophosphate, and collagen (C/GP/Co) to form an injectable, thermosensitive hydrogel to carry skeletal muscle satellite cells (SMSCs)."

Semi-synthetic Hydrogels

"Kim and others showed that C2C12 cellularized gelatin hydroxyphenylpropionic acid (GHPA) hydrogels supported cell proliferation and MHC was expressed indicating maturation into myofibers [132]."

"C2C12 cells were cultured on groove-ridge patterned GelMA–aligned CNTs, GelMA-random CNTs, and pristine GelMA hydrogels."

"Hong and others encapsulated C2C12 cells in GelMA hydrogels and investigated the effects of compression and compression with shear to simulate the effects of deep tissue injuries (DTIs)."

"C2C12 GelMA hydrogels cultured without the pillars ending up collapsing into a ball of cells by day 5 rather than forming multinucleated myotubes, thus showing that tension on the hydrogel is an important component of muscle development [133]."

"VML of the TA muscle in rats was created by removing a 6-mm diameter section and then implanting the HA + LMN containing minced muscle graft (MG), the HA + LMN hydrogel with no muscle tissue, or no repair to the defect."

Synthetic Hydrogels

"Browe and others combined PEGDA with AA to form a biocompatible, electroactive hydrogel to be used as an actuator while promoting mature skeletal muscle tissue."

"C2C12 cells survived and were metabolically active on 100% PEGDA, 1:4, 1:8, 1:12, and 1:16 PEGDA to AA hydrogels over a 10-day period, but the 1:4 PEGDA:AA hydrogel had significantly greater values."

"C2C12 cells were unable to adhere to PAAm hydrogels alone, but when GO was added, the amount of cells increased over 72 h. Conductivity also increased when GO was incorporated and even more so when reduced graphene oxide (rGO) was utilized."

"Villa and others combined poly(NIPAAm) with HEMA to culture C2C12 cells to form a detachable muscle sheet for implantation."

"C2C12 cells cultured on this hydrogel expressed desmin and a muscle cell sheet was successfully detached after 15 min."

Composite Hydrogels

"Mulyasasmita and others utilized protein-PEG hydrogels to deliver human induced pluripotent stem cell-derived endothelial cells (hiPSC-ECs) and vascular endothelial growth factor (VEGF) after an ischemic injury in the gastrocnemius muscle of non-obese diabetic severe combined immune deficient (NOD SCID) mice."

"After 2 weeks, the muscle tissue showed reduced necrosis and more myofibers compared with delivery of the hydrogel alone or PBS injection as a control [134]."

"Constantini and others 3D printed the PF solution containing C2C12 cells into aligned hydrogel fibers."

"The implants were explanted after 28 days, and histological analysis showed more aligned, striated, and MHC-positive myofibers compared with the cellularized bulk-hydrogel controls [135]."

"After 1 week in differentiating media, the C2C12 cells in 3D-printed hydrogel expressed higher levels of myogenin, α-sarcomeric actin, and MyoD compared with the 2D control [136]."

Conclusions and Future Directions

"Hydrogels are an attractive option when developing scaffolds for TE as they are an aqueous environment similar to the body and the amount of crosslinking can be tuned for mechanical properties similar to skeletal muscle."

"As skeletal muscle TE is a newer area of research, researchers are currently investigating a multitude of hydrogels to determine the best material, concentration of polymer, and crosslinking method that can best restore skeletal muscle function."

"Although this review focused on different compositions of hydrogels, mechanical, chemical, and electrical cues also play important roles in the successful development of muscle within or on the hydrogels."

"Future work should focus on mechanical property analysis and creating larger, more 3D scaffolds that form more mature, aligned skeletal muscle to better match the native skeletal muscle anatomy."

Effect of Photobiomodulation on Connective Tissue Remodeling and Regeneration of Skeletal Muscle in Elderly Rats [137]

This is a machine-generated summary of:

de Brito, Adriana; Alves, Agnelo Neves; Ribeiro, Beatriz Guimaraes; Barbosa, Daniel Victor D. Emilio; Magalhaes, Erick Moreno Ramos; Fernandes, Kristianne Porta Santos; Bussadori, Sandra Kalil; Goulardins, Juliana Barbosa; Mesquita-Ferrari, Raquel Agnelli: Effect of photobiomodulation on connective tissue remodeling and regeneration of skeletal muscle in elderly rats [137]

Published in: Lasers in Medical Science (2017)

Link to original: https://doi.org/10.1007/s10103-017-2392-6

If you want to cite the papers, please refer to the original.

For technical reasons we could not place the page where the original quote is coming from.

Abstract-Summary "The purpose of this study was to evaluate the effects of low-level laser therapy (LLLT) on morphological aspects, IL-6 and IL-1β expressions, as well as the distribution and organization of collagen in the tibialis anterior (TA) muscle of elderly rats submitted to cryoinjury."

"Male Wistar rats, aged 20 months, were distributed into three groups: (1) control animals not injured or treated with LLLT ($n = 5$), (2) cryoinjury without LLLT treatment ($n = 15$), and (3) cryoinjury treated with infrared LLLT ($n = 15$)."

"LLLT was applied to the TA 2 h after of the injury induction and consisted of daily applications until the sacrifice (1, 3, and 7 days)."

"Treatment induced a better collagen distribution and organization at 7 days in comparison to the untreated group ($p < 0.05$)."

"LLLT demonstrated a modulatory effect on the muscle repair process in elderly animals with regard to the collagen remodeling and morphological aspects of muscle tissue."

Introduction

"With the increase in life expectancy, it is essential to establish new therapeutic resources in clinical practice to improve the quality of the muscle repair process, motor quality, and injury recovery time in older adults [138]."

"When applied to injured muscle tissue, LLLT is able to reduce pain, edema, leukocyte influx, and myonecrosis as well as alter the expression of cytokines, inflammatory enzymes, and collagen remodeling, thereby accelerating the repair process [139–146]."

"The use of LLLT is well grounded in the literature due to its effect on inflammatory disorders, tissue regeneration, and skeletal muscle repair."

"The few studies that associate LLLT with skeletal muscle regeneration in older animals differ with regard to the dosimetric parameters employed [147, 148]."

"Purpose of the present study was to evaluate the effects of LLLT (780 nm, 40 mW, and 10 J/cm^2) on morphological aspects and the remodeling of connective tissue in the TA muscle of elderly rats submitted to cryoinjury."

Methods

"In each cut, five areas corresponding to 50% of the injury total area were photographed (magnification × 400) using a conventional light microscope (Zeiss Axioplan 2, Germany)."

"Total inflammatory cells, myonecrosis, blood vessels, and immature (new) muscle fibers were determined using the "cell counter" plug-in of the ImageJ software (National Institutes of Health, USA)."

"Dewaxing was performed with xylene, and the samples were immersed in alcohol, followed by incubation in a 3% hydrogen peroxide solution diluted in Tris-buffered saline (TBS) (pH 7.4)."

"Incubation was then performed with 3% goat serum (20 min), and the samples were blocked by immersion in citrate buffer solution (pH 6.0) for 20 min at 95 °C for antigen recovery."

"Image analysis was performed using the cell counter plug-in of the ImageJ program (free software, National Institutes of Health, Bethesda, MD, USA)."

Results

"At day 7, the group submitted to laser therapy exhibited a significant reduction in the amount of inflammatory infiltrate and myonecrosis compared to the untreated group ($p < 0.001$) and a significant increase occurred in the number of blood vessels and immature fibers."

"At days 1 and 3, the untreated and treated (LLLT) injury groups were similar with regard to collagen distribution, exhibiting fibers dispersed in the region of the inflammatory infiltrate and necrotic fibers in cell spaces."

"No differences were found in the amount of collagen in the untreated and treated groups at days 1, 3, and 7, but the group submitted to LLLT exhibited better collagen organization at day 7 compared to the other treated and untreated injury groups."

"An increase in the concentration of collagen was found in the untreated and treated injury groups compared to the control group at day 7 ($p < 0.05$)."

Discussion

"Pertille and others [148] used irradiation and treatment parameters different from the current study: infrared laser, wavelength 830 nm, output power 30 mW, and total energy 0.96 J; the treatment was initiated 24 h after injury induction, and only two points on the muscle were irradiated."

"The cryoinjury procedure has previously been performed by research groups, and many studies published in the literature have demonstrated that LLLT has an effect on the skeletal muscle repair process, despite the different dosimetric parameters employed [149, 150]."

"LLLT did not cause a significant difference in collagen area, as previously observed in young animals [151], with respect to the remodeling of the extracellular matrix during the muscle repair process in elderly rats."

Conclusion

"LLLT positively modulated the muscle repair process in elderly animals, reducing the amount of inflammatory infiltrate and myonecrosis as well as stimulating the formation of both new immature fibers and blood vessels."

"LLLT promoted improvements in the organization and remodeling of collagen fibers."

"It is important to emphasize that older animals responded more slowly to LLLT compared to data on young adult animals published in the literature."

The Effects of Neuregulin-1β on Intrafusal Muscle Fiber Formation in Neuromuscular Coculture of Dorsal Root Ganglion Explants and Skeletal Muscle Cells [152]

This is a machine-generated summary of:

Qiao, Yuan; Cong, Menglin; Li, Jianmin; Li, Hao; Li, Zhenzhong: The effects of neuregulin-1β on intrafusal muscle fiber formation in neuromuscular coculture of dorsal root ganglion explants and skeletal muscle cells [152]

Published in: Skeletal Muscle (2018)

Link to original: https://doi.org/10.1186/s13395-018-0175-9

If you want to cite the papers, please refer to the original.

For technical reasons we could not place the page where the original quote is coming from.

Abstract-Summary "The formation of IM fibers was observed in this coculture system after neuregulin-1β (NRG-1β) incubation."

"We found that NRG-1β promoted outgrowth of neurites and migration of neurons from the organotypic DRG explants and that this correlated with an induction of growth-associated protein 43 (GAP-43) expression."

"NRG-1β also increased the amount of nuclear bag fibers and nuclear chain fibers by elevating the proportion of tyrosine kinase receptor C (TrkC) phenotypic DRG neurons."

"These data imply that NRG-1β promoted neurite outgrowth and neuronal migration from the organotypic DRG explants and that this correlated with an induction of GAP-43 expression."

"The modulating effects of NRG-1β on TrkC DRG neuronal phenotype may link to promote IM fiber formation."

"The effects produced by NRG-1β in this neuromuscular coculture system provide new data for the therapeutic potential on IM fiber formation after muscle injury."

Background

"TrkC proprioceptive DRG neurons with their peripheral axons around IM fibers transmit information from muscle spindles to the spinal cord through central processes [153, 154]."

"Herndon and others confirmed that NRG-1 is required for the activation of ERK1/2 signaling pathway in response to induce the transcription factors in IM fibers of the muscle spindles [155]."

"How and to what extent NRG-1β influences the formation of IM fibers in the presence of both DRG neurons and developing muscle cells should be further evaluated."

"It is hypothesized that NRG-1β could induce neurite outgrowth of DRG neurons and the formation of IM fibers in a coculture system of DRG explants and dissociated muscle cells."

"The specific effects of NRG-1β and its downstream signaling on the formation of IM fibers and its involvement in TrkC phenotypic DRG neurons will be studied in this experiment."

Methods

"This coculture means organotypic DRG tissue explants and dissociated SKM cells would grow in the same well of the clusters and form neuromuscular junction (NMJ)-like structure in vitro."

"The neuromuscular coculture was prepared as follows: Each newly prepared DRG explant was plated into a well with 2-day-old SKM cell culture after confluent myoblast fusion has happened."

"The neuromuscular cocultures of organotypic DRG tissue explants and dissociated SKM cells were allowed to grow for an additional 4 days with media change every 2 days."

"Double fluorescence staining with neurofilament 200 (NF-200) for neuron and α-actin for SKM cell was used for determining the formation of nuclear bag or nuclear chain fibers."

"This coculture system of DRG neurons and SKM cells represents the formation of synaptic specializations, particularly muscle spindles, but the connective tissue capsule structures surrounding IM fibers could not be visualized."

Results

"We showed that NRG-1β (20 nmol/L) incubation elevated both mRNA and protein levels of GAP-43 suggesting the promotive effects of NRG-1β on GAP-43 expression in organotypic DRG tissue explants in this neuromuscular coculture system ($P < 0.001$)."

"We examined the effects of NRG-1β on TrkC expression in the neuromuscular coculture system of organotypic DRG tissue with dissociated SKM cells."

"We also showed that inhibiting each of the three downstream signaling pathways would block the effects of NRG-1β on TrkC in situ expression of DRG neurons in the neuromuscular cocultures of organotypic DRG tissue with dissociated SKM cells ($P < 0.05$ or $P < 0.001$)."

"We also showed that inhibiting each of the three downstream signaling pathways blocked the effects of NRG-1β on the elevation of the percentage of TrkC-positive neurons with their neurites on muscle fibers ($P < 0.05$ or $P < 0.001$)."

Discussion

"These results show that in addition to the effect of promoting the neurite outgrowth, NRG-1β promoted the formation of IM fibers with the sensory DRG neurons in neuromuscular cocultures."

"While the migrating neurons could target the muscle cells by the pathfinding route, the formation of IM fibers had more significance in these neuromuscular cocultures with DRG neurons, for the IM fibers are actually innervated by sensory neurons in vivo."

"Our results do not rule out that the nuclear bag and nuclear chain fibers could only be observed when the SKM is cocultured with sensory neurons, but they are consistent with the hypothesis that the formation of IM fibers critically depends on the growth environments, at least in part, for the existence of specific proprioceptive neurons and the trophic factor NRG-1β."

Conclusion

"NRG-1β promotes outgrowth of neurites and migration of neurons from the organotypic DRG explants, and this correlated with an induction of GAP-43 expression."

"Based on the promotion of neuronal outgrowth, NRG-1β induces IM fiber formation by modulating TrkC DRG neuronal phenotype in neuromuscular cocultures."

"The effects produced by NRG-1β in this neuromuscular coculture system provide new data for the therapeutic potential on IM fiber formation after muscle injury."

Skeletal Muscle Mechanics: Questions, Problems and Possible Solutions [156]

This is a machine-generated summary of:

Herzog, Walter: Skeletal muscle mechanics: questions, problems and possible solutions [156]

Published in: Journal of NeuroEngineering and Rehabilitation (2017)

Link to original: https://doi.org/10.1186/s12984-017-0310-6

If you want to cite the papers, please refer to the original.

For technical reasons we could not place the page where the original quote is coming from.

Abstract-Summary "Skeletal muscle mechanics have been studied ever since people have shown an interest in human movement."

"Experimental evidence suggests that our knowledge of the mechanisms of contraction is far from complete, and muscle properties and muscle function in human movement remain largely unknown."

"I am trying to identify some of the crucial challenges we are faced with in muscle mechanics, offer possible solutions to questions, and identify problems that might be worthwhile exploring in the future."

"It is my hope that this paper may serve as an inspiration for some, may challenge current beliefs in selected areas, tackle important problems in the area of muscle mechanics, physiology and movement control, and may guide and focus some of the thinking of future muscle mechanics research."

Background

"In order to produce half-sarcomere and sarcomere stability independent of sarcomere lengths, account for the experimentally observed residual force enhancement, and explain experimentally observed inconsistencies in the energetics and force trajectories in eccentric muscle contraction, a structural element connecting myosin with actin would be an elegant solution."

"When sarcomeres and single myofibrils are stretched in an activated preparation, force will continuously increase because of the increased stiffness in titin in active compared to passive muscle, thus providing positive stiffness at all lengths, including the descending limb of the force-length relationship and even when sarcomeres are pulled beyond actin-myosin filament overlap."

Conclusions

"Looking ahead to the next BANCOM meeting in 20 years from now (i.e., in 2036), I hope that the following problems and questions will have been solved in the three areas I discussed here."

"We will understand the mechanics of eccentric contractions in skeletal muscles much better than we do now."

"I anticipate that the molecular details and functions of titin (and possibly other structural proteins) in eccentric contractions are fully elucidated."

"We will know the mechanical properties and the functions of individual muscles for sub-maximal, dynamic conditions as occur in everyday human movements, and third, we will be able to quantify individual muscle forces in human movements reliably and accurately and will have solved the distribution problem in biomechanics and movement control."

Bibliography

1. Ahrens HE, Huettemeister J, Schmidt M, Kaether C, von Maltzahn J (2018) Klotho expression is a prerequisite for proper muscle stem cell function and regeneration of skeletal muscle. Skelet Mus. https://doi.org/10.1186/s13395-018-0166-x
2. Dumont NA, Wang YX, von Maltzahn J, Pasut A, Bentzinger CF, Brun CE, Rudnicki MA (2015) Dystrophin expression in muscle stem cells regulates their polarity and asymmetric division. Nat Med. 21:1455–1463
3. Price FD, von Maltzahn J, Bentzinger CF, Dumont NA, Yin H, Chang NC, Wilson DH, Frenette J, Rudnicki MA (2014) Inhibition of JAK-STAT signaling stimulates adult satellite cell function. Nat Med. 20:1174–1181
4. Sousa-Victor P, Garcia-Prat L, Serrano AL, Perdiguero E, Munoz-Canoves P (2015) Muscle stem cell aging: regulation and rejuvenation. Trends Endocrinol Metab. 26:287–296
5. Lukjanenko L, Jung MJ, Hegde N, Perruisseau-Carrier C, Migliavacca E, Rozo M, Karaz S, Jacot G, Schmidt M, Li L, Metairon S, Raymond F, Lee U, Sizzano F, Wilson DH, Dumont NA, Palini A, Fassler R, Steiner P, Descombes P, Rudnicki MA, Fan CM, von Maltzahn J, Feige JN, Bentzinger CF (2016) Loss of fibronectin from the aged stem cell niche affects the regenerative capacity of skeletal muscle in mice. Nat Med. 22:897–905
6. von Maltzahn J, Zinoviev R, Chang NC, Bentzinger CF, Rudnicki MA (2013) A truncated Wnt7a retains full biological activity in skeletal muscle. Nat Commun. 4:2869
7. von Maltzahn J, Chang NC, Bentzinger CF, Rudnicki MA (2012) Wnt signaling in myogenesis. Trends Cell Biol. 22:602–609
8. Brack AS, Conboy MJ, Roy S, Lee M, Kuo CJ, Keller C, Rando TA (2007) Increased Wnt signaling during aging alters muscle stem cell fate and increases fibrosis. Science. 317:807–810

9. Tian S, Chang S, Lu Y, Zhu J, Kong X (2021) The association between intracompartmental pressure and skeletal muscle recovery after tibial diaphyseal fractures: an ambispective cohort study. J Orthopaed Traumatol. https://doi.org/10.1186/s10195-021-00579-7
10. Donaldson J, Haddad B, Khan WS (2014) The pathophysiology, diagnosis and current management of acute compartment syndrome. Open Orthop J 8:185–193
11. Via AG, Oliva F, Spoliti M, Maffulli N (2015) Acute compartment syndrome. Musc Ligam Tendons J 5:18–22
12. Commean PK, Tuttle LJ, Hastings MK, Strube MJ, Mueller MJ (2011) Magnetic resonance imaging measurement reproducibility for calf muscle and adipose tissue volume. J Magn Reson Imaging 34:1285–1294
13. Flores DV, Mejia Gomez C, Estrada-Castrillon M, Smitaman E, Pathria MN (2018) MR imaging of muscle trauma: anatomy, biomechanics, pathophysiology, and imaging appearance. Radiographics 38:124–148
14. Laakman RW, Kaufman B, Han JS, Nelson AD, Clampitt M, O'Block AM, Haaga JR, Alfidi RJ (1985) MR imaging in patients with metallic implants. Radiology 157:711–714
15. Singh DR, Chin MS, Peh WC (2014) Artifacts in musculoskeletal MR imaging. Semin Musculoskelet Radiol 18:12–22
16. Rahman FA, Quadrilatero J (2021) Mitochondrial network remodeling: an important feature of myogenesis and skeletal muscle regeneration. Cell Mol Life Sci. https://doi.org/10.1007/s00018-021-03807-9
17. Wagatsuma A, Kotake N, Yamada S (2011) Muscle regeneration occurs to coincide with mitochondrial biogenesis. Mol Cell Biochem 349:139–147. https://doi.org/10.1007/s11010-010-0668-2
18. Nichenko AS, Southern WM, Tehrani KF et al (2020) Mitochondrial-specific autophagy linked to mitochondrial dysfunction following traumatic freeze injury in mice. Am J Physiol Cell Physiol 318:C242–C252. https://doi.org/10.1152/ajpcell.00123.2019
19. Chan JY, Kwong M, Lu R et al (1998) Targeted disruption of the ubiquitous CNC-bZIP transcription factor, Nrf-1, results in anemia and embryonic lethality in mice. EMBO J 17:1779–1787. https://doi.org/10.1093/emboj/17.6.1779
20. Ristevski S, O'Leary DA, Thornell AP et al (2004) The ETS transcription factor GABPalpha is essential for early embryogenesis. Mol Cell Biol 24:5844–5849. https://doi.org/10.1128/MCB.24.13.5844-5849.2004
21. Schreiber SN, Knutti D, Brogli K et al (2003) The transcriptional coactivator PGC-1 regulates the expression and activity of the orphan nuclear receptor estrogen-related receptor alpha (ERRalpha). J Biol Chem 278:9013–9018. https://doi.org/10.1074/jbc.M212923200
22. Schreiber SN, Emter R, Hock MB et al (2004) The estrogen-related receptor alpha (ERRalpha) functions in PPARgamma coactivator 1alpha (PGC-1alpha)-induced mitochondrial biogenesis. Proc Natl Acad Sci USA 101:6472–6477. https://doi.org/10.1073/pnas.0308686101
23. Cho Y, Hazen BC, Russell AP, Kralli A (2013) Peroxisome proliferator-activated receptor γ coactivator 1 (PGC-1)- and estrogen-related receptor (ERR)-induced regulator in muscle 1 (Perm1) is a tissue-specific regulator of oxidative capacity in skeletal muscle cells. J Biol Chem 288:25207–25218. https://doi.org/10.1074/jbc.M113.489674
24. Cho Y, Hazen BC, Gandra PG et al (2016) Perm1 enhances mitochondrial biogenesis, oxidative capacity, and fatigue resistance in adult skeletal muscle. FASEB J 30:674–687. https://doi.org/10.1096/fj.15-276360
25. Arany Z, Lebrasseur N, Morris C et al (2007) The transcriptional coactivator PGC-1beta drives the formation of oxidative type IIX fibers in skeletal muscle. Cell Metab 5:35–46. https://doi.org/10.1016/j.cmet.2006.12.003
26. Kamei Y, Ohizumi H, Fujitani Y et al (2003) PPARgamma coactivator 1beta/ERR ligand 1 is an ERR protein ligand, whose expression induces a high-energy expenditure and antagonizes obesity. Proc Natl Acad Sci USA 100:12378–12383. https://doi.org/10.1073/pnas.2135217100

27. Shao D, Liu Y, Liu X et al (2010) PGC-1 beta-regulated mitochondrial biogenesis and function in myotubes is mediated by NRF-1 and ERR alpha. Mitochondrion 10:516–527. https://doi.org/10.1016/j.mito.2010.05.012
28. Youle RJ, Karbowski M (2005) Mitochondrial fission in apoptosis. Nat Rev Mol Cell Biol 6:657–663. https://doi.org/10.1038/nrm1697
29. Chen H, Vermulst M, Wang YE et al (2010) Mitochondrial fusion is required for mtDNA stability in skeletal muscle and tolerance of mtDNA mutations. Cell 141:280–289. https://doi.org/10.1016/j.cell.2010.02.026
30. Ji W-K, Chakrabarti R, Fan X et al (2017) Receptor-mediated Drp1 oligomerization on endoplasmic reticulum. J Cell Biol 216:4123–4139. https://doi.org/10.1083/jcb.201610057
31. Bravo-San Pedro JM, Kroemer G, Galluzzi L (2017) Autophagy and mitophagy in cardiovascular disease. Circ Res 120:1812–1824. https://doi.org/10.1161/CIRCRESAHA.117.311082
32. Kerr JS, Adriaanse BA, Greig NH et al (2017) Mitophagy and Alzheimer's disease: cellular and molecular mechanisms. Trends Neurosci 40:151–166. https://doi.org/10.1016/j.tins.2017.01.002
33. Okatsu K, Oka T, Iguchi M et al (2012) PINK1 autophosphorylation upon membrane potential dissipation is essential for Parkin recruitment to damaged mitochondria. Nat Commun 3:1016. https://doi.org/10.1038/ncomms2016
34. Kondapalli C, Kazlauskaite A, Zhang N et al (2012) PINK1 is activated by mitochondrial membrane potential depolarization and stimulates Parkin E3 ligase activity by phosphorylating Serine 65. Open Biol 2:120080. https://doi.org/10.1098/rsob.120080
35. Rikka S, Quinsay MN, Thomas RL et al (2011) Bnip3 impairs mitochondrial bioenergetics and stimulates mitochondrial turnover. Cell Death Differ 18:721–731. https://doi.org/10.1038/cdd.2010.146
36. Saita S, Shirane M, Nakayama KI (2013) Selective escape of proteins from the mitochondria during mitophagy. Nat Commun 4:1410. https://doi.org/10.1038/ncomms2400
37. Jäger SS, Handschin CC, St-Pierre JJ, Spiegelman BM (2007) AMP-activated protein kinase (AMPK) action in skeletal muscle via direct phosphorylation of PGC-1α. Proc Natl Acad Sci 104:12017–12022. https://doi.org/10.1073/pnas.0705070104
38. Zhang Y, Uguccioni G, Ljubicic V et al (2014) Multiple signaling pathways regulate contractile activity mediated PGC-1α gene expression and activity in skeletal muscle cells. Physiol Rep 2:1–12. https://doi.org/10.14814/phy2.12008
39. Bo T, Yamamori T, Suzuki M et al (2018) Calmodulin-dependent protein kinase II (CaMKII) mediates radiation-induced mitochondrial fission by regulating the phosphorylation of dynamin-related protein 1 (Drp1) at serine 616. Biochem Biophys Res Commun 495:1601–1607. https://doi.org/10.1016/j.bbrc.2017.12.012
40. Touvier T, De Palma C, Rigamonti E et al (2015) Muscle-specific Drp1 overexpression impairs skeletal muscle growth via translational attenuation. Cell Death Dis 6:e1663. https://doi.org/10.1038/cddis.2014.595
41. Kraft CS, LeMoine CM, Lyons CN, Michaud D, Mueller CR, Moyes CD (2006) Control of mitochondrial biogenesis during myogenesis. Am J Physiol Cell Physiol. 290:C1119–C1127. https://doi.org/10.1152/ajpcell.00463.2005
42. Zhou Y, Sharma N, Dukes D, Myzithras MB, Gupta P, Khalil A, Kahn J, Ahlberg JS, Hayes DB, Franti M, Criswell T (2016) GDF11 treatment attenuates the recovery of skeletal muscle function after injury in older rats. AAPS J. https://doi.org/10.1208/s12248-016-0024-x
43. Loffredo FS, Steinhauser ML, Jay SM, Gannon J, Pancoast JR, Yalamanchi P et al (2013) Growth differentiation factor 11 is a circulating factor that reverses age-related cardiac hypertrophy. Cell. 153(4):828–839
44. Mendelsohn AR, Larrick JW (2014) Systemic factors mediate reversible age-associated brain dysfunction. Rejuvenat Res. 17(6):525–528
45. Egerman MA, Cadena SM, Gilbert JA, Meyer A, Nelson HN, Swalley SE et al (2015) GDF11 Increases with Age and Inhibits Skeletal Muscle Regeneration. Cell Metab. 22(1):164–174

46. Rodgers BD, Eldridge JA (2015) Reduced Circulating GDF11 Is Unlikely Responsible for Age-Dependent Changes in Mouse Heart, Muscle, and Brain. Endocrinology. 156(11):3885–3888
47. Myzithras M, Li H, Bigwarfe T, Waltz E, Gupta P, Low S et al (2016) Development of an ultra-sensitive Simoa assay to enable GDF11 detection: a comparison across bioanalytical platforms. Bioanalysis.
48. Criswell TL, Corona BT, Ward CL, Miller M, Patel M, Wang Z et al (2011) Compression-induced muscle injury in rats that mimics compartment syndrome in humans. Am J Pathol.
49. Zhou Y, Lovell D, Bethea M, Wang Z, Christ GJ, Soker S et al (2014) Age-dependent changes cooperatively impact skeletal muscle regeneration after compartment syndrome injury. Am J Pathol. 184(8):2225–2236
50. Sinha M, Jang YC, Oh J, Khong D, Wu EY, Manohar R et al (2014) Restoring systemic GDF11 levels reverses age-related dysfunction in mouse skeletal muscle. Science. 344(6184):649–652
51. Jacobsen NL, Morton AB, Segal SS (2023) Angiogenesis precedes myogenesis during regeneration following biopsy injury of skeletal muscle. Skelet Mus. https://doi.org/10.1186/s13395-023-00313-3
52. Hardy D, Besnard A, Latil M et al (2016) Comparative study of injury models for studying muscle regeneration in mice. PLoS ONE 11:e0147198. https://doi.org/10.1371/journal.pone.0147198
53. Morton AB, Norton CE, Jacobsen NL, Fernando CA, Cornelison DDW, Segal SS (2019) Barium chloride injures myofibers through calcium-induced proteolysis with fragmentation of motor nerves and microvessels. Skelet Muscle. 9(1):27
54. Tidball JG (2017) Regulation of muscle growth and regeneration by the immune system. Nat Rev Immunol. 17(3):165–178
55. Yin H, Price F, Rudnicki MA (2013) Satellite cells and the muscle stem cell niche. Physiol Rev. 93(1):23–67
56. Baghdadi MB, Tajbakhsh S (2018) Regulation and phylogeny of skeletal muscle regeneration. Dev Biol. 433(2):200–209
57. Juban G, Chazaud B (2021) Efferocytosis during skeletal muscle regeneration. Cells 10(12):3267
58. Carlson BM, Faulkner JA (1983) The regeneration of skeletal muscle fibers following injury: a review. Med Sci Sports Exerc. 15(3):187–198
59. Hansen-Smith FM, Carlson BM, Irwin KL (1980) Revascularization of the freely grafted extensor digitorum longus muscle in the rat. Am J Anat. 158(1):65–82
60. Grounds MD (2008) Complexity of extracellular matrix and skeletal muscle regeneration. In: Schiaffino S, Partridge T (eds) Skeletal muscle repair and regeneration. Springer, Dordrecht, pp 269–301
61. Sanes JR (2003) The basement membrane/basal lamina of skeletal muscle. J Biol Chem. 278(15):12601–12604
62. Webster MT, Manor U, Lippincott-Schwartz J, Fan CM (2016) Intravital imaging reveals ghost fibers as architectural units guiding myogenic progenitors during regeneration. Cell Stem Cell. 18(2):243–252
63. Senk A, Djonov V (2021) Collagen fibers provide guidance cues for capillary regrowth during regenerative angiogenesis in zebrafish. Sci Rep. 11(1):19520
64. Anderson SE, Han WM, Srinivasa V, Mohiuddin M, Ruehle MA, Moon JY et al (2019) Determination of a critical size threshold for volumetric muscle loss in the mouse quadriceps. Tissue Eng Part C Methods 25(2):59–70
65. Anderson JE (2022) Key concepts in muscle regeneration: muscle "cellular ecology" integrates a gestalt of cellular cross-talk, motility, and activity to remodel structure and restore function. Eur J Appl Physiol. 122(2):273–300
66. Christov C, Chretien F, Abou-Khalil R, Bassez G, Vallet G, Authier FJ et al (2007) Muscle satellite cells and endothelial cells: close neighbors and privileged partners. Mol Biol Cell. 18(4):1397–1409

67. Latroche C, Weiss-Gayet M, Muller L, Gitiaux C, Leblanc P, Liot S et al (2017) Coupling between myogenesis and angiogenesis during skeletal muscle regeneration is stimulated by restorative macrophages. Stem Cell Reports. 9(6):2018–2033
68. Biferali B, Proietti D, Mozzetta C, Madaro L (2019) Fibro–adipogenic progenitors cross-talk in skeletal muscle: the social network. Front Physiol. 10:1074
69. Chazaud B (2014) Macrophages: supportive cells for tissue repair and regeneration. Immunobiology. 219(3):172–178
70. Uezumi A, Ito T, Morikawa D, Shimizu N, Yoneda T, Segawa M et al (2011) Fibrosis and adipogenesis originate from a common mesenchymal progenitor in skeletal muscle. J Cell Sci. 124(Pt 21):3654–3664
71. Joe AW, Yi L, Natarajan A, Le Grand F, So L, Wang J et al (2010) Muscle injury activates resident fibro/adipogenic progenitors that facilitate myogenesis. Nat Cell Biol. 12(2):153–163
72. Collins BC, Kardon G (2021) It takes all kinds: heterogeneity among satellite cells and fibro-adipogenic progenitors during skeletal muscle regeneration. Development. 148(21):dev199861
73. Hardy D, Fefeu M, Besnard A, Briand D, Gasse P, Arenzana-Seisdedos F, Rocheteau P, Chrétien F (2019) Defective angiogenesis in CXCL12 mutant mice impairs skeletal muscle regeneration. Skelet Mus. https://doi.org/10.1186/s13395-019-0210-5
74. Zammit PS (2017) Function of the myogenic regulatory factors Myf5, MyoD, Myogenin and MRF4 in skeletal muscle, satellite cells and regenerative myogenesis. Semin Cell Dev Biol. 72:19–32
75. Bobadilla M, Sainz N, Abizanda G, Orbe J, Rodriguez JA, Páramo JA, Prósper F, Pérez-Ruiz A (2014) The CXCR4/SDF1 axis improves muscle regeneration through MMP-10 activity. Stem Cells Dev. 23:1417–1427
76. Verma M, Asakura Y, Murakonda BSR, Pengo T, Latroche C, Chazaud B et al (2018) Muscle satellite cell cross-talk with a vascular niche maintains quiescence via VEGF and notch signaling. Cell Stem Cell. 23(4):530–43 e9
77. Schmidt M, Schüler SC, Hüttner SS, von Eyss B, von Maltzahn J (2019) Adult stem cells at work: regenerating skeletal muscle. Cell Mol Life Sci. https://doi.org/10.1007/s00018-019-03093-6
78. Seale P, Sabourin LA, Girgis-Gabardo A, Mansouri A, Gruss P, Rudnicki MA (2000) Pax7 is required for the specification of myogenic satellite cells. Cell 102:777–786
79. Lepper C, Conway SJ, Fan CM (2009) Adult satellite cells and embryonic muscle progenitors have distinct genetic requirements. Nature 460:627–631
80. von Maltzahn J, Jones AE, Parks RJ, Rudnicki MA (2013) Pax7 is critical for the normal function of satellite cells in adult skeletal muscle. Proc Natl Acad Sci U S A. 110:16474–16479
81. Fry CS, Lee JD, Mula J, Kirby TJ, Jackson JR, Liu F, Yang L, Mendias CL, Dupont-Versteegden EE, McCarthy JJ, Peterson CA (2015) Inducible depletion of satellite cells in adult, sedentary mice impairs muscle regenerative capacity without affecting sarcopenia. Nat Med 21:76–80
82. McCarthy JJ, Mula J, Miyazaki M, Erfani R, Garrison K, Farooqui AB, Srikuea R, Lawson BA, Grimes B, Keller C, Van Zant G, Campbell KS, Esser KA, Dupont-Versteegden EE, Peterson CA (2011) Effective fiber hypertrophy in satellite cell-depleted skeletal muscle. Development 138:3657–3666
83. Chakkalakal J, Brack A (2012) Extrinsic regulation of satellite cell function and muscle regeneration capacity during aging. J Stem Cell Res Ther Suppl 11:001
84. Sousa-Victor P, Gutarra S, Garcia-Prat L, Rodriguez-Ubreva J, Ortet L, Ruiz-Bonilla V, Jardi M, Ballestar E, Gonzalez S, Serrano AL, Perdiguero E, Munoz-Canoves P (2014) Geriatric muscle stem cells switch reversible quiescence into senescence. Nature. 506:316–321
85. Tierney MT, Aydogdu T, Sala D, Malecova B, Gatto S, Puri PL, Latella L, Sacco A (2014) STAT3 signaling controls satellite cell expansion and skeletal muscle repair. Nat Med 20:1182–1186
86. Chakkalakal JV, Jones KM, Basson MA, Brack AS (2012) The aged niche disrupts muscle stem cell quiescence. Nature. 490:355–360

87. Schworer S, Becker F, Feller C, Baig AH, Kober U, Henze H, Kraus JM, Xin B, Lechel A, Lipka DB, Varghese CS, Schmidt M, Rohs R, Aebersold R, Medina KL, Kestler HA, Neri F, von Maltzahn J, Tumpel S, Rudolph KL (2016) Epigenetic stress responses induce muscle stem-cell ageing by Hoxa9 developmental signals. Nature. 540:428–432
88. Ahrens HE, Huettemeister J, Schmidt M, Kaether C, von Maltzahn J (2018) Klotho expression is a prerequisite for proper muscle stem cell function and regeneration of skeletal muscle. Skelet Muscle 8:20
89. Rozo M, Li L, Fan CM (2016) Targeting beta1-integrin signaling enhances regeneration in aged and dystrophic muscle in mice. Nat Med 22:889–896
90. Bentzinger CF, Wang YX, Dumont NA, Rudnicki MA (2013) Cellular dynamics in the muscle satellite cell niche. EMBO Rep. 14:1062–1072
91. Brack AS, Conboy IM, Conboy MJ, Shen J, Rando TA (2008) A temporal switch from notch to Wnt signaling in muscle stem cells is necessary for normal adult myogenesis. Cell Stem Cell. 2:50–59
92. Gunther S, Kim J, Kostin S, Lepper C, Fan CM, Braun T (2013) Myf5-positive satellite cells contribute to Pax7-dependent long-term maintenance of adult muscle stem cells. Cell Stem Cell 13:590–601
93. Relaix F, Rocancourt D, Mansouri A, Buckingham M (2005) A Pax3/Pax7-dependent population of skeletal muscle progenitor cells. Nature 435:948–953
94. Kuang S, Kuroda K, Le Grand F, Rudnicki MA (2007) Asymmetric self-renewal and commitment of satellite stem cells in muscle. Cell 129:999–1010
95. Wen Y, Bi P, Liu W, Asakura A, Keller C, Kuang S (2012) Constitutive Notch activation upregulates Pax7 and promotes the self-renewal of skeletal muscle satellite cells. Mol Cell Biol 32:2300–2311
96. Shinin V, Gayraud-Morel B, Tajbakhsh S (2009) Template DNA-strand co-segregation and asymmetric cell division in skeletal muscle stem cells. Methods Mol Biol 482:295–317
97. Conboy L, Seymour CM, Monopoli MP, O'Sullivan NC, Murphy KJ, Regan CM (2007) Notch signalling becomes transiently attenuated during long-term memory consolidation in adult Wistar rats. Neurobiol Learn Mem 88:342–351
98. Conboy IM, Rando TA (2002) The regulation of Notch signaling controls satellite cell activation and cell fate determination in postnatal myogenesis. Dev Cell. 3(3):397–409
99. Sahu A, Mamiya H, Shinde SN, Cheikhi A, Winter LL, Vo NV, Stolz D, Roginskaya V, Tang WY, St Croix C, Sanders LH, Franti M, Van Houten B, Rando TA, Barchowsky A, Ambrosio F (2018) Age-related declines in alpha-Klotho drive progenitor cell mitochondrial dysfunction and impaired muscle regeneration. Nat Commun 9:4859
100. Girardi F, Le Grand F (2018) Wnt signaling in skeletal muscle development and regeneration. Prog Mol Biol Transl Sci 153:157–179
101. Lacour F, Vezin E, Bentzinger CF, Sincennes MC, Giordani L, Ferry A, Mitchell R, Patel K, Rudnicki MA, Chaboissier MC, Chassot AA, Le Grand F (2017) R-spondin1 controls muscle cell fusion through dual regulation of antagonistic Wnt signaling pathways. Cell Rep 18:2320–2330
102. Huels DJ, Sansom OJ (2017) R-spondin is more than just Wnt's sidekick. Dev Cell 41:456–458
103. Rudolf A, Schirwis E, Giordani L, Parisi A, Lepper C, Taketo MM, Le Grand F (2016) Beta-catenin activation in muscle progenitor cells regulates tissue repair. Cell Rep 15:1277–1290
104. Mourikis P, Sambasivan R, Castel D, Rocheteau P, Bizzarro V, Tajbakhsh S (2012) A critical requirement for notch signaling in maintenance of the quiescent skeletal muscle stem cell state. Stem Cells 30:243–252
105. Pisconti A, Cornelison DD, Olguin HC, Antwine TL, Olwin BB (2010) Syndecan-3 and notch cooperate in regulating adult myogenesis. J Cell Biol 190:427–441
106. Bjornson CR, Cheung TH, Liu L, Tripathi PV, Steeper KM, Rando TA (2012) Notch signaling is necessary to maintain quiescence in adult muscle stem cells. Stem Cells 30:232–242

107. Verma M, Asakura Y, Murakonda BSR, Pengo T, Latroche C, Chazaud B, McLoon LK, Asakura A (2018) Muscle satellite cell cross-talk with a vascular niche maintains quiescence via VEGF and notch signaling. Cell Stem Cell 23(530–543):e9
108. Goel AJ, Rieder MK, Arnold HH, Radice GL, Krauss RS (2017) Niche cadherins control the quiescence-to-activation transition in muscle stem cells. Cell Rep 21:2236–2250
109. Mok GF, Lozano-Velasco E, Munsterberg A (2017) microRNAs in skeletal muscle development. Semin Cell Dev Biol 72:67–76
110. Wang J, Yang LZ, Zhang JS, Gong JX, Wang YH, Zhang CL, Chen H, Fang XT (2018) Effects of microRNAs on skeletal muscle development. Gene 668:107–113
111. Cheung TH, Quach NL, Charville GW, Liu L, Park L, Edalati A, Yoo B, Hoang P, Rando TA (2012) Maintenance of muscle stem-cell quiescence by microRNA-489. Nature 482:524–528
112. Ge Y, Sun Y, Chen J (2011) IGF-II is regulated by microRNA-125b in skeletal myogenesis. J Cell Biol 192:69–81
113. Cardinali B, Castellani L, Fasanaro P, Basso A, Alema S, Martelli F, Falcone G (2009) Microrna-221 and microrna-222 modulate differentiation and maturation of skeletal muscle cells. PLoS One 4:e7607
114. Feige P, Brun CE, Ritso M, Rudnicki MA (2018) Orienting muscle stem cells for regeneration in homeostasis. Aging Dis Cell Stem Cell 23:653–664
115. Mashinchian O, Pisconti A, Le Moal E, Bentzinger CF (2018) The muscle stem cell Niche in health and disease. Curr Top Dev Biol 126:23–65
116. Burridge K, Chrzanowska-Wodnicka M (1996) Focal adhesions, contractility, and signaling. Ann Rev Cell Dev Biol 12:463–519
117. Palecek SP, Loftus JC, Ginsberg MH, Lauffenburger DA, Horwitz AF (1997) Integrin-ligand binding properties govern cell migration speed through cell-substratum adhesiveness. Nature 385:537
118. Wu C, Keivenst VM, O'Toole TE, McDonald JA, Ginsberg MH (1995) Integrin activation and cytoskeletal interaction are essential for the assembly of a fibronectin matrix. Cell 83:715–724
119. Quach NL, Biressi S, Reichardt LF, Keller C, Rando TA (2009) Focal adhesion kinase signaling regulates the expression of caveolin 3 and β1 integrin, genes essential for normal myoblast fusion. Mol Biol Cell 20:3422–3435
120. Lee H-S, Bellin RM, Walker DL, Patel B, Powers P, Liu H, Garcia-Alvarez B, de Pereda JM, Liddington RC, Volkmann N, Hanein D, Critchley DR, Robson RM (2004) Characterization of an actin-binding site within the Talin FERM domain. J Mol Biol 343:771–784
121. Tadokoro S, Shattil SJ, Eto K, Tai V, Liddington RC, de Pereda JM, Ginsberg MH, Calderwood DA (2003) Talin binding to integrin β tails: a final common step in integrin activation. Science 302:103
122. Tidball JG (1995) Inflammatory cell response to acute muscle injury. Med Sci Sports Exerc 27:1022–1032
123. Lewis FC, Cottle BJ, Shone V, Marazzi G, Sassoon D, Tseng CCS, Dankers PYW, Chamuleau SAJ, Nadal-Ginard B, Ellison-Hughes GM (2017) Transplantation of allogeneic PW1(pos)/Pax7(neg) interstitial cells enhance endogenous repair of injured porcine skeletal muscle. JACC Basic Transl Sci 2:717–736
124. Xiao W, Liu Y, Chen P (2016) Macrophage depletion impairs skeletal muscle regeneration: the roles of pro-fibrotic factors, inflammation, and oxidative stress. Inflammation. https://doi.org/10.1007/s10753-016-0438-8
125. Nozaki M, Li Y, Zhu J, Ambrosio F, Uehara K, Fu FH, Huard J (2008) Improved muscle healing after contusion injury by the inhibitory effect of suramin on myostatin, a negative regulator of muscle growth. Am J Sports Med 36(12):2354–2362
126. Tidball JG, Villalta SA (2010) Regulatory interactions between muscle and the immune system during muscle regeneration. Am J Physiol Regul Integr Comp Physiol. 298(5):R1173–R1187
127. Summan M, Warren GL, Mercer RR, Chapman R, Hulderman T, Van Rooijen N, Simeonova PP (2006) Macrophages and skeletal muscle regeneration: a clodronate-containing liposome

depletion study. American Journal of Physiology. Regulatory, Integrative and Comparative Physiology 290(6):R1488–R1495

128. Fischer KM, Scott TE, Browe DP, McGaughey TA, Wood C, Wolyniak MJ, Freeman JW (2020) Hydrogels for skeletal muscle regeneration. Regener Eng Trans Med. https://doi.org/10.1007/s40883-019-00146-x
129. Scott JB, Ward CL, Corona BT, Deschenes MR, Harrison BS, Saul JM et al (2017) Achieving acetylcholine receptor clustering in tissue-engineered skeletal muscle constructs in vitro through a materials-directed Agrin delivery approach. Front Pharmacol.:7. https://doi.org/10.3389/fphar.2016.00508
130. Baker HB, Passipieri JA, Siriwardane M, Ellenburg MD, Vadhavkar M, Bergman CR et al (2017) Cell and growth factor-loaded keratin hydrogels for treatment of volumetric muscle loss in a mouse model. Tissue Eng A 23(11–12):572. https://doi.org/10.1089/ten.tea.2016.0457
131. Pollot BE, Rathbone CR, Wenke JC, Guda T (2018) Natural polymeric hydrogel evaluation for skeletal muscle tissue engineering. J Biomed Mater Res B Appl Biomater. 106(2):672–679. https://doi.org/10.1002/jbm.b.33859
132. Kim MJ, Shin YC, Lee JH, Jun SW, Kim C-S, Lee Y et al (2016) Multiphoton imaging of myogenic differentiation in gelatin-based hydrogels as tissue engineering scaffolds. Biomater Res. 20:2. https://doi.org/10.1186/s40824-016-0050-x
133. Agrawal G, Aung A, Varghese S (2017) Skeletal muscle-on-a-chip: an in vitro model to evaluate tissue formation and injury. Lab Chip. 17(20):3447–3461. https://doi.org/10.1039/c7lc00512a
134. Mulyasasmita W, Cai L, Dewi RE, Jha A, Ullmann SD, Luong RH et al (2014) Avidity-controlled hydrogels for injectable co-delivery of induced pluripotent stem cell-derived endothelial cells and growth factors. J Contr Rel 191:71–81. https://doi.org/10.1016/j.jconrel.2014.05.015
135. Costantini M, Testa S, Mozetic P, Barbetta A, Fuoco C, Fornetti E et al (2017) Microfluidic-enhanced 3D bioprinting of aligned myoblast-laden hydrogels leads to functionally organized myofibers in vitro and in vivo. Biomaterials. 131:98–110. https://doi.org/10.1016/j.biomaterials.2017.03.026
136. Mozetic P, Giannitelli SM, Gori M, Trombetta M, Rainer A (2017) Engineering muscle cell alignment through 3D bioprinting. J Biomed Mater Res A. 105(9):2582–2588. https://doi.org/10.1002/jbm.a.36117
137. de Brito A, Alves AN, Ribeiro BG, Barbosa DVDE, Magalhaes EMR, Fernandes KPS, Bussadori SK, Goulardins JB, Mesquita-Ferrari RA (2017) Effect of photobiomodulation on connective tissue remodeling and regeneration of skeletal muscle in elderly rats. Lasers Med Sci. https://doi.org/10.1007/s10103-017-2392-6
138. Sobrian SK, Walters E (2014) Enhanced satellite cell activity in aging skeletal muscle after manual acupuncture-induced injury. Chin Med 5:22–33
139. Dourado DM, Favero S, Baranauskas V, da Cruz-Hofling MA (2003) Effects of the Ga-As laser irradiation on myonecrosis caused by Bothrops moojeni snake venom. Lasers Surg Med 33(5):352–357
140. Barbosa AM, Villaverde AB, Guimaraes-Souza L, Ribeiro W, Cogo JC, Zamuner SR (2008) Effect of low-level laser therapy in the inflammatory response induced by Bothrops jararacussu snake venom. Toxicon 51(7):1236–1244
141. Barbosa AM, Villaverde AB, Sousa LG, Munin E, Fernandez CM, Cogo JC, Zamuner SR (2009) Effect of low-level laser therapy in the myonecrosis induced by Bothrops jararacussu snake venom. Photomed Laser Surg 27(4):591–597
142. Mesquita-Ferrari RA, Martins MD, Silva JA Jr, da Silva TD, Piovesan RF, Pavesi VC, Bussadori SK, Fernandes KP (2011) Effects of low-level laser therapy on expression of TNF-α and TGF-β in skeletal muscle during the repair process. Lasers Med Sci 26(3):335–340
143. De Souza TO, Mesquita DA, Ferrari RA, Dos Santos PD, Correa L, Bussadori SK, Fernandes KP, Martins MD (2011) Phototherapy with low-level laser affects the remodeling of types I and III collagen in skeletal muscle repair. Lasers Med Sci 26(6):803–814

144. Baptista J, Martins MD, Pavesi VC, Bussadori SK, Fernandes KP, Pinto Junior D dos S, Ferrari RA (2011) Influence of laser photobiomodulation on collagen IV during skeletal muscle tissue remodeling after injury in rats. Photomed Laser Surg 29(1):11–17
145. Fernandes KP, Alves AN, Nunes FD, Souza NH, Silva JA Jr, Bussadori SK, Ferrari RA (2013) Effect of photobiomodulation on expression of IL-1β in skeletal muscle following acute injury. Lasers Med Sci 28(3):1043–1046
146. Alves AN, Fernandes KP, Deana AM, Bussadori SK, Mesquita-Ferrari RA (2014) Effects of low-level laser therapy on skeletal muscle repair: a systematic review. Am J Phys Med Rehabil 93(12):1073–1085
147. Vatansever F, Rodrigues NC, Assis LL, Peviani SS, Durigan JL, Moreira FM, Hamblin MR, Parizotto NA (2012) Low intensity laser therapy accelerates muscle regeneration in aged rats. Photonics Lasers Med 1(4):287–297
148. Pertille A, Macedo AB, Oliveira CP (2012) Evaluation of muscle regeneration in aged animals after treatment with low-level laser therapy. Rev Bras Fisioter 16(6):495–501
149. Rodrigues NC, Brunelli R, Abreu DC, Fernandes K, Parizotto NA, Renno AC (2014) Morphological aspects and Cox-2 expression after exposure to 780-nm laser therapy in injured skeletal muscle: an in vivo study. Braz J PhysTher 18(5):395–401
150. Pires D, Xavier M, Araújo T, Silva JA Jr, Aimbire F, Albertini R (2011) Low-level laser therapy (LLLT; 780 nm) acts differently on mRNA expression of anti- and pro-inflammatory mediators in an experimental model of collagenase-induced tendinitis in rat. Lasers Med Sci 26:85–94
151. Alves AN, Fernandes KP, Melo CA, Yamaguchi RY, França CM, Teixeira DF, Bussadori SK, Nunes FD, Mesquita-Ferrari RA (2013) Modulating effect of low level-laser therapy on fibrosis in the repair process of the tibialis anterior muscle in rats. Lasers Med Sci 29(2):813–821
152. Qiao Y, Cong M, Li J, Li H, Li Z (2018) The effects of neuregulin-1β on intrafusal muscle fiber formation in neuromuscular coculture of dorsal root ganglion explants and skeletal muscle cells. Skelet Mus. https://doi.org/10.1186/s13395-018-0175-9
153. Liu H, Lu J, He Y, Yuan B, Li Y, Li X (2014) Insulin-like growth factor-1 prevents dorsal root ganglion neuronal tyrosine kinase receptor expression alterations induced by dideoxycytidine in vitro. Cell Mol Neurobiol. 34:183–194
154. Ogihara Y, Masuda T, Ozaki S, Yoshikawa M, Shiga T (2016) Runx3-regulated expression of two Ntrk3 transcript variants in dorsal root ganglion neurons. Dev Neurobiol. 76:313–322
155. Herndon CA, Ankenbruck N, Fromm L (2014) The Erk MAP kinase pathway is activated at muscle spindles and is required for induction of the muscle spindle-specific gene Egr3 by neuregulin1. J Neurosci Res. 92:174–184
156. Herzog W (2017) Skeletal muscle mechanics: questions, problems and possible solutions. J Neuro Eng Rehab. https://doi.org/10.1186/s12984-017-0310-6
157. Vracko R, Benditt EP (1972) Basal lamina: the scaffold for orderly cell replacement. Observations on regeneration of injured skeletal muscle fibers and capillaries. J Cell Biol. 55(2):406–419
158. Rueda P, Richart A, Récalde A, Gasse P, Vilar J, Guérin C, Lortat-Jacob H, Vieira P, Baleux FC, Chretien F et al (2012) Homeostatic and tissue reparation defaults in mice carrying selective genetic invalidation of CXCL12/proteoglycan interactions. Circulation. 126:1882–1895

The Effects of Exercise on Skeletal Muscle

8

Christopher Myers

8.1 Introduction by the Editor

This chapter investigates the profound effects of exercise on skeletal muscle. In its many forms, exercise plays an indisputable role in influencing skeletal muscle's structure, function, and metabolic properties. This chapter unravels the multifaceted effects of exercise on the muscle and how regular physical activity can optimize muscle health and systemic metabolism.

Exercise improves skeletal structure and function in aged and diseased states. This chapter explores how exercise affects heart failure patients' skeletal muscle abnormalities and central nervous system function. Additionally, this chapter discusses how exercise and age affect vasocontraction in resistance vessels.

In addition, we examine how exercise regimens can be customized to achieve specific outcomes. For instance, a program with an emphasis on resistance exercise would be more suited to individuals seeking to build muscle mass before surgery to quicken recovery. A targeted mix-modality training regime can target gene expression, HSPs, and anti-inflammatory responses.

The final section of this chapter addresses the role of exercise on nutrition and insulin sensitivity.

Exercise has myriad effects on skeletal muscle and overall health. By examining the impacts on gene expression, recovery, and nutritional effects, this chapter underscores the importance of exercise as a powerful tool for promoting muscle health and systemic metabolic wellness.

The profound effects of exercise on skeletal muscle encompass a broad spectrum of physiological adaptations critical for overall health and well-being. Regular

C. Myers (✉)
3 Sports Science, 3 Sports Science LLC, Akron, OH, United States of America
e-mail: chris@3sportsscience.com

C. Myers (ed.), *Skeletal Muscle Physiology*,
https://doi.org/10.1007/978-3-031-47065-3_8

physical activity triggers a cascade of molecular and cellular responses within skeletal muscle fibers, enhancing strength, endurance, and metabolic efficiency. As discussed in various chapters of this book, these adaptations are mediated through intricate signaling pathways that promote protein synthesis, mitochondrial biogenesis, and the remodeling of muscle architecture. Moreover, exercise has been shown to play a pivotal role in lessening the decline in muscle function associated with aging and chronic diseases by counteracting muscle atrophy and fostering resilience against metabolic disorders. This dynamic interplay between exercise and skeletal muscle underscores the importance of physical activity as a cornerstone of preventive healthcare, with far-reaching implications for improving quality of life and extending health span. This chapter presents recent research into the effects of exercise on skeletal muscle.

A critical study explored in this chapter investigated the impact of a 5-week resistance training regimen on Hsp70 and inflammatory cytokines in adipose tissue and skeletal muscles of diabetic and healthy rats. Researchers applied a ladder-climbing exercise protocol targeting the flexor hallucis longus (FHL) and soleus muscles. Findings revealed that resistance training reduced Hsp70 protein levels in the FHL muscle across diabetic and healthy rats without affecting the soleus muscle. Additionally, a significant reduction in inflammatory cytokines within the fast skeletal muscle of diabetic rats occurred alongside an increase in these markers within adipose tissue, which also saw an increase in Hsp70. These changes were positively correlated, suggesting a link between inflammatory cytokines and Hsp70 protein levels in muscle and adipose tissue. Importantly, resistance training preserved muscle mass in diabetic rats, pointing towards its efficacy in maintaining skeletal muscle health amid diabetes by modulating inflammatory and stress-related factors. The alterations in Hsp70 protein expression and inflammatory markers between skeletal muscle and adipose tissue highlight the differential impacts of resistance training. This study underscores the potential of resistance training in diabetes management through its influence on muscle and adipose tissue at the molecular level, advocating for further investigation into the underlying mechanisms.

The progressive loss of skeletal muscle mass, strength, and function with aging, a condition known as sarcopenia, presents a significant challenge to health systems worldwide. This decline in muscle health not only impairs the quality of life among the elderly but also predisposes individuals to a range of metabolic disorders, including obesity and type 2 diabetes (T2D). Intriguingly, the skeletal muscle is not merely a bystander in aging and metabolic regulation but plays a central role in maintaining systemic metabolic homeostasis. It serves as the primary site for glucose disposal. It is intricately involved in the body's overall metabolic regulation, highlighting the critical importance of preserving muscle health to combat metabolic diseases.

A study discussed in this chapter examined the effects of short-term endurance and strength exercise on Slc2a4+/− mice, a model characterized by hyperglycemia, hyperinsulinemia, and glucose intolerance, to assess the molecular regulation of skeletal muscle. Results highlighted improvements in skeletal muscle glucose uptake and metabolism, which were linked to significant transcriptomic changes, particularly with endurance training, mitochondrial activity enhancements, and markers of unfolded protein response (UPRmt). Experiments further suggested that insulin or glucose levels could influence these mitochondrial adaptations in skeletal muscle. While endurance exercise notably impacted transcriptome and mitochondrial activity, strength exercise predominantly influenced post-translational mechanisms and protein synthesis. Both exercise modalities effectively mitigate hyperglycemia and hyperinsulinemia's impacts in this model. Key findings included the elevation of mitochondrial and UPRmt markers in endurance-exercised mice and the identification of Sirtuin 3 as a significant gene affected by strength training, known for its role in muscle insulin signaling and fatty acid oxidation. These insights underscore the importance of exercise in managing conditions preceding obesity and type 2 diabetes, advocating for a tailored approach to physical exercise prescriptions in future healthcare strategies.

Additionally, recent research has illuminated the profound changes in skeletal muscle with aging, revealing extensive transcriptional transformations that impact muscle structure, function, and metabolic processes. Studies utilizing animal models, such as the Slc2a4+/− mice, have provided valuable insights into the metabolic challenges associated with hyperinsulinemia and hyperglycemia, conditions that precede the onset of T2D and obesity. These models have emphasized the significance of understanding the molecular underpinnings of muscle adaptation to physical activity and dietary interventions.

Physical exercise emerges as a potent modulator of muscle health, with endurance and strength training exercises showing distinct benefits for skeletal muscle. Endurance training, for instance, has been linked to improvements in mitochondrial activity and the unfolded protein response in muscle, suggesting a role in enhancing muscle metabolic function and resilience. On the other hand, strength training exercises have been associated with post-translational mechanisms and protein synthesis, underscoring the complexity of exercise-induced adaptations in muscle tissue. Moreover, interventions such as very-low-energy diets (VLEDs) and low-energy diets (LEDs) combined with exercise have been explored. Still, they are very inconsistent in their potential to mitigate sarcopenia and improve metabolic health, highlighting the interplay between diet, exercise, and muscle physiology.

As we delve deeper into the molecular responses of skeletal muscle to aging, exercise, and dietary interventions, it becomes increasingly clear that a multifaceted approach is necessary to preserve muscle health and combat metabolic diseases. Understanding the specific contributions of different exercise and dietary strategies to muscle adaptation and metabolic regulation will be crucial in developing targeted interventions to enhance muscle health across the lifespan. This comprehensive body of research expands our knowledge of muscle biology. It paves the way for innovative strategies to improve health outcomes in aging populations and those at risk of metabolic disorders.

8.2 Machine Generated Summaries

Disclaimer: The summaries in this chapter were generated from Springer Nature publications using extractive AI auto-summarization: An extraction-based summarizer aims to identify the most important sentences of a text using an algorithm and uses those original sentences to create the auto-summary (unlike generative AI). As the constituted sentences are machine selected, they may not fully reflect the body of the work, so we strongly advise that the original content is read and cited. The auto generated summaries were curated by the editor to meet Springer Nature publication standards. To cite this content, please refer to the original papers.

Machine generated keywords: exercise, training, resistance, exercise training, weight, resistance exercise, insulin, aerobic, aerobic exercise, endurance, physical exercise, participant, patient, glucose, age.

Effects of Training Status on PDH Regulation in Human Skeletal Muscle During Exercise [1]

This is a machine-generated summary of:

Gudiksen, Anders; Bertholdt, Lærke; Stankiewicz, Tomasz; Tybirk, Jonas; Plomgaard, Peter; Bangsbo, Jens; Pilegaard, Henriette: Effects of training status on PDH regulation in human skeletal muscle during exercise [1]

Published in: Pflügers Archiv – European Journal of Physiology (2017)

Link to original: https://doi.org/10.1007/s00424-017-2019-6

If you want to cite the papers, please refer to the original.

For technical reasons we could not place the page where the original quote is coming from.

Abstract-Summary "The purpose of this study was to investigate the impact of training state on post-translational regulation of PDHa activity during submaximal and exhaustive exercise." "Exercising at the same relative intensity led to similar muscle PDH activation in untrained and trained subjects, whereas PDHa activity at exhaustion was higher ($P < 0.05$) in trained than untrained."

"Exercise induced similar PDH dephosphorylation in untrained and trained subjects, while PDH acetylation was increased ($P < 0.05$) only in trained subjects."

"PDHa activity and PDH dephosphorylation were well adjusted to the relative exercise intensity during submaximal exercise."

"Higher PDHa activity in trained than untrained at exhaustion seemed related to differences in glycogen utilization rather than differences in PDH phosphorylation and acetylation state, although site-specific contributions cannot be ruled out."

Introduction

"Several studies have reported increased skeletal muscle PDH-E1α protein content with exercise training in both humans and rodents [2–4]."

"8 weeks of endurance exercise training increased PDK2, but not PDK4 protein levels, while also increasing total PDH and PDK activity in human skeletal muscle [3]."

"The impact of the adaptive changes associated with increased skeletal muscle oxidative capacity on post-translational modifications of PDH during steady state and exhaustive exercise remains to be thoroughly studied in human skeletal muscle."

"Studies have indicated an attenuation of IL-6 release from skeletal muscle with increased training state, which may influence exercise-induced PDH regulation in skeletal muscle [5, 6]."

"The aim of the present study was therefore to investigate the impact of training state on post-translational regulation of PDHa activity in human skeletal muscle during submaximal exercise at the same relative exercise intensity as well as during abrupt changes in intensity and during exhaustive exercise."

Methods

"Citrate synthase (CS) activity and oxidative phosphorylation (OXPHOS) complex protein content were measured in skeletal muscle biopsies as markers of oxidative capacity after the intervention to establish two non-overlapping populations of trained and untrained individuals."

"Muscle glycogen content was determined on 0.8–1.5 mg freeze-dried, dissected muscle tissue as previously described by boiling samples for 2 h in 1 M HCl and determining glycogen as glycosyl units [7]."

"Citrate synthase (CS) activity was determined colorimetrically on the homogenate at baseline and after addition of oxaloacetate and normalized to protein content measured in the homogenate samples using the bicinchoninic acid method (Thermo Fischer Scientific, USA) [8]."

"PDHa activity was determined after homogenizing 10–15 mg wet weight muscle tissue in ice-cold buffer containing sucrose, KCl, MgCl2, EGTA, Tris-HCl, NaF, DCA, and Triton X-100 for ~50 s using a motor-driven glass homogenizer and pestle (Kimble-Kontes, NJ, USA) and immediately snap frozen in liquid nitrogen as previously described [9–11]."

Results

"The muscle G-6-P concentration was higher ($P < 0.05$) at 50% IPPO and 65% IPPO and Exh than at Pre in the untrained subject, and was higher ($P < 0.05$) at 65% IPPO and Exh than before exercise in the trained subjects."

"Muscle PDHa activity was higher ($P < 0.05$) at all time points during exercise than Pre in both untrained and trained subjects."

"PDH^{Ser232} phosphorylation in muscle was lower ($P < 0.05$) at all time points during exercise than at Pre in both untrained and trained subjects."

"Muscle AMPK^{Thr172} phosphorylation was in untrained subjects higher ($P < 0.05$) at all time points during exercise than Pre and in trained subjects higher ($P < 0.05$) at Exh than Pre."

"Muscle STAT3^{Tyr705} phosphorylation in untrained subjects was higher ($P < 0.05$) at 65% IPPO and at Exh than Pre, while there was no effect of exercise in trained subjects."

Discussion

"The main findings of the present study are that exercising at the same relative intensity at steady state and with a short-term increase in intensity elicited similar PDHa activity in skeletal muscle from untrained and trained subjects, whereas PDHa activity was higher in trained than untrained at exhaustion."

"The finding that skeletal muscle PDHa activity was similar in untrained and trained subjects when exercising at 50% IPPO indicates that PDHa activity depends on the relative rather than the absolute exercise intensity."

"The observation that the muscle acetyl-CoA concentration also was similar in trained and untrained subjects at 50% IPPO may suggest a similar flux through PDH and hence equal carbohydrate oxidation, although the trained subjects exercised at a higher absolute intensity."

"The observation that muscle PDHa activity and acetyl-CoA were higher in the trained than untrained at exhaustion is in accordance with the augmented PDH-E1α protein content and hence capacity for PDH activation and indicates elevated carbohydrate oxidation in the trained subjects at exhaustion."

Exercise Training in Heart Failure Patients: Effects on Skeletal Muscle Abnormalities and Sympathetic Nervous Activity—A Literature Review [12]

This is a machine-generated summary of:

Iliopoulos, Fotios; Mazis, Nicolas: Exercise training in heart failure patients: effects on skeletal muscle abnormalities and sympathetic nervous activity—a literature review [12]

Published in: Sport Sciences for Health (2018)

Link to original: https://doi.org/10.1007/s11332-018-0442-5

If you want to cite the papers, please refer to the original.

For technical reasons we could not place the page where the original quote is coming from.

Abstract-Summary "Patients suffering from heart failure exhibit fatigue resulting in reduced exercise tolerance and thus, decreased functional capacity and quality of life." "Evidence supporting that cardiac function is poorly correlated with the exercise capacity, led to investigations into peripheral abnormalities, such as impaired function and oxidative capacity of the skeletal muscle, and increased activation of the sympathetic nervous system."

"There is ample evidence demonstrating that improvements in muscle metabolism and sympathetic nervous activity, which are closely connected with exercise capacity, lead to lower rates of hospitalization and improvements in quality of life."

Introduction

"Although the peak cardiac output is restored with pharmacological agents or even heart transplantation, it is proved that exercise intolerance and fatigue remain in HF patients [13, 14]."

"It should be noticed that skeletal myopathy and impaired SNA are more evident in patients with HF and reduced ejection fraction (systolic HF), while concerning the patients with preserved ejection fraction (diastolic HF) there is evidence for impairments at the microvascular system only."

"There are investigations demonstrating that in HF with preserved ejection fraction, impaired microvascular function leads to reduced vasodilation during exercise [15, 16]."

"The purpose of this literature review is to evaluate the current evidence for all modes of ET (AET, RET, combination) in improving abnormal skeletal muscle features and SNA, as well as, the impact on exercise capacity and quality of life, only in humans with systolic HF."

Methods

"Only studies from 2000 to 2017 that examined adult patients with systolic heart failure were included."

"Search, a complication of words and terms was used, to find the most relevant articles for the review."

"Such key words were: exercise training, heart failure, skeletal muscle, skeletal myopathy, exercise capacity and sympathetic nerve activity."

"Analyzing the articles led to a final total of 30 studies for inclusion in the present review."

"The studies examined the effects of AET, RET or a combination of both, on the skeletal muscle impairments and the sympathetic nervous activity in patients suffering from HF."

Discussion

"The effects of ET on the impaired metabolism of the skeletal muscle in HF patients have been studied extensively."

"An amount of six studies have examined the effects of ET on the aerobic/oxidative capacity of the skeletal muscles in a total of 86 HF patients."

"The literature provides evidence supporting that ET has diverse positive effects on metabolic properties of the skeletal muscle in HF patients, which lead on improved oxidative capacity, exercise tolerance and fatigue, as well."

"Studies applied only RET or AET combined with RET, as many HF patients lack of muscle strength, resulting in inability to perform even daily tasks and thus, the application of RET on exercise programs in HF patients was emphasized [17, 18]."

"A considerable amount of studies has investigated the beneficial effects of ET on inflammatory markers of skeletal muscle in HF patients."

Conclusion

"Studies concerning AET have revealed better outcomes in favor of intrinsic muscle alterations and peripheral factors, whereas, studies regarding RET have demonstrated beneficial effects on skeletal muscle function and strength."

"Since there is a strong correlation between these factors and prognosis of HF, and pharmacological and surgical interventions fail to improve considerably the functional capacity and the clinical characteristics of the disease, ET seems to be a physical intervention of great importance for HF patients."

"Although a small proportion of trials have been conducted examining the effects of ET together with standard medications, further studies evaluating the outcomes of ET in conjunction with remedies will be of great importance to recognize potential therapeutic results of combined physical and pharmacological therapy."

"The effects of RET on metabolic properties of the skeletal muscle in HF should be investigated more in the future."

Effects of Aging and Exercise Training on the Dynamics of Vasoconstriction in Skeletal Muscle Resistance Vessels [19]

This is a machine-generated summary of:

Gittemeier, Elizabeth M.; Ericson, Tyler; Ghosh, Payal; Copp, Steven W.; Opoku-Acheampong, Alexander B.; Behnke, Bradley J.: Effects of aging and exercise training on the dynamics of vasoconstriction in skeletal muscle resistance vessels [19]

Published in: European Journal of Applied Physiology (2017)

Link to original: https://doi.org/10.1007/s00421-017-3541-0

If you want to cite the papers, please refer to the original.

For technical reasons we could not place the page where the original quote is coming from.

Abstract-Summary "It is unknown whether aging or exercise training affect the dynamics of arteriolar vasoconstriction." "We hypothesized that old age will slow, and exercise training will speed, the dynamics of skeletal muscle arteriolar vasoconstriction in resistance vessels of aged rats."

"Old age blunted all kinetic parameters (i.e., time delay, time constant) resulting in vasoconstriction taking ~3 times as long to reach a steady state (SS) versus younger counterparts for NE (aged-sed: 15.6 ± 6.0 versus young-sed: 4.6 ± 0.5 s; $P < 0.05$) with a similar time course to caffeine."

"Exercise training resulted in a similar time to SS between age groups for NE (aged-ET: 6.8 ± 1.6 versus young-ET: 7.0 ± 0.6 s) and caffeine (aged-ET: 7.8 ± 0.6 versus young-ET: 8.6 ± 1.0 s)."

"Further, exercise training speeds the dynamics of constriction to both NE and caffeine with old age."

Introduction

"We assessed the time course of vasoconstriction to norepinephrine, the principal neurotransmitter of the sympathetic nervous system (SNS), in the resistance vasculature of the red portion of the gastrocnemius muscle."

"We hypothesize that the dynamics of vasoconstriction in resistance vessels of the red gastrocnemius muscle will be blunted (i.e., slower) with aging to both norepinephrine and caffeine."

"A secondary purpose of this study was to determine the effects of exercise training on the dynamics of vasoconstriction in the red portion of the gastrocnemius muscle."

"Given the crucial role of the peripheral resistance vessels in the maintenance of vascular resistance, this study will determine if the rate of vasoconstriction in skeletal muscle arterioles is impacted by aging and, if so, whether endurance exercise training can improve such constrictor dynamics."

Methods

"Spontaneous tone was determined according to the following formula: where Db is the initial baseline inner diameter before the addition of a pharmacological agonist (i.e., norepinephrine or caffeine), and Dm is the maximal intraluminal diameter obtained in calcium-free PSS as described below."

"Once spontaneous tone was displayed, the vessels were incubated in calcium-free PSS for 30 min (2 × 15 min washes) followed by exposure to caffeine (20 mM; as previously described [20]) with images recorded as described above for off-line analysis."

"Upon completion of the dynamics study, all vessels were incubated in calcium-free PSS solution for 60 min in the presence of 10^{-4} M sodium nitroprusside to determine maximal passive diameter [21]."

"A two-way ANOVA was used to determine differences in vessel characteristics, vessel dynamics, and muscle citrate synthase activity data between the young and old sedentary and exercise-trained groups."

Discussion

"There are several potential mechanisms that may slow the dynamics of arterial constriction to norepinephrine with old age including reduced responsiveness to α-adrenergic agonism, altered calcium release and handling and/or increased vascular stiffness."

"Although we cannot extrapolate the current findings to muscle comprising of different fiber types, in a previous study we observed similar age-related blunting in the dilation dynamics of the soleus and red portion of the gastrocnemius [22], which provided a mechanism for the slower transition of blood flow [23] and microvascular oxygen pressures in a mixed fiber-type muscle (spinotrapezius) to contractions with aging [24]."

"As we would not expect a differential effect of aging on vascular stiffness on the small arteries of the red portion of the gastrocnemius muscle versus those from surrounding muscles (i.e., soleus and white portion of the gastrocnemius muscle), it is unlikely that an increased stiffness significantly affects contractile dynamics with old age as observed herein [25]."

"In soleus muscle arterioles with old age, exercise training has been demonstrated to attenuate α-adrenergic [26] and enhance myogenic [27] -induced vasoconstriction."

Limitations

"We chose to look at a maximal dose of norepinephrine or caffeine as, from preliminary studies, these doses induce a similar magnitude of vasoconstriction between age groups, similar to that observed in arterioles from other locomotory muscle [25]."

"The aged animals from the current study, regardless of training status, would expectedly have greater whole body adiposity versus conditioned matched younger counterparts, as demonstrated for similar age/exercise-trained groups by Mazzeo and Horvath [28]."

"Further, some studies suggest perivascular adipose tissue can blunt agonist-induced vasoconstriction, which could ultimately slow contractile dynamics [29]."

"We investigated resistance vessels (~200 μm) within the muscle that were naturally devoid of perivascular adipose tissue, making this mechanism unlikely."

"We cannot rule out the effect of circulating adipokines (e.g., leptin, adiponectin) with aging and/or adiposity as contributing to the age and/or exercised-trained effects on contractile dynamics."

Conclusions

"The results of this study clearly demonstrate that the time course of vasoconstriction, and therefore the ability to rapidly increase vascular resistance, is affected by both aging and exercise training."

"Old age significantly slowed the time course of vasoconstriction to both norepinephrine and caffeine, whereas exercise training mitigated the age-related blunted contractile dynamics (i.e., training resulted in a faster vasoconstriction) to both these agonists."

"In the young rats, exercise training slowed the time course of vasoconstriction."

"These results may explain why initial orthostatic intolerance is worse with increasing age and is improved in the elderly with exercise training but is exacerbated in young healthy subjects with exercise training."

Effectiveness of Early Exercise on Reducing Skeletal Muscle Loss During Preoperative Neoadjuvant Chemotherapy for Esophageal Cancer [30]

This is a machine-generated summary of:

Ikeda, Tomohiro; Noma, Kazuhiro; Maeda, Naoaki; Tanabe, Shunsuke; Sakamoto, Yoko; Katayama, Yoshimi; Shirakawa, Yasuhiro; Fujiwara, Toshiyoshi; Senda, Masuo: Effectiveness of early exercise on reducing skeletal muscle loss during preoperative neoadjuvant chemotherapy for esophageal cancer [30]

Published in: Surgery Today (2022)

Link to original: https://doi.org/10.1007/s00595-021-02449-5

If you want to cite the papers, please refer to the original.

For technical reasons we could not place the page where the original quote is coming from.

Abstract-Summary "To investigate if early exercise can help prevent skeletal muscle loss and improve the clinical outcomes of esophageal cancer patients receiving preoperative neoadjuvant chemotherapy (NAC)." "We analyzed the effect of early exercise on the risk of skeletal muscle loss (defined as > 2.98%) during NAC and the subsequent clinical outcomes."

"Skeletal muscle loss occurred in 34% and 67% of the early and late exercise groups, respectively ($p < 0.001$)."

"Multivariate analysis revealed that early exercise reduced the risk of skeletal muscle loss (OR = 0.25, 95% CI 0.09–0.65, $p = 0.006$)."

"Our results suggest that early exercise reduces the risk of both skeletal muscle loss during NAC and subsequent surgical site infection in patients with esophageal cancer."

Introduction

"Skeletal muscle loss is a common adverse event during NAC and found by several studies to be associated with poor prognosis [31–33]."

"Skeletal muscle loss in patients with cancer is associated not only with prognosis, but also with decreased physical function [34] and healthcare resource burden [35]."

"We hypothesized that early exercise would improve the outcomes of patients with esophageal cancer during NAC; however, there are few reports on the prevention of skeletal muscle loss during NAC in patients with esophageal cancer."

"The present study aimed to examine the effect of early exercise on skeletal muscle loss during NAC in patients with esophageal cancer."

Methods

"The exclusion criteria were as follows: missing computed tomography (CT) imaging data either before or after NAC, missing data about exercise during NAC, a diagnosis of dementia, treatment other than DCF for the tumor, surgical feasibility unknown before NAC, and stage T4b esophageal cancer."

"We selected early exercise as a factor that may help minimize skeletal muscle loss."

"We used three models, the Crude Model, the Adjusted Model 1, and the Adjusted Model 2, to analyze the effect of early exercise on the risk of skeletal muscle loss (> 2.98%)."

"For the Adjusted Model 1, we selected and adjusted confounding factors, namely, age, sex, stage, CCI, PS, malnutrition (baseline), diarrhea, FN, and RECIST, from the baseline data."

"For the Adjusted Model 2, we selected and adjusted the confounding factors; namely, age, sex, stage, CCI, PS, malnutrition (baseline), malnutrition (week 3), diarrhea, FN, and RECIST, from the baseline data and clinical characteristics."

Results

"The early exercise group comprised 71 (65%) patients (median age = 65.4 years, mean BMI = 21.1 kg/m^2) who started participating in exercise therapy 1 day before Day 1 of NAC."

"The patients in the late exercise group started their exercise therapy 7 days after Day 1 of NAC."

"The proportion of patients with malnutrition in week 3 in the early and late exercise groups was 42% and 28%, respectively."

"Anorexia was observed in 46% of the early exercise group patients vs. 72% of the late exercise group patients, with skeletal muscle loss observed in 34% vs. 67% of patients in the early and late exercise groups, respectively."

"The skeletal muscle loss rate was significantly lower in the early exercise group than that in the late exercise group."

Discussion

"The results of this study, which examined the association between early exercise during NAC and skeletal muscle loss in patients with esophageal cancer, suggest that early exercise helps maintain skeletal muscle mass."

"Although the results of our study were contradictory to those reported by Xu and others [36], our exercise program, which included resistance training with physical activity (walking), may have been effective in preventing skeletal muscle loss."

"Early exercise during NAC may have helped prevent skeletal muscle loss and reduced the rate of wound infection."

"Although the number of days of exercise attendance was significantly higher in the early exercise group, there was no significant association between skeletal muscle loss and exercise attendance between the two groups (Online Resource 3)."

"Further clinical trials are required to establish the effectiveness of exercise therapy on skeletal muscle loss during NAC in patients with esophageal cancer."

Conclusion

"This study investigated the effect of early exercise on skeletal muscle loss during NAC in patients with esophageal cancer and found that early exercise reduced the risk of skeletal muscle loss and SSI."

"Further studies are required to investigate the effects of exercise during NAC on short- and long-term clinical outcomes."

Effects of Acute Aerobic and Concurrent Exercise on Skeletal Muscle Metabolic Enzymes in Untrained Men [37]

This is a machine-generated summary of:

Solfest, Jessica S.; Nie, Yaohui; Weiss, Jessica A.; Garner, Ron T.; Kuang, Shihuan; Stout, Julianne; Gavin, Timothy P.: Effects of acute aerobic and concurrent exercise on skeletal muscle metabolic enzymes in untrained men [37]

Published in: Sport Sciences for Health (2019)

Link to original: https://doi.org/10.1007/s11332-019-00547-z

If you want to cite the papers, please refer to the original.

For technical reasons we could not place the page where the original quote is coming from.

Abstract-Summary "Vastus lateralis biopsies were taken prior to and 1 h post AEx and A+REx for the measurement of phosphofructokinase (PFK), 3-l-hydroxyacyl CoA dehydrogenase (β-HAD), succinate dehydrogenase (SDH) and CS." Introduction

"Moderate intensity, chronic aerobic exercise (AEx) training is well known to increase the maximal activity of several skeletal muscle mitochondrial enzymes including 3-l-hydroxyacyl-CoA dehydrogenase (β-HAD), citrate synthase (CS), and succinate dehydrogenase (SDH) [38]."

"In rodents, acute exercise appears to either decrease (CS, SDH, PFK, β-HAD) or not change (β-HAD) muscle enzyme activities accessed immediately post--exercise [39–42]."

"Whether skeletal muscle enzyme activity is increased in response to more traditional (mode, duration, intensity) acute exercise in humans is unknown."

"The addition of resistance exercise to aerobic exercise appears be additive on the skeletal muscle enzyme responses to training."

"In the current report, we hypothesized that traditional, systemic aerobic exercise increases the activity of skeletal muscle PFK, β-HAD, and SDH as well as CS and the addition of acute resistance exercise enhances the response to aerobic exercise."

Materials and Methods

"The 1-h post-exercise time point was chosen in an effort to replicate the post-exercise biopsy time previously employed when a 50% increase in CS activity was reported [43]."

"Following total protein determination, homogenized samples were frozen and kept at − 80 °C until enzyme analysis."

"Analysis of enzyme activity was performed according to previously reported procedures for PFK [44], β-HAD [45], CS [46], and SDH [46]."

"Because enzyme activity is measured in the presence of excess reagents, increases in activity could be the result of increased protein content or allosteric regulation."

"Exercise training-induced increases in muscle enzyme activity reflect increases in enzyme protein content [38, 47]."

"Western blotting was performed to evaluate if changes in muscle enzyme activity following the acute exercise bout could be attributed to alterations in protein content."

"Student's t test was used to analyze aerobic exercise-induced differences in muscle enzyme activity between LOW and NORM."

Results

"When expressed relative to total tubulin, phosphorylation of ACC tended ($p = 0.06$) to be increased following AEx and was significantly increased by A + REx compared to PRE."

"When expressed relative to total ACC, there was a trend toward an increase with exercise ($p = 0.07$)."

"There was a relationship following AEx between p-ACC and SDH activity ($r = 0.65$; $p = 0.02$)."

Discussion

"In the current report, we investigated if acute, moderate intensity, cycle exercise increases skeletal muscle PFK, β-HAD, CS, and SDH activity and if the addition of fatiguing acute knee extension resistance exercise alters the response in sedentary men."

"Increases in enzyme activity are routinely used to quantify skeletal muscle responses to exercise training [38]."

"The current results suggest there are individuals with exercise-induced increases in muscle enzyme activity at 1 h post-exercise."

"To address this, we attempted to measure phosphorylated Ser/Thr on PFK, β-HAD, CS, and SDH in our samples, hoping to link changes in muscle enzyme activities with exercise-induced allosteric modifications of the enzymes."

"Combined with the lack of increases in muscle enzyme activity, the results suggest AMPK likely is not responsible for exercise-induced PFK, β-HAD, CS, and SDH activity changes."

"Consistent with this, there was no difference in ACC phosphorylation between NORM and LOW suggesting that exercise-induced changes in muscle enzyme activity are not due to activation by AMPK."

Effects of 10 Weeks of Regular Running Exercise with and Without Parallel PDTC Treatment on Expression of Genes Encoding Sarcomere-Associated Proteins in Murine Skeletal Muscle [48]

This is a machine-generated summary of:

Schmitt, Angelika; Haug, Anne-Lena; Schlegel, Franziska; Fragasso, Annunziata; Munz, Barbara: Effects of 10 weeks of regular running exercise with and without parallel PDTC treatment on expression of genes encoding sarcomere-associated proteins in murine skeletal muscle [48]

Published in: Cell Stress and Chaperones (2018)

Link to original: https://doi.org/10.1007/s12192-018-0914-1

If you want to cite the papers, please refer to the original.

For technical reasons we could not place the page where the original quote is coming from.

Abstract-Summary "Physical exercise can induce various adaptation reactions in skeletal muscle tissue, such as sarcomere remodeling." "We studied skeletal muscle adaptation to running exercise in a murine model system, with and without parallel treatment with the NFκB-inhibitory, anti-oxidant and anti-inflammatory drug pyrrolidine dithiocarbamate (PDTC)."

"In control mice, 10 weeks of uphill (15° incline) treadmill running for 60 min thrice a week at a final speed of 14 m/min had differential, but only minor effects on many genes encoding molecular chaperones for sarcomere proteins, and/or factors involved in the degradation of the latter."

"PDTC treatment modulated gene expression patterns as well, both in sedentary and exercising mice; however, most of these effects were also modest and there was little effect of PDTC treatment on exercise-induced changes in gene expression."

"Our data suggest that moderate-intensity treadmill running, with or without parallel PDTC treatment, had little effect on the expression of genes encoding sarcomere components and sarcomere-associated factors in murine skeletal muscle tissue."

Introduction

"A chronic, low-grade inflammatory situation induces the activation of skeletal muscle NFκB, which triggers upregulation of the UPS and other factors involved in the degradation of sarcomeric proteins (for review, see Bowen and others [49])."

"Against this background, it is not surprising that at least acute physical exercise can lead to upregulation of specific components of the protein-degrading machinery, mainly the UPS, the E3 ubiquitin ligases MuRF1 and atrogin-1, and the calpains (for review, see [50, 51])."

"Against this background, it is tempting to speculate whether exercise-induced inflammation and NFκB activation might also be involved in long-term skeletal muscle remodeling, specifically de novo folding of newly synthetized sarcomere proteins and their incorporation into the sarcomeric unit."

"Here, we analyzed skeletal muscle gene expression patterns, specifically genes encoding sarcomere components and sarcomere-associated factors, in response to a 10-week uphill running protocol, with and without parallel treatment with the NFκB inhibitor pyrrolidine dithiocarbamate (PDTC), in mice."

Experimental Procedures

"They were run for 1 h on Mondays, Wednesdays, and Fridays every week, at an initial speed of 2 m/min and 0° incline, with gradually increasing speed and incline (3 m/min and 5° per week), until animals ran at a final speed of 14 m/min and 15° incline from the sixth week onwards."

"Forty-eight hours after the last training session, mice were sacrificed, skeletal muscle tissue was dissected, snap-frozen in liquid nitrogen, and stored at −80 °C or in RNAlater® (Thermo Fisher Scientific, Waltham, MA, USA) for RNA or protein isolation."

"Of the total RNA, 500 ng was used for reverse transcription using M-MLV reverse transcriptase and random primers (Promega, Madison, WI, USA) in a total volume of 20 μL. Briefly, RNA samples were denatured for 5 min at 70 °C, and after adding reverse transcriptase and random primers, the mix was incubated at room temperature for 10 min, followed by 60 min at 50 °C."

Results

"We analyzed expression of metabolism-associated genes in samples derived from two different types of muscles (Q and G) of all four animal groups."

"These data indicate that there were moderate, but no major effects on the expression of fiber type-specific MyH genes in the analyzed muscle tissues."

"We analyzed expression of a broad variety of genes encoding either sarcomere proteins or molecular chaperones assisting in their proper folding, trafficking, and incorporation into sarcomeric structures."

"We determined expression of the genes encoding the key E3 ubiquitin ligases MuRF1 and atrogin-1, both of which have been implicated in the degradation of sarcomere proteins during skeletal muscle atrophy, but also in exercise adaptation."

"We analyzed expression of genes encoding components of the calpain protease system, which is also involved in the degradation of sarcomere proteins."

Discussion

"Despite the known role of the calpain system in exercise-induced skeletal muscle remodeling (for review, see Murphy [51, 52]), we did not find significant changes with regard to the expression of genes encoding different calpains or their regulators."

"The effects of exercise on the expression of genes encoding sarcomere--associated molecular chaperones were also minor, mostly non-significant, variable, and very much dependent on the type of muscle analyzed, indicating that there is no major difference between "steady state" levels of skeletal muscle remodeling between trained and untrained individuals."

"In response to PDTC treatment, there was no consistent pattern of gene regulation, suggesting that basal levels of skeletal muscle remodeling, both in trained and in untrained individuals, are not highly dependent on NFκB activity."

"Whereas PDTC alone exerted persistent and in some cases profound effects on skeletal muscle gene expression patterns, there was little effect of this factor on exercise-induced changes in gene expression."

Effect of Resistance Exercise Training on Expression of Hsp70 and Inflammatory Cytokines in Skeletal Muscle and Adipose Tissue of STZ-Induced Diabetic Rats [53]

This is a machine-generated summary of:

Molanouri Shamsi, M.; Mahdavi, M.; Quinn, L. S.; Gharakhanlou, R.; Isanegad, A.: Effect of resistance exercise training on expression of Hsp70 and inflammatory cytokines in skeletal muscle and adipose tissue of STZ-induced diabetic rats [53]

Published in: Cell Stress and Chaperones (2016)

Link to original: https://doi.org/10.1007/s12192-016-0703-7

If you want to cite the papers, please refer to the original.

For technical reasons we could not place the page where the original quote is coming from.

Abstract-Summary "Resistance training can maintain muscle mass by changing both inflammatory cytokines and stress factors in adipose tissue and skeletal muscle." "The purpose of this study was to determine the effects of a 5-week ladder climbing resistance training program on the expression of Hsp70 and inflammatory cytokines in adipose tissue and fast-twitch flexor hallucis longus (FHL) and slow-twitch soleus muscles in healthy and streptozotocin-induced diabetic rats."

"Hsp70 protein levels were decreased in normal and diabetic rats by resistance training in the FHL, but not soleus muscle."

"Significant positive correlations between inflammatory cytokines in adipose tissue and skeletal muscles with Hsp70 protein levels were observed."

"We found that in diabetic rats, resistance training decreased inflammatory cytokines and Hsp70 protein levels in fast skeletal muscle, increased adipose tissue inflammatory cytokines and Hsp70, and preserved FHL muscle mass."

"These results suggest that resistance training can maintain skeletal muscle mass in diabetes by changing inflammatory cytokines and stress factors such as Hsp70 in skeletal muscle and adipose tissue."

Introduction

"Inflammatory cytokines and stress factors may also induce skeletal muscle atrophy in type 1 diabetes [54]."

"The effects of resistance training on inflammatory factors in adipose tissue and skeletal muscle in relation to Hsp70 levels have not been studied."

"We hypothesized that resistance training would maintain muscle mass and modulate both inflammatory cytokines and stress factors such as Hsp70 in adipose tissue and skeletal muscle of diabetic subjects."

"The protein levels of Hsp70 and the inflammatory cytokines IL-6, TNF-α, and IL-1β in adipose tissue of healthy and streptozotocin (STZ)-induced diabetic rats subjected to resistance exercise training were assessed."

"Possible relationships between inflammatory cytokines and Hsp70 in adipose tissue and skeletal muscle were assessed in the current study."

"Our findings suggest that changes in inflammatory cytokine and stress factor levels in skeletal muscle and adipose tissue are involved in skeletal muscle adaptation following resistance training in diabetes."

Methods

"Rats in the T and DT groups were trained using a ladder-climbing protocol that specifically targets the FHL muscle, with progressively larger weight loads attached to the tail as previously described [55]."

"Rats were familiarized with the exercise for 3 days, 48 h before STZ injection, and exercise training was initiated after STZ injection."

"Animals from the T and DT groups were exercised with five sets of four repetitions each with a 60-s rest interval between the reps and 3 min between the sets per session."

"Maximal loads that rats could bear prior to training were 120 % of body mass in the DT group and 150 % of body mass in the T group."

"TNF-α, IL-6, IL-1β, and Hsp70 protein levels in tissue and serum were determined in duplicate using commercially available rat ELISA kits (DuoSet ELISA, R&D Systems, Minneapolis, MN)."

Results

"Resistance training decreased Hsp70 protein in FHL muscle of diabetic rats about 45 % compared to control diabetic rats."

"The effects of diabetes and resistance training on protein levels of IL-6, TNF-α, and IL-1β in the FHL and soleus muscles were assessed."

"Higher concentrations of IL-6 and IL-1β were observed in diabetic FHL and soleus muscle compared to healthy rats."

"Training decreased muscle IL-6 levels in both the soleus and FHL muscles of diabetic rats (training × diabetes interaction, $P < 0.001$)."

Discussion

"This study was undertaken to test the hypothesis that resistance training would maintain muscle mass, as well as change inflammatory cytokine and Hsp70 levels in adipose tissue and skeletal muscle in diabetic subjects."

"Hsp70 protein expression exhibited divergent changes in skeletal muscles and adipose tissue following resistance training."

"Hsp70 protein expression was decreased in fast-twitch skeletal muscle following resistance training."

"Diabetes and resistance training increased Hsp70 and inflammatory cytokine levels in adipose tissue."

"Resistance training decreased Hsp70 and IL-6 protein levels in fast-twitch skeletal muscle."

"Our results propose that resistance training maintain the skeletal muscle mass in diabetes by changing inflammatory cytokines and stress factors such as Hsp70 in skeletal muscle and adipose tissue."

"Further studies are in order to determine the mechanism controlling Hsp70 and inflammatory cytokines in adipose tissue and skeletal muscle following resistance training."

Resistance Exercise Training-Induced Skeletal Muscle Strength Provides Protective Effects on High-Fat-Diet-Induced Metabolic Stress in Mice [56]

This is a machine-generated summary of:

Kim, Hye Jin; Kim, Youn Ju; Kim, Il Yong; Seong, Je Kyung: Resistance exercise training-induced skeletal muscle strength provides protective effects on high-fat-diet-induced metabolic stress in mice [56]

Published in: Laboratory Animal Research (2022)

Link to original: https://doi.org/10.1186/s42826-022-00145-0

If you want to cite the papers, please refer to the original.

For technical reasons we could not place the page where the original quote is coming from.

Abstract-Summary "Resistance exercise training is known to improve metabolic disorders, such as obesity and type2 diabetes." "We investigated whether the beneficial effects of resistance exercise training persisted even after the discontinuation of training with high-fat diet (HFD)-induced metabolic stress."

"We further evaluated whether the improvement in skeletal muscle strength and endurance by training were correlated with improved metabolism."

"Resistance exercise-trained mice had a lean phenotype and counteracted diet-induced obesity and glucose tolerance, even after exercise cessation."

"These results have strong implications for the preventive effect of resistance exercise-induced metabolic improvement by enhancing skeletal muscle strength rather than endurance."

Background

"Aerobic exercise has been extensively studied as a mechanism for anti-obesity effects, including improved insulin sensitivity."

"Aerobic exercise training improves cardio-respiratory capacity, which mobilizes energy using a lot of oxygen, and is recognized as the most effective type of exercise to improve obesity without increasing skeletal muscle mass and strength (<Author>, <year>, J, N, Engl, Med., 376:<first-page>-<last-page>)."

"It has been extensively researched as a model mechanism for anti-obesity effect, including improved insulin sensitivity by aerobic exercise training."

"According to a recent studies, resistance training is not limited to simply increasing muscle mass and strength but is also an important type of exercise for improving obesity and metabolic diseases via modulation of metabolic-related signaling transduction [57–60]."

"Studies on the physiological and molecular mechanisms of resistance exercise to improve systemic metabolism, including skeletal muscle, liver, and adipose tissue metabolism, are being conducted."

Results

"Gastrocnemius muscle weight was significantly increased in the resistance exercise-trained group."

"These results show that climbing exercise-induced enhancement of skeletal muscle strength and skeletal muscle endurance capacities were not blunted by training cessation with HFD induction."

"These results suggest that resistance exercise training can regulate glucose homeostasis despite training cessation and HFD induction."

"These results indicate that resistance climbing exercise training specifically stimulates muscle fiber-type gene expression changes only in the triceps."

"These results suggest that the metabolic response to resistance exercise training differs depending on the type of muscle and is not related to changes in muscle fiber type."

Discussion

"Resistance exercise enhances neuromuscular adaptations that increase muscle strength or endurance without significantly changing the oxygen availability."

"Our study investigated the preventive effects of resistance exercise on body weight, gonadal fat mass, and glucose tolerance and determined whether the beneficial effects of exercise were associated with skeletal muscle strength and endurance in mice."

"We observed that grip strength and hanging time were significantly increased in resistance exercise-trained mice, despite exposure to HFD for 1 week after exercise discontinuation."

"Although we did not study the molecular mechanisms underlying the preventive effects of resistance training, we analyzed whether resistance training enhanced skeletal muscle function in relation to the body or gonadal fat weight."

"Our data also showed that PGC-1α protein expression and GLUT4 expression increased in gastrocnemius muscle after resistance exercise, despite exposure to HFD for 1 week after exercise was discontinued."

Conclusions

"The mice-climbing resistance exercise improved metabolic parameters and increased muscle strength and endurance."

"Muscle strength had a higher correlation with body fat and glucose tolerance than muscle endurance."

"These effects persisted even under HFD-induced metabolic stress after cessation of exercise."

"Further research is needed to determine whether a continuous high-fat diet after cessation of exercise has a lasting protective effect."

Methods

"After 1 week of acclimatization, the mice were randomly divided into two groups and maintained on a normal chow diet (NCD, NIH-31, Zeigler, PA, USA)."

"Mice that had fasted for 16 h were injected intraperitoneally (IP) with D-glucose (1 g kg^{-1}/body weight) (G8270, Sigma-Aldrich, MO, USA)."

"The slides were washed once with PBS and probed with anti-GLUT4 antibody (PA5-23052, Invitrogen) at a dilution of 1:500 overnight in 3% BSA in PBS at 4 °C."

"The slides were then washed three times for 5 min each in 0.05% Tween 20 in PBS, after which they were incubated with Alexa 488-conjugated goat anti-rabbit IgG secondary antibody (A11008, Invitrogen) diluted 1:200 in PBS containing 3% BSA for 30 min at room temperature."

"The membranes were then washed three times for 5 min each in TBST, after which they were incubated for 1 h with anti-rabbit or mouse IgG horseradish peroxidase-linked secondary antibody (1:5000) (Cell Signaling, USA)."

The Effects of Very Low Energy Diets and Low Energy Diets with Exercise Training on Skeletal Muscle Mass: A Narrative Review [61]

This is a machine-generated summary of:

Ardavani, Arash; Aziz, Hariz; Smith, Ken; Atherton, Philip J.; Phillips, Bethan E.; Idris, Iskandar: The Effects of Very Low Energy Diets and Low Energy Diets with Exercise Training on Skeletal Muscle Mass: A Narrative Review [61]

Published in: Advances in Therapy (2020)

Link to original: https://doi.org/10.1007/s12325-020-01562-0

If you want to cite the papers, please refer to the original.

For technical reasons we could not place the page where the original quote is coming from.

Abstract-Summary "Very-low-energy diets (VLEDs) have been recognised as a viable strategy for improving the extent of weight loss and cardio-metabolic outcomes in people who are either overweight or obese." "Concerns exist regarding the reductions in lean body mass (LBM) during VLEDs, particularly in vulnerable demographic groups, such as middle-aged and older adults already prone to developing sarcopenia."

"This narrative review explores the potential benefits and limitations of exercise and/or protein supplementation for LBM retention during VLEDs based on the available evidence."

"Current studies suggest that both protein supplementation and exercise training may result in improved LBM retention (and skeletal muscle function) during VLEDs."

Digital Features

"This article is published with digital features, including a summary slide, to facilitate understanding of the article."

"To view digital features for this article go to https://doi.org/10.6084/m9.figshare.13187495 ."

Very-Low-Energy Diets

"The observed rate of weight loss in VLEDs and LEDs in multiple studies is up to 2.5 kg (kg) per week during the first 4 to 6 weeks of intervention [62]."

"At least three studies have demonstrated that in insulin-dependent T2DM patients, a reduction in their insulin administration is observed following VLED interventions [63]."

"This study demonstrated that VLED resulted in an observed remission of T2DM (defined as HbA1c < 6.5% after withdrawal of all anti-diabetic medications for at least 2 months) in 46% of participants in the intervention cohort following 1 year of intervention compared with only 4% in the control group (OR 19.7, 95% Confidence Interval (CI) 7.8–49.8) [64]."

"The DiRECT study provided direct evidence of T2DM remission via VLED with an associated loss of weight, although the proportion of patients experiencing these benefits was less following the first year of intervention."

Skeletal Muscle: Maintenance and Metabolic Functions

"To the recognised role of skeletal muscle in facilitating locomotion, posture and strength, it is also implicated in numerous metabolic processes essential for overall health [65–67]."

"As the primary site of both insulin-dependent and -independent glucose disposal, skeletal muscle is especially important in the context of metabolic diseases—in particular, T2DM and obesity [68, 69]."

"Low skeletal muscle mass and subsequent low function (such as grip strength) have been linked to mortality, frailty and increased insulin resistance, with the latter being an important risk factor for adverse cardiovascular outcomes and metabolic dysfunction [69, 70]."

"Despite the crucial roles of skeletal muscle outlined above, some degree of muscle atrophy is inevitable (yet partially modifiable) with advancing age, a phenomenon known as sarcopenia [71, 72]."

"Nutrition is essential for regulating muscle mass and function, whereby mitigating skeletal muscle atrophy whilst on a VLED is particularly important for middle-aged and older adults who are either pre-sarcopenic or are already suffering from clinically-defined sarcopenia."

Impact of Protein on LBM Retention in Vleds

"Although additional protein supplementation has an evidence base that suggests some protective effect for LBM preservation in sarcopenic patients, the utilisation of this strategy should be patient-tailored and past medical history or susceptibility to renal or bone metabolism abnormalities should be considered [73, 74]."

"An increase of dietary protein to between 1.7–2.3 g/kg/day has been shown to result in improvements in LBM retention [75, 76]."

"The current evidence concerning the effects of additional protein supplementation in the context of VLED has to date yielded conflicting results."

"In 2018, Larsen and others conducted an RCT assessing the effects of additional daily protein supplementation (0.4 g/kg/day) with 30 min of aerobic exercise training (AET) five times per week in 29 patients aged between 18 to 55 years who were either overweight or obese over a 4-week duration [77]."

Impact of Exercise on LBM Retention in VLEDS

"Hemysfield and others conducted a 6-week RCT assessing the body composition and metabolic outcomes of 12 women between the ages of 25–45 years with obesity and discovered that the AET group, which consisted of a progressively increasing exercise volume (up to 5.6 km of walking), observed less reduction in LBM relative to the controls (AET group: $-$ 2.2 ± 0.8 kg, control group: 2.6 ± 0.6 kg, $p < 0.001$) [78]."

"The above findings were not however observed in another non-randomised interventional study published in 1988 by Phinney and others This study assessed the extent of weight loss in 12 women with obesity between the ages of 22–39 years assigned to diet-only or diet-plus-AET groups [79]."

Exercise-Nutrition Interactions in LEDS

"It was determined that 30 g protein elicited approximately double the MPS rate response when combined with exercise compared with 15 g (16%) [80]."

"The findings of Areta and colleagues are in accordance with the work of Pasiakos and others (2013), who reported that an increase in protein intake (double [1.6 g/kg/day] or triple [2.4 g/kg/day] the recommended daily allowance [RDA] [0.8 g/kg/day]) in a short-term LED state (−40% maintenance intake) in combination with RET and AET resulted in preservation of LBM [81]."

"Although the literature is suggestive of a LBM retention enhancement with protein nutrition in the context of VLEDs, further studies are required to both replicate these observed outcomes and investigate whether AET and/or HIIT exercise resembles the effect of RET when combined with additional protein supplementation in VLEDs."

Conclusion

"Inconsistent evidence currently exists concerning the capacity for RET to serve as a retainer of LBM when applied concurrently with a VLED, with different studies demonstrating opposing outcomes."

"Current evidence suggests that the addition of AET to a VLED demonstrates either an improvement in LBM retention or no difference compared to VLED-only interventions."

"No conclusive evidence currently exists concerning the interaction between HIIT and VLEDs."

Effects of Short-Term Endurance and Strength Exercise in the Molecular Regulation of Skeletal Muscle in Hyperinsulinemic and Hyperglycemic Slc2a4$^{+/-}$ Mice [82]

This is a machine-generated summary of:

Muñoz, Vitor Rosetto; Botezelli, José Diego; Gaspar, Rafael Calais; da Rocha, Alisson L.; Vieira, Renan Fudoli Lins; Crisol, Barbara Moreira; Braga, Renata Rosseto; Severino, Matheus Brandemarte; Nakandakari, Susana Castelo Branco Ramos; Antunes, Gabriel Calheiros; Brunetto, Sérgio Q.; Ramos, Celso D.; Velloso, Lício Augusto; Simabuco, Fernando Moreira; de Moura, Leandro Pereira; da Silva, Adelino Sanchez Ramos; Ropelle, Eduardo Rochete; Cintra, Dennys Esper; Pauli, José Rodrigo: Effects of short-term endurance and strength exercise in the molecular regulation of skeletal muscle in hyperinsulinemic and hyperglycemic Slc2a4$^{+/-}$ mice [82]

Published in: Cellular and Molecular Life Sciences (2023)

Link to original: https://doi.org/10.1007/s00018-023-04771-2

If you want to cite the papers, please refer to the original.

For technical reasons we could not place the page where the original quote is coming from.

Abstract-Summary "Physical exercise stimulates skeletal muscle glucose uptake, improving whole-body glucose homeostasis." "When Slc2a4$^{+/-}$ mice were submitted to the endurance or strength training protocol, improvements were observed in the skeletal muscle glucose uptake and glucose metabolism, associated with broad transcriptomic modulation, that was, in part, related to mitochondrial adaptations."

"The endurance training, but not the strength protocol, was effective in improving skeletal muscle mitochondrial activity and unfolded protein response markers (UPRmt)."

"Experiments with C2C12 cells indicated that insulin or glucose levels could contribute to these mitochondrial adaptations in skeletal muscle."

"While endurance exercise plays an important role in transcriptome and mitochondrial activity, strength exercise mostly affects post-translational mechanisms and protein synthesis in skeletal muscle."

"The performance of both types of physical exercise proved to be a very effective way to mitigate the impacts of hyperglycemia and hyperinsulinemia in the $Slc2a4^{+/-}$ mouse model."

Introduction

"We proposed to study the initial responses to endurance and strength physical exercise in the heterozygous Slc2a4 mouse line ($Slc2a4^{+/-}$), that harbors GLUT4 content, accompanied by hyperglycemia, hyperinsulinemia, insulin resistance, and glucose intolerance, but without the presence of increased body adiposity."

"The vast majority of studies that have explored the impacts of physical exercise on glucose metabolism have done so in models of diet-induced or genetically modified obesity in which animals show increased adiposity [83, 84]."

"Other studies that investigated the effects of exercise on skeletal muscle metabolism were conducted with lean animals and, therefore, without changes in circulating insulin and glucose levels."

"The findings will contribute to the understanding of the impacts of aerobic and strength exercise on conditions of hyperglycemia and hyperinsulinemia, which are manifestations that may precede the development of obesity and T2D, with both types of exercise having specific responses that contribute to glycemic homeostasis at the beginning of a physical training program."

Methods

"After a 48 h washout, the mice performed 5 days of endurance exercise training (60 min running at 60% of exhaustion velocity) followed by a 16 h rest period to perform the glucose tolerance test (GTT)."

"On the next day, the mice were submitted to another session of endurance exercise training followed by performance of the insulin tolerance test (ITT) 16 h after the last exercise session."

"The animals performed 5 days of strength exercise training followed by a 16 h rest period to complete the GTT."

"On the next day, the mice were submitted to another session of strength exercise training followed by a 16 h rest to perform the ITT."

"On the day of the tissue collection (after 7 days of exercise, 16 h after the final exercise session, and with a 6 h fasting period), before the anesthesia injection, the blood from the tail was collected to measure the fasting glycemia, centrifuged (3000 rpm, 4 °C, for 5 min) to collect the serum, and stored at − 80 °C for future analysis."

Results

"The physiological changes observed in the Slc2a4$^{+/-}$ mice are partly reflected by alterations in the skeletal muscle insulin signaling pathway."

"The adaptations from the short-term exercise training protocol were reflected in improved skeletal muscle glucose uptake."

"As gene ontogeny terms indicated mitochondrial adaptations as important components of the muscle response to PA in Slc2a4$^{+/-}$ mice, we investigated mitochondrial modulations in the skeletal muscle of Slc2a4$^{+/-}$ mice, including mitochondrial unfolded protein response markers (UPRmt)."

"Because our model is hyperglycemic and hyperinsulinemic, we correlated the fasting glucose and insulin levels of the Slc2a4$^{+/-}$ mice with OXPHOS and UPRmt-related genes in skeletal muscle."

Discussion

"The most significant gene improved by the strength training protocol, Sirtuin 3 (Sirt3), is a well-established deacetylase protein, playing a role in skeletal muscle insulin signaling and mitochondrial fatty acid oxidation [85, 86]."

"The molecular functions related to the upregulated genes observed in both exercise training protocols led us to look at mitochondrial and UPRmt markers in the skeletal muscle of sedentary Slc2a4$^{+/-}$ mice and C2C12 cells submitted to different challenges."

"In aged mice presenting mitochondrial defects, 4 weeks of aerobic treadmill exercise provided the following adaptations in skeletal muscle compared to aged control mice: a increased VDAC; b increased mitochondrial-encoded genes (mt-ND1, mt-CytB, and mt-D-loop); c increased mitonuclear imbalance (MTCO1/ATP5a ratio); and d increased UPRmt makers (HSP60, LONP1, and YME1L1) [87]."

"In skeletal muscle, we observed increased mitochondrial and UPRmt markers (Tfam, VDAC, CV-ATP5A, CIII-UQCRC2, HSP60, and YME1L1) in the endurance exercised Slc2a4$^{+/-}$ mice compared to sedentary Slc2a4$^{+/-}$ mice."

Conclusion

"The findings of the present study contribute to the prospect of a personalized and targeted prescription in the future of physical exercise."

"Both short-term exercise models efficiently improved glycemic homeostasis and overcame the metabolic challenges from Slc2a4$^{+/-}$ mice."

"Performing both types of physical exercise (aerobic and resistance) proved to be a very effective way to mitigate the impacts of hyperglycemia and hyperinsulinemia (metabolic alterations of an insidious and elusive nature) that precede the onset of diabetes."

Skeletal Muscle in Aged Mice Reveals Extensive Transformation of Muscle Gene Expression [88]

This is a machine-generated summary of:

Lin, I-Hsuan; Chang, Junn-Liang; Hua, Kate; Huang, Wan-Chen; Hsu, Ming-Ta; Chen, Yi-Fan: Skeletal muscle in aged mice reveals extensive transformation of muscle gene expression [88]

Published in: BMC Genomic Data (2018)
Link to original: https://doi.org/10.1186/s12863-018-0660-5

If you want to cite the papers, please refer to the original.

For technical reasons we could not place the page where the original quote is coming from.

Abstract-Summary "Aging leads to decreased skeletal muscle function in mammals and is associated with a progressive loss of muscle mass, quality and strength." "Age-related muscle loss (sarcopenia) is an important health problem associated with the aged population."

"We investigated the alteration of genome-wide transcription in mouse skeletal muscle tissue (rectus femoris muscle) during aging using a high-throughput sequencing technique."

"Analysis revealed significant transcriptional changes between skeletal muscles of mice at 3 (young group) and 24 (old group) months of age."

Background

"Activation and overexpression of SIRT1 reduces the oxidative stress and inflammation associated with ameliorating diseases, such as vascular endothelial disorders, neurodegenerative diseases, as well as skeletal muscle aging [89–92]."

"Skeletal muscle has been classified into several fiber types depending on the expression of myosin isoforms, twitch duration, shortening velocity, endurance, and metabolic mechanisms used for energy production."

"Sarcopenia, a term for the loss of skeletal muscle mass, strength, and function during the aging process [93], is a severe muscle disorder that impairs the quality of life in human society."

"Aging-induced sarcopenia has been characterized by the preferential loss of fast-twitch muscle fibers [94]."

"Many possible mechanisms have been reported in aged skeletal muscle, such as the increase of apoptosis, decrease in the number of satellite cells and regeneration rate, and dysfunction of autophagic and lysosomal degradation [95–97]."

Methods

"At least four mice from each age group were sacrificed and the rectus femoris muscle was harvested from both legs of each mouse."

"All measurements were done using RNA samples prepared from the rectus femoris muscles from four mice at young (3-month old) and old (24-month old) age."

"The sequencing library for mRNA-seq was prepared using TruSeq RNA Sample Preparation Kit v2 (Illumina) according to manufacturer's instructions."

"The difference in gene expression levels was calculated as the log2 fold changes of genes between young and old skeletal muscle samples."

"In our RNA-seq dataset, there were 114 over-expressed and 634 under-expressed genes in 24-month old mice compared to the 3-month old mice."

"IHC staining was performed using paraffin-embedded skeletal muscle sections (4 μm)."

"We performed ultrastructural observation using 4 mice of each age (3-month and 24-month of age), 5 blocks of each mouse, and 3 grids of each block."

Results

"Genes associated with type I muscle fibers, Tnnt1 (Troponin T1, slow skeletal type) and Atp2a2 (ATPase sarcoplasmic/endoplasmic reticulum Ca2+ transporting 2), were up-regulated in 24-month-old muscle (2.53- and 1.93- fold respectively), indicating a fast-to-slow muscle fiber transition during aging."

"The cardiac-specific Ryr2 gene for mediating the release of sarcoplasmic calcium ion, and Trpc3 (Transient receptor potential cation channel, subfamily C, member 3), involved in forming calcium ion channels in both skeletal and cardiac muscles were also up-regulated in aged skeletal muscles (4.69- and 2.73-fold respectively)."

"The mitochondrial translational activator encoded by Mss51 (Mitochondrial Translational Activator, as known as Zmynd17) is a fast-twitch fiber gene and was strongly up-regulated (4.66-fold) in aged muscle."

"The Lpin1 (Lipin 1) gene, which encodes phosphatidic acid phosphohydrolase was found to be down-regulated (1.53-fold) in 24-month-old muscle."

Discussion

"Comparative transcriptome analysis between young and aged muscles revealed an extensive and complex transcriptional programming of genes associated with muscle structural proteins, NMJ, energy metabolism, stress response, polyamine biosynthesis, and immune functions."

"Several type I muscle fiber genes were found to be up-regulated, and there is a switch to intermediate fast-twitch muscle fiber during aging, which employs both aerobic metabolism/respiration as well as glycolysis as an energy source."

"Several genes related to macrophage, T lymphocytes, B lymphocytes, and complement system were silenced or strongly down-regulated in aged skeletal muscles."

"Several genes, which participate in the pathological mechanisms for muscle disorders, had expression changes in aged skeletal muscle."

"The Nrap (Nebulin-related-anchoring protein) and Xirp1 (Xin actin-binding repeat containing 1) genes, which are expressed in the myotendinous junction (MTJ) of skeletal muscle, were up-regulated (1.69- and 2.53-fold respectively) in aged muscle."

Conclusions

"Notable evidence in worms, fruit flies, mice, and humans suggest that skeletal muscle plays an important role in the systemic regulation of aging."

"Potential biomarkers of aging through the endocrine system will be of clinical relevance and is a critical issue that needs our attention."

Bibliography

1. Gudiksen A, Bertholdt L, Stankiewicz T, Tybirk J, Plomgaard P, Bangsbo J, Pilegaard H (2017) Effects of training status on PDH regulation in human skeletal muscle during exercise. Pflügers Archiv Eur J Physiol. https://doi.org/10.1007/s00424-017-2019-6
2. Bienso RS, Olesen J, Gliemann L, Schmidt JF, Matzen MS, Wojtaszewski JF, Hellsten Y, Pilegaard H (2015) Effects of exercise training on regulation of skeletal muscle glucose metabolism in elderly men. J Gerontol A Biol Sci Med Sci 70:866–872
3. LeBlanc PJ, Peters SJ, Tunstall RJ, Cameron-Smith D, Heigenhauser GJ (2004) Effects of aerobic training on pyruvate dehydrogenase and pyruvate dehydrogenase kinase in human skeletal muscle. J Physiol 557:559–570
4. Ringholm S, Olesen J, Pedersen JT, Brandt CT, Halling JF, Hellsten Y, Prats C, Pilegaard H (2013) Effect of lifelong resveratrol supplementation and exercise training on skeletal muscle oxidative capacity in aging mice; impact of PGC-1alpha. Exp Gerontol 48:1311–1318
5. Fischer CP, Plomgaard P, Hansen AK, Pilegaard H, Saltin B, Pedersen BK (2004) Endurance training reduces the contraction-induced interleukin-6 mRNA expression in human skeletal muscle. Am J Physiol Endocrinol Metab 287:E1189–E1194
6. Ronsen O, Holm K, Staff H, Opstad PK, Pedersen BK, Bahr R (2001) No effect of seasonal variation in training load on immuno-endocrine responses to acute exhaustive exercise. Scand J Med Sci Sports 11:141–148
7. Passonneau JV, Lauderdale VR (1974) A comparison of three methods of glycogen measurement in tissues. Anal Biochem 60:405–412
8. Lowry OH, Passonneau JV (1972) A flexible system of enzymatic analysis. Academic Press, New York
9. Cederblad G, Carlin JI, Constantin-Teodosiu D, Harper P, Hultman E (1990) Radioisotopic assays of CoASH and carnitine and their acetylated forms in human skeletal muscle. Anal Biochem 185:274–278
10. Constantin-Teodosiu D, Cederblad G, Hultman E (1991) A sensitive radioisotopic assay of pyruvate dehydrogenase complex in human muscle tissue. Anal Biochem 198:347–351
11. Pilegaard H, Birk JB, Sacchetti M, Mourtzakis M, Hardie DG, Stewart G, Neufer PD, Saltin B, van Hall G, Wojtaszewski JF (2006) PDH-E1alpha dephosphorylation and activation in human skeletal muscle during exercise: effect of intralipid infusion. Diabetes 55:3020–3027

12. Iliopoulos F, Mazis N (2018) Exercise training in heart failure patients: effects on skeletal muscle abnormalities and sympathetic nervous activity—a literature review. Sport Sci Health. https://doi.org/10.1007/s11332-018-0442-5
13. Haykowsky MJ, Kouba EJ, Brubaker PH et al (2014) Skeletal muscle composition and its relation to exercise intolerance in older patients with heart failure and preserved ejection fraction. Am J Cardiol 113:1211–1216
14. Piepoli MF (2006) Exercise training in heart failure. Curr Heart Fail Rep 3:189–196
15. Borlaug BA, Melenovsky V, Russell SD et al (2006) Impaired chronotropic and vasodilator reserves limit exercise capacity in patients with heart failure and a preserved ejection fraction. Circulation 114:2138–2147
16. Kishimoto S, Kajikawa M, Maruhashi T et al (2017) Endothelial dysfunction and abnormal vascular structure are simultaneously present in patients with heart failure with preserved ejection fraction. Int J Cardiol 231:181–187
17. Werber-Zion G, Goldhammer E, Shaar A et al (2004) Left ventricular function during strength testing and resistance exercise in patients with left ventricular dysfunction. J Cardiopulm Rehabil 24:100–109
18. Braith RW, Stewart KJ (2006) Resistance exercise training: its role in the prevention of cardiovascular disease. Circulation 113:2642–2650
19. Gittemeier EM, Ericson T, Ghosh P, Copp SW, Opoku-Acheampong AB, Behnke BJ (2017) Effects of aging and exercise training on the dynamics of vasoconstriction in skeletal muscle resistance vessels. Eur J Appl Physiol. https://doi.org/10.1007/s00421-017-3541-0
20. Colleran PN, Behnke BJ, Wilkerson MK, Donato AJ, Delp MD (2008) Simulated microgravity alters rat mesenteric artery vasoconstrictor dynamics through an intracellular Ca(2+) release mechanism. Am J Physiol Regul Integr Comp Physiol 294(5):R1577–R1585. https://doi.org/10.1152/ajpregu.00084.2008
21. Muller-Delp JM, Spier SA, Ramsey MW, Delp MD (2002b) Ageing impairs endothelium-dependent vasodilation in rat skeletal muscle arterioles. Am J Physiol Heart Circ Physiol 283(4):H1662–H1672. https://doi.org/10.1152/ajpheart.00004.2002
22. Behnke BJ, Delp MD (2010) Ageing blunts the dynamics of vasodilation in isolated skeletal muscle resistance vessels. J Appl Physiol 108(1):14–20. https://doi.org/10.1152/japplphysiol.00970.2009
23. Copp SW, Ferreira LF, Herspring KF, Musch TI, Poole DC (2009) The effects of ageing on capillary hemodynamics in contracting rat spinotrapezius muscle. Microvasc Res 77(2):113–119. https://doi.org/10.1016/j.mvr.2008.11.001
24. McCullough DJ, Davis RT 3rd, Dominguez JM 2nd, Stabley JN, Bruells CS (1985) Behnke BJ (2011) Effects of ageing and exercise training on spinotrapezius muscle microvascular PO2 dynamics and vasomotor control. J Appl Physiol 110(3):695–704. https://doi.org/10.1152/japplphysiol.01084.2010
25. Muller-Delp J, Spier SA, Ramsey MW, Lesniewski LA, Papadopoulos A, Humphrey JD, Delp MD (2002a) Effects of ageing on vasoconstrictor and mechanical properties of rat skeletal muscle arterioles. Am J Physiol Heart Circ Physiol 282(5):H1843–H1854. https://doi.org/10.1152/ajpheart.00666.2001
26. Donato AJ, Lesniewski LA, Delp MD (2007) Ageing and exercise training alter adrenergic vasomotor responses of rat skeletal muscle arterioles. J Physiol 579(Pt 1):115–125. https://doi.org/10.1113/jphysiol.2006.120055
27. Ghosh P, Mora Solis FR, Dominguez JM 2nd, Spier SA, Donato AJ, Delp MD, Muller-Delp JM (2015) Exercise training reverses ageing-induced impairment of myogenic constriction in skeletal muscle arterioles. J Appl Physiol 118(7):904–911. https://doi.org/10.1152/japplphysiol.00277.2014
28. Mazzeo RS, Horvath SM (1986) Effects of training on weight, food intake, and body composition in ageing rats. Am J Clin Nutr 44(6):732–738
29. Gollasch M (2012) Vasodilator signals from perivascular adipose tissue. Br J Pharmacol 165(3):633–642. https://doi.org/10.1111/j.1476-5381.2011.01430.x

30. Ikeda T, Noma K, Maeda N, Tanabe S, Sakamoto Y, Katayama Y, Shirakawa Y, Fujiwara T, Senda M (2022) Effectiveness of early exercise on reducing skeletal muscle loss during preoperative neoadjuvant chemotherapy for esophageal cancer. Surgery Today. https://doi.org/10.1007/s00595-021-02449-5
31. Fujihata S, Ogawa R, Nakaya S, Hayakawa S, Okubo T, Sagawa H et al (2021) The impact of skeletal muscle wasting during neoadjuvant chemotherapy on postoperative anastomotic leakage in patients with esophageal cancer. Esophagus. 18:258–266
32. Järvinen T, Ilonen I, Kauppi J, Salo J, Räsänen J (2018) Loss of skeletal muscle mass during neoadjuvant treatments correlates with worse prognosis in esophageal cancer: a retrospective cohort study. World J Surg Oncol. 16:27
33. Kamitani N, Migita K, Matsumoto S, Wakatsuki K, Kunishige T, Nakade H et al (2019) Association of skeletal muscle loss with the long-term outcomes of esophageal cancer patients treated with neoadjuvant chemotherapy. Surg Today. 49:1022–1028
34. Naito T, Okayama T, Aoyama T, Ohashi T, Masuda Y, Kimura M et al (2017) Skeletal muscle depletion during chemotherapy has a large impact on physical function in elderly Japanese patients with advanced non-small-cell lung cancer. BMC Cancer. 17:571
35. Naito T, Okayama T, Aoyama T, Ohashi T, Masuda Y, Kimura M et al (2017) Unfavorable impact of cancer cachexia on activity of daily living and need for inpatient care in elderly patients with advanced non-small-cell lung cancer in Japan: a prospective longitudinal observational study. BMC Cancer. 17:800
36. Xu Y-J, Cheng JC-H, Lee J-M, Huang P-M, Huang G-H, Chen CC-H (2015) A walk-and-eat intervention improves outcomes for patients with esophageal cancer undergoing neoadjuvant chemoradiotherapy. Oncologist. 20:1216–1222
37. Solfest JS, Nie Y, Weiss JA, Garner RT, Kuang S, Stout J, Gavin TP (2019) Effects of acute aerobic and concurrent exercise on skeletal muscle metabolic enzymes in untrained men. Sport Sci Health. https://doi.org/10.1007/s11332-019-00547-z
38. Henriksson J, Reitman JS (1976) Quantitative measures of enzyme activities in type I and type II muscle fibres of man after training. Acta Physiol Scand 97(3):392–397. https://doi.org/10.1111/j.1748-1716.1976.tb10279.x
39. Dohm GL, Huston RL, Askew EW, Fleshood HL (1973) Effects of exercise, training, and diet on muscle citric acid cycle enzyme activity. Can J Biochem 51(6):849–854
40. Ji LL, Stratman FW, Lardy HA (1988) Enzymatic down regulation with exercise in rat skeletal muscle. Arch Biochem Biophys 263(1):137–149
41. Lawler JM, Powers SK, Visser T, Van Dijk H, Kordus MJ, Ji LL (1993) Acute exercise and skeletal muscle antioxidant and metabolic enzymes: effects of fiber type and age. Am J Physiol 265(6 Pt 2):R1344–R1350. https://doi.org/10.1152/ajpregu.1993.265.6.R1344
42. Gollnick PD, Bertocci LA, Kelso TB, Witt EH, Hodgson DR (1990) The effect of high-intensity exercise on the respiratory capacity of skeletal muscle. Pflugers Arch 415(4):407–413
43. Leek BT, Mudaliar SR, Henry R, Mathieu-Costello O, Richardson RS (2001) Effect of acute exercise on citrate synthase activity in untrained and trained human skeletal muscle. Am J Physiol Regul Integr Comp Physiol 280(2):R441–R447
44. Shan C, Elf S, Ji Q, Kang HB, Zhou L, Hitosugi T, Jin L, Lin R, Zhang L, Seo JH, Xie J, Tucker M, Gu TL, Sudderth J, Jiang L, DeBerardinis RJ, Wu S, Li Y, Mao H, Chen PR, Wang D, Chen GZ, Lonial S, Arellano ML, Khoury HJ, Khuri FR, Lee BH, Brat DJ, Ye K, Boggon TJ, He C, Kang S, Fan J, Chen J (2014) Lysine acetylation activates 6-phosphogluconate dehydrogenase to promote tumor growth. Mol Cell 55(4):552–565. https://doi.org/10.1016/j.molcel.2014.06.020
45. Jaworowski A, Porter MM, Holmback AM, Downham D, Lexell J (2002) Enzyme activities in the tibialis anterior muscle of young moderately active men and women: relationship with body composition, muscle cross-sectional area and fibre type composition. Acta Physiol Scand 176(3):215–225. https://doi.org/10.1046/j.1365-201X.2002.t01-2-01004.x
46. Spinazzi M, Casarin A, Pertegato V, Salviati L, Angelini C (2012) Assessment of mitochondrial respiratory chain enzymatic activities on tissues and cultured cells. Nat Protoc 7(6):1235–1246. https://doi.org/10.1038/nprot.2012.058

47. Simoneau JA, Lortie G, Boulay MR, Marcotte M, Thibault MC, Bouchard C (1987) Effects of two high-intensity intermittent training programs interspaced by detraining on human skeletal muscle and performance. Eur J Appl Physiol Occup Physiol 56(5):516–521
48. Schmitt A, Haug A-L, Schlegel F, Fragasso A, Munz B (2018) Effects of 10 weeks of regular running exercise with and without parallel PDTC treatment on expression of genes encoding sarcomere-associated proteins in murine skeletal muscle. Cell Stress Chaperon. https://doi.org/10.1007/s12192-018-0914-1
49. Bowen TS, Schuler G, Adams V (2015) Skeletal muscle wasting in cachexia and sarcopenia: molecular pathophysiology and impact of exercise training. J Cachexia Sarcopenia Muscle 6:197–207
50. Murton AJ, Constantin D, Greenhaff PL (2008) The involvement of the ubiquitin proteasome system in human skeletal muscle remodelling and atrophy. Biochim Biophys Acta 1782:730–743
51. Pasiakos SM, Carbone JW (2014) Assessment of skeletal muscle proteolysis and the regulatory response to nutrition and exercise. IUBMB Life 66:478–484
52. Murphy RM (2010) Calpains, skeletal muscle function and exercise. Clin Exp Pharmacol Physiol 37:385–391
53. Molanouri Shamsi M, Mahdavi M, Quinn LS, Gharakhanlou R, Isanegad A (2016) Effect of resistance exercise training on expression of Hsp70 and inflammatory cytokines in skeletal muscle and adipose tissue of STZ-induced diabetic rats. Cell Stress Chaperon. https://doi.org/10.1007/s12192-016-0703-7
54. Ono T, Takada S, Kinugawa S, Tsutsui H (2015) Curcumin ameliorates skeletal muscle atrophy in type 1 diabetic mice by inhibiting protein ubiquitination. Exp Physiol 100(9):1052–1063
55. Lee S, Barton ER, Sweeney HL, Farrar RP (2004) Viral expression of insulin-like growth factor-I enhances muscle hypertrophy in resistance-trained rats. J Appl Physiol 96(3):1097–1104
56. Kim HJ, Kim YJ, Kim IY, Seong JK (2022) Resistance exercise training-induced skeletal muscle strength provides protective effects on high-fat-diet-induced metabolic stress in mice. Laboratory. Anim Res. https://doi.org/10.1186/s42826-022-00145-0
57. Westcott WL (2012) Resistance training is medicine: effects of strength training on health. Curr Sports Med Rep. 11(4):209–216
58. Strasser B, Schobersberger W (2011) Evidence for resistance training as a treatment therapy in obesity. J Obes. 2011:482564
59. Sayer AA, Syddall HE, Dennison EM, Martin HJ, Phillips DI, Cooper C et al (2007) Grip strength and the metabolic syndrome: findings from the hertfordshire cohort study. QJM 100(11):707–713
60. Winett RA, Carpinelli RN (2001) Potential health-related benefits of resistance training. Prev Med. 33(5):503–513
61. Ardavani A, Aziz H, Smith K, Atherton PJ, Phillips BE, Idris I (2020) The effects of very low energy diets and low energy diets with exercise training on skeletal muscle mass: a narrative review. Adv Ther. https://doi.org/10.1007/s12325-020-01562-0
62. Saris WH (2001) Very-low-calorie diets and sustained weight loss. Obes Res. 9(Suppl 4):295S–301S
63. Sellahewa L, Khan C, Lakkunarajah S et al (2017) A systematic review of evidence on the use of very low calorie diets in people with diabetes. Curr Diabetes Rev. 13:35–46
64. Lean MEJ, Leslie WS, Barnes AC et al (2017) Primary care-led weight management for remission of type 2 diabetes (DiRECT): an open-label, cluster-randomised trial. Lancet. 391:541–551
65. Clausen T (1986) Regulation of active Na + -K + transport in skeletal muscle. Physiol Rev. 66:542–580
66. Sinacore DR, Gulve EA (1993) The role of skeletal muscle in glucose transport, glucose homeostasis, and insulin resistance: implications for physical therapy. Phys Ther. 73:878–891
67. Straznicky NE, Lambert EA, Grima MT et al (2012) The effects of dietary weight loss with or without exercise training on liver enzymes in obese metabolic syndrome subjects. Diabetes Obes Metab. 14:139–148

68. Baron AD, Brechtel G, Wallace P et al (1988) Rates and tissue sites of non-insulin- and insulin-mediated glucose uptake in humans. Am J Physiol. 255:E769–E774
69. Bouzakri K, Koistinen HA, Zierath JR (2005) Molecular mechanisms of skeletal muscle insulin resistance in type 2 diabetes. Curr Diabetes Rev. 1:167–174
70. Chaston TB, Dixon JB, O'Brien PE (2007) Changes in fat-free mass during significant weight loss: a systematic review. Int J Obes. 31:743–750
71. Sayer AA, Syddall H, Martin H et al (2008) The developmental origins of sarcopenia. J Nutr Heal Aging. 12:427
72. Cruz-Jentoft AJ, Bahat G, Bauer J et al (2019) Sarcopenia: revised European consensus on definition and diagnosis. Age Ageing. 48:16–31
73. Esmeijer K, Geleijnse JM, de Fijter JW, Kromhout D, Hoogeveen EK (2020) Dietary protein intake and kidney function decline after myocardial infarction: the Alpha Omega Cohort. Nephrol Dial Transp. 35(1):106–115
74. Delimaris I (2013) Adverse effects associated with protein intake above the recommended dietary allowance for adults. ISRN Nutr 2013:126929
75. Barrows K, Snook JT (1987) Effect of a high-protein, very-low-calorie diet on body composition and anthropometric parameters of obese middle-aged women. Am J Clin Nutr. 45:381–390
76. Longland TM, Oikawa SY, Mitchell CJ et al (2016) Higher compared with lower dietary protein during an energy deficit combined with intense exercise promotes greater lean mass gain and fat mass loss: aa randomized trial. Am J Clin Nutr. 103:738–746
77. Larsen AE, Bibby BM, Hansen M (2018) Effect of a whey protein supplement on preservation of fat free mass in overweight and obese individuals on an energy restricted very low caloric diet. Nutrients. https://doi.org/10.3390/nu10121918
78. Heymsfield SB, Casper K, Hearn J et al (1989) Rate of weight loss during underfeeding: relation to level of physical activity. Metabolism. 38:215–223
79. Phinney SD, LaGrange BM, O'Connell M et al (1988) Effects of aerobic exercise on energy expenditure and nitrogen balance during very low calorie dieting. Metabolism. 37:758–765
80. Areta JL, Burke LM, Camera DM et al (2014) Reduced resting skeletal muscle protein synthesis is rescued by resistance exercise and protein ingestion following short-term energy deficit. Am J Physiol Endocrinol Metab. 306:E989–E997
81. Pasiakos SM, Cao JJ, Margolis LM et al (2013) Effects of high-protein diets on fat-free mass and muscle protein synthesis following weight loss: a randomized controlled trial. FASEB J. 27:3837–3847
82. Muñoz VR, Botezelli JD, Gaspar RC, da Rocha AL, Vieira RFL, Crisol BM, Braga RR, Severino MB, Nakandakari SCBR, Antunes GC, Brunetto SQ, Ramos CD, Velloso LA, Simabuco FM, de Moura LP, da Silva ASR, Ropelle ER, Cintra DE, Pauli JR (2023) Effects of short-term endurance and strength exercise in the molecular regulation of skeletal muscle in hyperinsulinemic and hyperglycemic Slc2a4$^{+/-}$ mice. Cell Mol Life Sci. https://doi.org/10.1007/s00018-023-04771-2
83. Hoene M, Kappler L, Kollipara L, Hu C, Irmler M, Bleher D et al (2021) Exercise prevents fatty liver by modifying the compensatory response of mitochondrial metabolism to excess substrate availability. Mol Metab 54:101359. https://doi.org/10.1016/j.molmet.2021.101359
84. Fredrickson G, Barrow F, Dietsche K, Parthiban P, Khan S, Robert S et al (2021) Exercise of high intensity ameliorates hepatic inflammation and the progression of NASH. Mol Metab. https://doi.org/10.1016/j.molmet.2021.101270
85. Hirschey MD, Shimazu T, Goetzman E, Jing E, Schwer B, Lombard DB, Grueter CA, Harris C, Biddinger S, Ilkayeva OR, Stevens RD, Li Y, Saha AK, Ruderman NB, Bain JR, Newgard CB, Farese RV Jr, Alt FW, Kahn CR, Verdin E (2010) SIRT3 regulates mitochondrial fatty-acid oxidation by reversible enzyme deacetylation. Nature 464(7285):121–125. https://doi.org/10.1038/nature08778
86. Jing E, O'Neill BT, Rardin MJ, Kleinridders A, Ilkeyeva OR, Ussar S, Bain JR, Lee KY, Verdin EM, Newgard CB, Gibson BW, Kahn CR (2013) Sirt3 regulates metabolic flexibility of skeletal muscle through reversible enzymatic deacetylation. Diabetes 62:3404–3417

87. Cordeiro AV, Brícola RS, Braga RR, Lenhare L, Silva VRR, Anaruma CP et al (2020) Aerobic exercise training induces the mitonuclear imbalance and UPRmt in the skeletal muscle of aged mice. J Gerontol Ser A 75(12):2258–2261. https://doi.org/10.1093/gerona/glaa059
88. Lin I-H, Chang J-L, Hua K, Huang W-C, Hsu M-T, Chen Y-F (2018) Skeletal muscle in aged mice reveals extensive transformation of muscle gene expression. BMC Genom Data 19. https://doi.org/10.1186/s12863-018-0660-5
89. Kim D, Nguyen MD, Dobbin MM, Fischer A, Sananbenesi F, Rodgers JT, Delalle I, Baur JA, Sui G, Armour SM, Puigserver P, Sinclair DA, Tsai LH (2007) SIRT1 deacetylase protects against neurodegeneration in models for Alzheimer's disease and amyotrophic lateral sclerosis. EMBO J. 26(13):3169–3179. https://doi.org/10.1038/sj.emboj.7601758
90. Donmez G, Arun A, Chung CY, McLean PJ, Lindquist S, Guarente L (2012) SIRT1 protects against alpha-synuclein aggregation by activating molecular chaperones. J Neurosci. 32(1):124–132. https://doi.org/10.1523/JNEUROSCI.3442-11.2012
91. Gano LB, Donato AJ, Pasha HM, Hearon CM Jr, Sindler AL, Seals DR (2014) The SIRT1 activator SRT1720 reverses vascular endothelial dysfunction, excessive superoxide production, and inflammation with aging in mice. Am J Physiol Heart Circ Physiol. 307(12):H1754–H1763. https://doi.org/10.1152/ajpheart.00377.2014
92. Pardo PS, Boriek AM (2011) The physiological roles of Sirt1 in skeletal muscle. Aging (Albany NY). 3(4):430–437. https://doi.org/10.18632/aging.100312
93. Cruz-Jentoft AJ, Baeyens JP, Bauer JM, Boirie Y, Cederholm T, Landi F, Martin FC, Michel JP, Rolland Y, Schneider SM, Topinkova E, Vandewoude M, Zamboni M, European Working Group on Sarcopenia in Older P (2010) Sarcopenia: European consensus on definition and diagnosis: report of the European working group on sarcopenia in older people. Age Ageing. 39(4):412–423. https://doi.org/10.1093/ageing/afq034
94. Nilwik R, Snijders T, Leenders M, Groen BB, van Kranenburg J, Verdijk LB, van Loon LJ (2013) The decline in skeletal muscle mass with aging is mainly attributed to a reduction in type II muscle fiber size. Exp Gerontol. 48(5):492–498. https://doi.org/10.1016/j.exger.2013.02.012
95. Chakkalakal JV, Jones KM, Basson MA, Brack AS (2012) The aged niche disrupts muscle stem cell quiescence. Nature. 490:355–360
96. Marzetti E, Leeuwenburgh C (2006) Skeletal muscle apoptosis, sarcopenia and frailty at old age. Exp Gerontol. 41(12):1234–1238. https://doi.org/10.1016/j.exger.2006.08.011
97. Shefer G, Rauner G, Yablonka-Reuveni Z, Benayahu D (2010) Reduced satellite cell numbers and myogenic capacity in aging can be alleviated by endurance exercise. PLoS One. 5(10):e13307. https://doi.org/10.1371/journal.pone.0013307
98. Okita K, Kinugawa S, Tsutsui H (2013) Exercise intolerance in chronic heart failure–skeletal muscle dysfunction and potential therapies. Circ J 77:293–300
99. Middlekauff HR, Vigna C, Verity MA et al (2012) Abnormalities of calcium handling proteins in skeletal muscle mirror those of the heart in humans with heart failure: a shared mechanism? J Card Fail 18:724–733
100. Szentesi P, Bekedam MA, Van Beek-Harmsen BJ et al (2005) Depression of force production and ATPase activity in different types of human skeletal muscle fibers from patients with chronic heart failure. J Appl Physiol 99:2189–2195
101. Lunde PK, Sjaastad I, Schiotz Thorud HM et al (2001) Skeletal muscle disorders in heart failure. Acta Physiol Scand 171:277–294
102. Vescovo G, Volterrani M, Zennaro R et al (2000) Apoptosis in the skeletal muscle of patients with heart failure: investigation of clinical and biochemical changes. Heart 84:431–437
103. Middlekauff HR (2010) Making the case for skeletal myopathy as the major limitation of exercise capacity in heart failure. Circ Heart Fail 3:537–546
104. Lambert CP, Evans WJ (2005) Adaptations to aerobic and resistance exercise in the elderly. Rev Endocr Metab Disord. 6(2):137–143
105. Villareal DT, Aguirre L, Gurney AB, Waters DL, Sinacore DR, Colombo E et al (2017) Aerobic or resistance exercise, or both, in dieting obese older adults. N Engl J Med. 376(20):1943–1955

Central and Peripheral Fatigue 9

Christopher Myers

9.1 Introduction by the Editor

Skeletal muscle fatigue, characterized by a temporary decline in the ability to generate force or power output, encompasses peripheral and central components, each with distinct mechanisms and implications for physical performance. Peripheral fatigue originates in the muscle and is attributed to factors such as energy depletion, accumulation of metabolic byproducts, and impaired calcium handling, which directly affect muscle contraction efficiency. On the other hand, central fatigue involves a reduction in voluntary activation of muscles, stemming from alterations in neurotransmitter levels and neural drive from the central nervous system. Understanding the intricate balance between central and peripheral fatigue is crucial for developing strategies to enhance athletic performance, manage fatigue in clinical populations, and improve overall functional capacity. This dual perspective on muscle fatigue broadens our comprehension of the limits of human performance but also opens avenues for targeted interventions to mitigate fatigue's impact.

In Chap. 7, low-level laser therapy (LLLT) was discussed, and how it improves skeletal muscle recovery after injury was discussed. This chapter highlights a study by Leal Junior et al. investigating the effects of 830 nm low-level laser therapy (LLLT) on exercise-induced skeletal muscle fatigue in humans, revealing significant findings that could impact future therapeutic strategies. By administering active LLLT or a placebo to the biceps humeri muscle before exercise, researchers found that subjects who received active LLLT performed a higher mean number of repetitions and exhibited no significant difference in blood lactate levels compared to the placebo group, despite an increase in lactate levels post-exercise. These results suggest that 830 nm LLLT can delay the onset of muscle fatigue during high-intensity

C. Myers (✉)
3 Sports Science, 3 Sports Science LLC, Akron, OH, United States of America
e-mail: chris@3sportsscience.com

C. Myers (ed.), *Skeletal Muscle Physiology*,
https://doi.org/10.1007/978-3-031-47065-3_9

exercises without altering lactate accumulation. This research underscores the potential of LLLT as a non-invasive method to enhance athletic performance and possibly aid in muscle recovery and rehabilitation, marking an essential step in exploring LLLT's role in managing and mitigating skeletal muscle fatigue.

The relationship between resistance exercise and skeletal muscle adaptation represents a fundamental aspect of exercise physiology, particularly concerning how much physical activity can influence muscle morphology and function. This chapter highlights a study by Drummond et al. on the impact of resistance exercise volume on myogenic regulatory factors (MRFs), specifically myoD and myogenin, in the rodent quadriceps muscle, providing insightful findings into this complex interplay. This research found that low-volume resistance exercise (LV) significantly increased myogenin mRNA and protein expression in red quadriceps muscle within 6–12 h post-exercise. Interestingly, doubling the exercise volume (high-volume resistance exercise, HV) delayed and diminished myogenin and myoD expression response. This suggests that increased exercise volume does not proportionally enhance MRF expression, possibly due to excessive muscle damage or injury from the heightened exertion. This research highlights the importance of optimizing resistance exercise volume to effectively stimulate muscle adaptation without inducing counterproductive stress responses.

Under mechanical and metabolic stress, resistance exercise initiates a cascade of molecular and cellular responses within skeletal muscle, leading to adaptations such as hypertrophy and enhanced muscle function. Central to these adaptations are MRFs, which include myoD and myogenin, transcription factors that play pivotal roles in muscle cell differentiation and repair. Previous studies have established the upregulation of MRFs following resistance exercise, suggesting their contribution to exercise-induced muscle remodeling.

The study by Drummond et al. explicitly addresses how resistance exercise volume affects the expression of myoD and myogenin within muscle tissue. Utilizing a rodent model of resistance exercise, the research demonstrates that increased exercise volume does not proportionally enhance MRF expression. Indeed, in the red quadriceps muscle, where significant elevations in myogenin mRNA and protein were observed following low-volume exercise, doubling the exercise volume resulted in a delayed and blunted response. This finding is particularly intriguing, suggesting an optimal exercise volume exists beyond which the beneficial effects on MRF expression may not be realized, potentially due to excessive muscle damage or an overwhelmed reparative response.

Furthermore, this chapter delves into two studies investigating the role of specific channels and compounds in mitigating muscle fatigue, offering insights into potential therapeutic strategies. The first study, conducted by Pritschow et al., showcases that TRPV4 channels, part of the transient receptor potential (TRP) family, are present in mouse skeletal muscle and can influence calcium influx and muscle fatigue. Activation of TRPV4 channels was found to significantly attenuate muscle fatigue, highlighting a novel pathway for modulating muscle function. The second study by Sánchez-Duarte et al. investigates nicorandil's effects on slow skeletal muscle fibers in chickens, revealing its dual function as a mitoKATP channel opener and a nitric oxide (NO) donor. Nicorandil enhanced post-fatigue muscle tension, modulated the glutathione redox state, reduced lipid peroxidation, and increased NO production, indicating a comprehensive approach to combating muscle fatigue. Together, these studies underscore the complexity of muscle fatigue mechanisms and suggest innovative avenues for intervention through channel activation and pharmacological agents.

Other compounds of significant interest are taurine and β-alanine. The research conducted by Horvath et al. explores the impact of taurine and β-alanine supplementation on muscle function and taurine transporter (TauT) protein expression in mdx mice, a model for Duchenne Muscular Dystrophy (DMD). Their findings reveal that while supplementation alters muscle taurine content, it does not affect TauT protein levels in wild-type and mdx mice. Interestingly, β-alanine supplementation increased muscle fatigue resistance, suggesting potential therapeutic benefits. This study highlights the complexity of muscle adaptation. It indicates that taurine and β-alanine could offer benefits beyond their effects on TauT protein, especially in improving muscle function and resistance to fatigue in conditions resembling DMD.

Exploring non-invasive interventions for enhancing skeletal muscle recovery and adaptation has unveiled significant insights, mainly through low-level laser therapy (LLLT) and resistance exercise. Studies highlighted in this chapter present compelling evidence on the efficacy of 830 nm LLLT in delaying muscle fatigue without affecting lactate accumulation and the nuanced relationship between resistance exercise volume and myogenic regulatory factors (MRFs) expression in skeletal muscle. These findings highlight the potential of LLLT as a tool for improving athletic performance and rehabilitation processes. Moreover, research into the modulation of muscle function and fatigue through TRPV4 channel activation and the effects of nicorandil and dietary supplementation with taurine and β-alanine suggests innovative approaches to combat muscle fatigue and enhance recovery. These studies collectively underscore the intricate mechanisms underlying muscle adaptation to physical stressors and the potential of targeted therapies to optimize muscle health and function, marking significant advancements in exercise physiology and therapeutic strategies.

9.2 Machine Generated Summaries

Disclaimer: The summaries in this chapter were generated from Springer Nature publications using extractive AI auto-summarization: An extraction-based summarizer aims to identify the most important sentences of a text using an algorithm and uses those original sentences to create the auto-summary (unlike generative AI). As the constituted sentences are machine selected, they may not fully reflect the body of the work, so we strongly advise that the original content is read and cited. The auto generated summaries were curated by the editor to meet Springer Nature publication standards. To cite this content, please refer to the original papers.

Machine generated keywords: fatigue, channel, exercise, muscle fatigue, fiber, tension, slow skeletal, peak, wildtype, atp, resistance, contractile protein, influx, reactive, resistance exercise.

Effect of 830 nm Low-Level Laser Therapy in Exercise-Induced Skeletal Muscle Fatigue in Humans [1]

This is a machine-generated summary of:

Leal Junior, Ernesto Cesar Pinto; Lopes-Martins, Rodrigo Álvaro Brandão; Vanin, Adriane Aver; Baroni, Bruno Manfredini; Grosselli, Douglas; De Marchi, Thiago; Iversen, Vegard V.; Bjordal, Jan Magnus: Effect of 830 nm low-level laser therapy in exercise-induced skeletal muscle fatigue in humans [1]

Published in: Lasers in Medical Science (2008)

Link to original: https://doi.org/10.1007/s10103-008-0592-9

If you want to cite the papers, please refer to the original.

For technical reasons we could not place the page where the original quote is coming from.

Abstract-Summary "Active LLLT (830 nm wavelength, 100 mW output, spot size 0.0028 cm^2, 200 s total irradiation time) or an identical placebo LLLT was delivered to four points on the biceps humeri muscle immediately before exercises."

"After active LLLT the mean number of repetitions was significantly higher than after placebo irradiation [mean difference 4.5, standard deviation (SD) ± 6.0, P = 0.042], the blood lactate levels increased after exercises, but there was no significant difference between the treatments."

"We concluded that 830 nm LLLT can delay the onset of skeletal muscle fatigue in high-intensity exercises, in spite of increased blood lactate levels."

Introduction

"Skeletal muscle fatigue is a common experience in daily life, although the mechanisms of action, development or prevention are not fully understood."

"In clinical settings, skeletal muscle fatigue may also subsequently contribute to the development of muscle pain [2]."

"Skeletal muscle fatigue is a novel area of low-level laser therapy (LLLT) research, and the optimal parameters of application are not fully understood."

"In clinical settings, LLLT has been used in the treatment of muscle pain, and some positive findings have been found for neck muscle pain [3] and conditions like fibromyalgia [4], which may be related to skeletal muscle fatigue."

"We decided to investigate whether an infrared wavelength (830 nm) would have similar effects of LLLT on skeletal muscle fatigue, and whether there were differences in effects between active LLLT and placebo LLLT in the same persons in a crossover design."

Methods

"For participants drawing lot A, active LLLT was given at the first session and placebo LLLT was given at the second session."

"For participants drawing lot B, placebo LLLT was given at the first session and active LLLT at the second session."

"After 2 weeks of familiarization with the exercises, they were given a maximal contraction force test that consisted of one single repetition of flexion to 90° from full extension of the elbow, so that we could evaluate the strength of the biceps muscle while the subject was seated in the Scott chair."

"Each subject received two different treatment sessions, 1 week apart, with either active LLLT or placebo LLLT before the exercise tests."

"Immediately after LLLT, the participants were repositioned and started on the fatigue exercise test within 60 s. In order to measure blood lactate concentrations, following aseptic cleaning of the finger, a qualified nurse, who was unaware of the group allocation, took one blood sample before and another 3 min after the exercises had been completed."

Results

"The mean number of repetitions performed in the exercise test was 25.60 (± 6.15) when the subjects received placebo LLLT, and 30.10 (± 8.08) when the subjects received active LLLT before the exercise fatigue tests ($P = 0.042$)."

"The time taken to perform the exercise tests was slightly longer after active LLLT (37.15 ± 6.45 s) than after placebo LLLT (34.34 ± 6.77 s), but this difference did not reach statistical significance ($P = 0.096$)."

"There was no significant difference in the change in blood lactate levels from pre-exercise tests to post-exercise tests ($P = 0.200$) between active LLLT, at 1.35 mmol l^{-1} (± 0.90 mmol l^{-1}), or placebo LLLT, at 1.78 mmol l^{-1} (± 1.10 mmol l^{-1})."

Discussion

"In this small trial we found some indication that LLLT might be able to delay the onset of a skeletal muscle fatigue response during high-intensity exercises, by significantly increasing the number of repetitions after pre-treatment with active LLLT in comparison with placebo LLLT."

"We find this unlikely, because previous animal studies have shown that infrared LLLT can prevent ischemic muscle injuries by reducing the release of reactive oxygen species (ROS) and creatine phosphokinase activity, while increasing levels of antioxidants and heat shock proteins [5, 6]."

"Our findings of a delayed onset of skeletal muscle fatigue are consistent with previous findings in an animal study and must be considered as a first step in a novel area of LLLT research where many questions are, as yet, unanswered [7]."

"Clinical studies have previously demonstrated that active LLLT with a greater dose of 8.9 J increased post-exercise microcirculation, but did not reduce spontaneous pain, in the masseter muscle [8]."

Conclusion

"Blood lactate concentrations were similar immediately after the muscle performance test, but we cannot rule out that the accumulation of lactate may have been slightly delayed by LLLT."

"The clinical impact of our findings may be limited by the experimental model, which only measured the immediate effects on muscle performance within 2 min of irradiation."

"The study may open another clinical avenue to treatment of musculoskeletal conditions where muscle fatigue is a precursor of pain."

Glibenclamide Increases Post-Fatigue Tension in Slow Skeletal Muscle Fibers of the Chicken [9]

This is a machine-generated summary of:

Andrade, Felipa; Trujillo, Xóchitl; Sánchez-Pastor, Enrique; Montoya-Pérez, Rocío; Saavedra-Molina, Alfredo; Ortiz-Mesina, Mónica; Huerta, Miguel: Glibenclamide increases post-fatigue tension in slow skeletal muscle fibers of the chicken [9]

Published in: Journal of Comparative Physiology B (2010)

Link to original: https://doi.org/10.1007/s00360-010-0527-1

If you want to cite the papers, please refer to the original.

For technical reasons we could not place the page where the original quote is coming from.

Abstract-Summary "We investigated the action of glibenclamide (an anti-diabetic sulphonylurea that acts on K_{ATP} channels) on fatigued slow skeletal muscle, studying twitch and tetanus tension after inducing the muscle to fatigue by continuous electrical stimulation."

"Our results showed that glibenclamide (150 μM) increased post-fatigue twitch tension by about 25% with respect to the fatigued condition ($P < 0.05$)."

"Glibenclamide (150 μM) increased post-fatigue tetanic tension (83.61 ± 15.7% in peak tension, and 85.0 ± 19.0% in tension-time integral, P = 0.02, and 0.04, respectively; n = 3)."

"After muscle pre-incubation with glibenclamide (150 μM), tension of caffeine-evoked contractures increased (6.5 ± 1.5% in maximal tension, and 5.9 ± 3.8% in tension-time integral, P < 0.05)."

"These results suggest a possible role of K_{ATP} channels in the fatigue process, since glibenclamide increases twitch and tetanus tension in fatigued slow muscle of the chicken and during metabolic inhibition, possibly by increasing intracellular calcium."

Introduction

"During fatigue, intracellular ATP decreases leading to a reduction in ATP hydrolysis in order to protect the cells, thus K_{ATP} channels are sensors of ATP and ADP intracellular rate [10]."

"The activation of K_{ATP} channels succeeds when the cells are compromised in ATP-demanding functions such as fatigue or metabolic inhibition [11]."

"Some authors have explored the role of K_{ATP} channels in skeletal muscle using agonists of these channels, suggesting that their activation is involved in the diminution of force when muscle is fatigued [12–16]."

"In the present paper, the effects of glibenclamide on twitch and tetanus tension of fatigued slow skeletal muscle fibers of the chicken, as well as its action during metabolic inhibition (low ATP), were studied."

Methods

"The vertebral cord was pinned at the bottom of the experimental chamber and 3.0 surgical silk threads were tied around the humerus bone to attach muscle bundle (1–2 mm thick) distal end to a force transducer (Grass FT03, West Warwick, RI, USA) by means of a lightweight wire."

"Each protocol consisted of a period with initial stimulation to induce fatigue (about 30% of initial tension), after which glibenclamide was applied to the bath for 5 min, maintaining the stimulation."

"After drug washout, electrical stimulation continued to record tension for 15 min for post-fatigue analysis."

"Contractures in the ALD muscle were induced by applying 8 mM caffeine."

"Protocol consisted of 5 min exposure to caffeine contracture followed by a resting period of 30 min including 5 min of glibenclamide exposure."

"5 min caffeine (6 mM) contracture was recorded in the presence of glibenclamide."

Results

"These results show that glibenclamide, a K_{ATP} channels blocker, increases tension of low frequency fatigued muscle fibers."

"Trace 4 shows tetanic tension recorded after glibenclamide washout, in which the muscle was further fatigued."

"In the presence of glibenclamide tetanic tension increased to 83.61 ± 15.71% in peak tension and to 85.06 ± 19.00% in tension-time integral (P = 0.0267, P = 0.0484 respectively; n = 3)."

"When glibenclamide (150 μM) was added, peak tension increased to 30.58 ± 5.00% with respect to the fatigued state, and tension-time integral was 28.96 ± 5.64% (n = 3)."

"The increase in caffeine contracture observed in the presence of glibenclamide was 6.55 ± 1.57% in maximal tension and 5.93 ± 3.86% in tension-time integral ($P < 0.05$; n = 3)."

Discussion

"There are some reports indicating the presence of K_{ATP} channels in the membranes of mammalian and avian fast skeletal muscle fibers, and it has been suggested that this kind of channels may have a functional role in muscular fatigue."

"The effects of glibenclamide could be due to an involvement of this drug in one of the steps involved in excitation contraction coupling and in the modulation of intracellular Ca^{2+} during fatigue in slow skeletal muscle fibers."

"Glibenclamide was found to avoid the diminution in tension caused by cyanide suggesting that K_{ATP} channel activation is involved in the action mechanism causing muscle tension reduction during metabolic inhibition."

"Indirect evidence of possible K_{ATP} channel involvement or the stopping of excitation–contraction coupling in the reduction of tension in slow skeletal muscle fibers of the chicken produced by fatigue or metabolic inhibition processes is presented in this paper."

Functional TRPV4 Channels Are Expressed in Mouse Skeletal Muscle and Can Modulate Resting Ca^{2+} Influx and Muscle Fatigue [17]

This is a machine-generated summary of:

Pritschow, Bernd W.; Lange, Thom; Kasch, Joachim; Kunert-Keil, Christiane; Liedtke, Wolfgang; Brinkmeier, Heinrich: Functional TRPV4 channels are expressed in mouse skeletal muscle and can modulate resting Ca^{2+} influx and muscle fatigue [17]

Published in: Pflügers Archiv – European Journal of Physiology (2010)

Link to original: https://doi.org/10.1007/s00424-010-0883-4

If you want to cite the papers, please refer to the original.

For technical reasons we could not place the page where the original quote is coming from.

Abstract-Summary "Candidate channels that may control Ca^{2+} influx into muscle fibers are the STIM proteins, Orai, and the members of the transient receptor potential (TRP) family of cation channels."

"We show that TRPV4, an osmo-sensitive cation channel of the vanilloid subfamily of TRP channels is functionally expressed in mouse skeletal muscle."

"Western blot analysis showed the presence of TRPV4-specific bands at about 85 and 100 kDa in all tested muscles."

"Using the manganese quench technique, we studied the resting influx of divalent cations into isolated wild-type muscle fibers."

"The specific TRPV4-channel activator 4α-phorbol-12,13-didecanoate (4α-PDD) stimulated resting influx by about 60% only in wild-type fibers."

"When soleus muscles were stimulated with a fatigue protocol, muscle fatigue was significantly attenuated in the presence of 4α-PDD."

"We conclude that TRPV4 is functionally expressed in mouse skeletal muscle and that TRPV4 activation modulates resting Ca^{2+} influx and muscle fatigue."

Introduction

"The underlying Ca^{2+} influx pathways are not well characterized for skeletal muscle fibers, and may include various mechanisms and macromolecular complexes containing ion channels [18–21]."

"Stretch-activated channels [22], calcium leak channels [23] mechanosensitive channels [24], insulin-like growth factor (IGF-1) activated channels [25, 26] and store-operated channels [18] are regarded as candidates for influx pathways of Ca^{2+} into muscle fibers."

"Physiological and pharmacological data indicated that stretch-activated channels and store-operated channels belong to the same channel population in wild-type skeletal muscle or share common constituents [19]."

"We have recently shown that members of the TRP (transient receptor potential) channel family of cation channels are expressed in mouse skeletal muscle [27], and that some members of this superfamily are localized in the sarcolemma [28], among them TRPC6, TRPV4, and TRPM7."

"The present study was designed to investigate the role of TRPV4 in skeletal muscle."

"We addressed this question in this study, and investigated specifically whether TRPV4 activity can modulate Ca^{2+} homeostasis and contractile function of skeletal muscle."

Materials and Methods

"Muscles for contraction experiments and for fiber isolation were rapidly removed weighted and transferred into the appropriate buffer solutions."

"Force measurements on whole soleus muscles were performed in a Krebs–Henseleit bathing solution that was continuously bubbled with 95% O_2 and 5% CO_2."

"Skeletal muscle fibers were enzymatically dissociated from interosseus muscles in a solution containing 1.5 mg/ml collagenase (Serva, Heidelberg, Germany), 2 mg/ml dispase (Roche, Basel, Switzerland) and 1% streptomycin/penicillin dissolved in a mixture of DMEM/Ham's F12 medium (1:1)."

"Muscle fibers isolated from interosseus muscles were loaded with the membrane permeable Ca^{2+} indicator Fura-2/AM (2 μM) for 45 min in the standard external solution."

"Mn^{2+} was added to the external solution to reach a final concentration of 0.5 mM. The fluorescence (emitted light of 490–530 nm) after excitation at 360 nm, the isosbestic point of Fura-2, was recorded every 10 s. Mn^{2+} enters the cell via the same routes as Ca^{2+} and quenches the Fura-2 fluorescence."

Results

"Such fibers were used to test whether TRPV4 activity can contribute to the influx of divalent cations into muscle fibers."

"The effect of 4α-PDD on divalent cation influx led us to test whether the drug could induce changes in $[Ca^{2+}]_i$ in muscle fibers."

"To study a possible contribution of TRPV4 on muscular force production, we used isolated soleus muscles, and stimulated them under isometric conditions."

"To induce muscle fatigue, we used a protocol consisting of 200 tetanic stimulations (at 50 Hz) of 500-ms duration applied every 2 s. Under these conditions, the maximum isometric force dropped within 250 s by 70%."

"Within 30 min, the muscles recovered nearly completely from such a stimulation protocol and another fatigue protocol could be applied."

"During prolonged and repeated stimulation, Ca^{2+} influx via TRPV4 attenuates muscle fatigue."

Discussion

"This finding supports the view that TRPV4 channels are localized in the sarcolemma of muscle fibers and can be functionally activated."

"Although a moderately increased concentration of 4α-PDD (5 μM) was needed to stimulate divalent cation influx, 1 μM was sufficient to attenuate muscle fatigue."

"This conclusion led us to interrogate muscle fatigue as a function of TRPV4."

"Fatigue involves changes in neuronal activity in the central nervous system in vivo and depends on several peripheral cellular mechanisms in skeletal muscle."

"We were able to show that TRPV4 activation by 4α-PDD could antagonize muscle fatigue during repetitive tetanic stimuli."

"We show novel evidence that TRPV4 is functional in skeletal muscle fibers and that its activation could play a role in counteracting muscle fatigue."

Nicorandil Improves Post-Fatigue Tension in Slow Skeletal Muscle Fibers by Modulating Glutathione Redox State [29]

This is a machine-generated summary of:

Sánchez-Duarte, E.; Trujillo, X.; Cortés-Rojo, C.; Saavedra-Molina, A.; Camargo, G.; Hernández, L.; Huerta, M.; Montoya-Pérez, R.: Nicorandil improves post-fatigue tension in slow skeletal muscle fibers by modulating glutathione redox state [29]

Published in: Journal of Bioenergetics and Biomembranes (2017)
Link to original: https://doi.org/10.1007/s10863-016-9692-6

If you want to cite the papers, please refer to the original.

For technical reasons we could not place the page where the original quote is coming from.

Abstract-Summary "K^+ channel openers like nicorandil has been recognized for their ability to protect skeletal muscle from ischemia-reperfusion injury, however, the effects of nicorandil on fatigue in slow skeletal muscle fibers has not been explored, being the aim of this study."

"Nicorandil (10 μM), improved the muscle function reversing fatigue as increased post-fatigue tension in the peak and total tension significantly with respect to the fatigued condition."

"Nicorandil also decreased lipid peroxidation and maintained both reduced glutathione (GSH) levels and an elevated GSH/GSSG ratio, whereas total glutathione (TGSH) remained unaltered during post-fatigue tension."

"NO production, measured through nitrite concentrations was significantly increased with nicorandil during post-fatigue tension; this increase remained unaltered in the presence of nicorandil plus L-NAME, nonetheless, this effect was reversed with nicorandil plus MPG."

"These results suggest that nicorandil improves the muscle function reversing fatigue in slow skeletal muscle fibers of chicken through its effects not only as a mitoK_{ATP} channel opener but also as NO donor and as an antioxidant."

Introduction

"In skeletal muscle, the protective effects of mitoK_{ATP} channels openers have been suggested that these channels are an important myoprotective target during muscle metabolic stress and ischemic processes [30–33]."

"A physiological role of mitoK_{ATP} channels on muscle fatigue has been proposed by García and others [34], since it was found that in mammal fast skeletal muscle fibers, the onset of fatigue has been delayed with diazoxide, a selective opener of mitoK_{ATP} channels."

"It has been demonstrated that nicorandil protects ischemic myocardium through its direct effect on mitoK_{ATP} channels and via a NO-PKG-dependent pathway, which was associated to the activity of this drug like a nitric oxide (NO) donor, which in turn, was attributed to its structure constituted by an organic nitrate and a nicotinamide group of [35, 36]."

"We hypothesized that nicorandil can protect skeletal muscle and likely have an effect on recovery from fatigue by decreasing oxidative stress and improving NO levels."

Materials and Methods

"Thiobarbituric acid reactive substances (TBARS) were used as a marker of lipid peroxidation in homogenates of ALD muscle using the method described by Ratajczak-Wrona and others [37], with minor modifications."

"Protein concentration of ALD muscle homogenates was measured by the Biuret method [38], using bovine serum albumin as the protein standard."

"Homogenates of ALD muscle samples were treated with 5% (v/v) sulfosalicylic acid and centrifuged at 7800 g for 10 min to remove denatured proteins; the supernatant was used for measurements."

"Nitrite accumulation was assayed in muscle homogenates according to the method described by Miranda and others [39], with minor modifications."

"ALD muscle samples were treated with Griess reagent, consisting of N-(1-Naphthyl) ethylenediamine dihydrochloride (NEDD, 75 μl) and sulfanilamide (SULF, 75 μL), together with HCl (150 μl) and VCl_3 (150 μl), and then mixed and incubated at 37 °C for 30 min."

Results

"After bathing the muscle with nicorandil (10 μM) the post-fatigue tension increased by 23.68 ± 6.61% in the peak tension and 31.04 ± 5.57% in total tension, related to fatigue tension."

"When a high concentration of nicorandil was applied to muscle (30 μM), the post-fatigue tension had an increment of 17.35 ± 2.51% in the peak, and 21.75 ± 5.89% in total tension."

"The effect of nicorandil at 10 μM was the concentration at which was observed the major increase in the post-fatigue tension or in reversing muscle fatigue."

"This effect was reversed in the presence of nicorandil, as GSH levels were 40% higher and a decrease of the same magnitude was observed for GSSG when compared respect to the fatigued condition."

Discussion

"Our results suggest that MPG could prevent the beneficial effect of nicorandil in post fatigue tension, not by scavenging ROS but rather by altering the redox state of thiol groups in the channel involved in its opening, which means that the oxidation of these putative cysteines is important for channel opening though low level of ROS without oxidatively impairing other structures or processes sensitive to oxidative stress like oxidative phosphorylation or the electron transport chain."

"Altogether these results indicate that the beneficial effect mediated by nicorandil by preserving adequate muscle function by reversing fatigue, it is carried out through its several functions not only as an mitoK_{ATP} channel opener but also as a NO donor and antioxidant, being the first study exploring these properties during fatigue in slow skeletal muscle fibers."

The Effect of Taurine and β-Alanine Supplementation on Taurine Transporter Protein and Fatigue Resistance in Skeletal Muscle from mdx Mice [40]

This is a machine-generated summary of:

Horvath, Deanna M.; Murphy, Robyn M.; Mollica, Janelle P.; Hayes, Alan; Goodman, Craig A.: The effect of taurine and β-alanine supplementation on taurine transporter protein and fatigue resistance in skeletal muscle from mdx mice [40]

Published in: Amino Acids (2016)

Link to original: https://doi.org/10.1007/s00726-016-2292-2

If you want to cite the papers, please refer to the original.

For technical reasons we could not place the page where the original quote is coming from.

Abstract-Summary "This study investigated the effect of taurine and β-alanine supplementation on muscle function and muscle taurine transporter (TauT) protein expression in mdx mice."

"Wild-type (WT) and mdx mice (5 months) were supplemented with taurine or β-alanine for 4 weeks, after which in vitro contractile properties, fatigue resistance and force recovery, and the expression of the TauT protein and proteins involved in excitation–contraction (E–C) coupling were examined in fast-twitch muscle."

"Supplementation with taurine and β-alanine increased and reduced taurine content, respectively, in muscle from WT and mdx mice but had no effect of TauT protein."

"These findings suggest that TauT protein expression is relatively insensitive to changes in muscle taurine content in WT and mdx mice, and that taurine and β-alanine supplementation may be viable therapeutic strategies to improve fatigue resistance of dystrophic skeletal muscle."

Introduction

"Corticosteroid treatment, the most common pharmacological therapy for DMD, results in an improvement in mdx muscle function that is associated with an increase in muscle taurine content [41]."

"These findings suggest that taurine supplementation may lead to improved Ca^{2+} handling and force generation in mdx fast-twitch muscle."

"This study found reduced TauT protein at ~3 and 6 weeks of age in mdx mice; however, neither studies have examined TauT protein in older mdx mice (e.g. 6 months) nor whether taurine supplementation alters muscle TauT content in mdx muscle [42]."

"The aims of this study were to examine the effect of taurine supplementation on muscle taurine content, TauT protein, ex vivo contractile properties, fatigue resistance and force recovery after fatiguing contractions, in isolated fast-twitch mdx muscle."

Materials and Methods

"Mice (n = 8–12/group) were anesthetised with an intraperitoneal injection of 40–50 mg/kg of Nembutal (pentobarbitone sodium, Rhone Merieux, QLD, Australia) and the extensor digitorum longus (EDL) muscle was rapidly removed and placed in Krebs–Ringer solution containing (in mM) 118 NaCl, 1.18; $MgSO_4$; 4.75 KCl; 1.0 Na_2HPO_4; 2.5 $CaCl_2$; 24.0 $NaHCO_3$; and 11.0 glucose, pH 7.4 and bubbled with 95 % O_2–5 % CO_2 (BOC Gases, Preston, Victoria, Australia) at 30 °C."

"Equal amounts of whole muscle homogenates (~20 ug muscle mass) were loaded onto Criterion TGX 4–15 % Stain-Free gradient gels (Bio-Rad, Australia) and total protein separated at 200 volts (V) for 60 min."

"Following separation of proteins, gels were imaged for total protein on a Criterion Stain Free Imager (Bio-Rad, Australia) and then the proteins were wet-transferred from the gels onto a nitrocellulose membrane for 60 min at 100 V (140 mM glycine, 37 mM Tris base, and 20 % v/v methanol)."

Results

"Taurine and β-alanine supplementation had no effect on any of these parameters in either mdx or WT EDL muscles."

"β-Alanine supplementation had no effect on P_0 or SP_0 produced by EDL muscles from WT mice."

"β-Alanine supplementation had no effect on P_0 or SP_0 produced by EDL muscles from mdx mice."

"While taurine supplementation had no effect on fatigue in the WT EDL muscles, β-alanine supplementation resulted in significantly greater force production at the end of stimulation (i.e. decreased fatigue) in WT muscles."

Discussion

"This is the first study to examine the effects of prolonged taurine and β-alanine supplementation on skeletal muscle taurine content, ex vivo contractile properties, fatigue resistance and force recovery in muscles from mdx mice."

"We provide a novel comparison of skeletal muscle TauT protein abundance between mdx and WT mice at 6 months of age with and without taurine or β-alanine supplementation."

"We also demonstrated that, in spite of a ~21 % supplementation-induced increase in taurine content, TauT protein content was not altered in muscle from WT or mdx mice."

"We found that β-alanine modestly reduced taurine content in muscles from WT and mdx mice with no significant change in TauT protein."

"Taurine, but not β-alanine, supplementation significantly decreased both body mass and EDL muscle mass in mdx mice, with no effect in WT animals."

Conclusion

"We found that neither taurine nor β-alanine, supplementation altered muscle TauT protein content in EDL muscles from either mdx or WT mice."

"β-Alanine supplementation did, however, improve fatigue resistance in WT and mdx muscles."

"There were no significant differences in CSQ1, RyR1, DHPR or SERCA1 protein abundances in muscles from mdx and WT mice before or after supplementation with taurine or β-alanine."

Skeletal Muscle Myofibrillar Protein Oxidation and Exercise Capacity in Heart Failure [43]

This is a machine-generated summary of:

Vescovo, Giorgio; Ravara, Barbara; Dalla Libera, Luciano: Skeletal muscle myofibrillar protein oxidation and exercise capacity in heart failure [43]

Published in: Basic Research in Cardiology (2007)

Link to original: https://doi.org/10.1007/s00395-007-0692-x

If you want to cite the papers, please refer to the original.

For technical reasons we could not place the page where the original quote is coming from.

Abstract-Summary "We have previously demonstrated in a model of heart failure, the monocrotaline treated rat, that oxidation of skeletal muscle actin, tropomyosin and myosin produces a reduction of contractile efficiency, which may further depress muscle function and exercise capacity."

"To investigate the presence of oxidized myofibrillar proteins in skeletal muscle of CHF patients by means of the Oxyblot technique and to correlate it with exercise capacity."

"We have analyzed skeletal muscle biopsies taken from six patients with class III–IV NYHA CHF and four control patients (peak VO_2 12.8 ± 1.9 vs. 29.7 ± 1.7 ml/kg/min, $p < 0.0001$)."

"In the skeletal muscle of CHF patient there was a much higher level of myofibrillar protein oxidation as expressed by the Oxyblot/Red Ponceau (Oxy/RP) ratio as compared to controls (2.1 ± 0.3 vs. 1.02 ± 0.09, $p < 0.0001$)."

"Higher levels of muscle oxidation were found in patients with lower exercise capacity with an inverse correlation between Oxyblot and VO_2 values ($r^2 = 0.83$)."

Introduction

"Heart failure (CHF) is a clinical syndrome characterized by decreased exercise capacity with early appearance of dyspnoea and fatigue [44] and by high mortality."

"We have previously demonstrated in a well-known model of CHF, the monocrotaline treated rat, that skeletal muscle protein oxidation may cause further deterioration of muscle efficiency [45]."

"We suggested that contractile proteins oxidation may play a role in further deteriorating muscle efficiency and therefore in limiting exercise capacity in patients with heart failure and we envisaged the potential of using drugs for the treatment of CHF that can prevent myofibrillar oxidation."

"We have investigated the degree of contractile proteins oxidation on skeletal muscle microbiopsies taken from the quadriceps femoris of six CHF patients and four controls and we correlated the Oxyblot/Red Ponceau ratio with exercise capacity measured as peak VO_2."

Patients and Methods

"The suspension was centrifuged at 12,000 g for 10 min and the pellet containing myofibrillar proteins was resuspended in 200 µl of 0.01% TFA and assayed for protein concentration [45]."

"Homogenates of control patients muscles were incubated with 1 mM, H_2O_2 and ferrous sulphate for 10 min at room temperature."

"The principle of oxyblot procedure is to detect carbonyl groups, deriving from protein oxidation, in the protein side chains."

"The Oxyblot Oxidized Protein Detection Kit was purchased from Chemicon."

"One-dimensional electrophoresis of the resuspended myofibrillar proteins, was carried out on 10% SDS/polyacrylamide gel after 2,4-dinitro phenyl hydrazine derivatization."

"To quantify the amount of oxidation and allow the comparison between the various samples we defined the oxidative index (Oxy/RP) [the ratio between densitometric values of the oxyblot bands (oxidation level) and those stained with Red Ponceau (protein content)]."

"To identify oxidized protein parallel gels were run."

Results

"Using specific monoclonal antibodies against myosin heavy chains, actin and tropomyosin, we demonstrated that oxidation occurs at level of these sarcomeric contractile proteins (Panel C, D, E)."

"The oxidation pattern observed in patients is very similar to that obtained from an experimental CHF rat (the monocrotaline treated rat)."

"The level of oxidation is very low both in Controls and in the normal rat soleus muscle."

"The obtained pattern after oxidation is similar to that seen in CHF patients."

Discussion

"In a model of CHF, the monocrotaline treated rat, we have demonstrated in the soleus muscle the presence of oxidized myofibrillar proteins and the impairement of contraction in the whole muscle."

"We suggest that protein oxidation in the skeletal muscle is related to the high oxidative stress, iNOS production and free radicals formation which have been described in CHF [46]."

"It is in fact possible that drugs with anti-oxidative properties could improve muscle performance and exercise capacity."

"Although the number of patients of this study is small, we think that the finding of protein oxidation in the skeletal muscle of patients with CHF and the correlation with severity of exercise capacity limitation, is fairly new and could be the starting point for further studies, including also the effect of treatment with beta-blockers or other drugs capable to block oxidative stress."

Skeletal Muscle Phenotyping of Hippo Gene-Mutated Mice Reveals that Lats1 Deletion Increases the Percentage of Type I Muscle Fibers [47]

This is a machine-generated summary of:

Nezhad, Fakhreddin Yaghoob; Riermeier, Annett; Schönfelder, Martin; Becker, Lore; de Angelis, Martin Hrabě; Wackerhage, Henning: Skeletal muscle phenotyping of Hippo gene-mutated mice reveals that Lats1 deletion increases the percentage of type I muscle fibers [47].

Published in: Transgenic Research (2022)

Link to original: https://doi.org/10.1007/s11248-021-00293-4

If you want to cite the papers, please refer to the original.

For technical reasons we could not place the page where the original quote is coming from.

Abstract-Summary "We aimed to phenotype the hindlimb muscles of Hippo gene-mutated $Lats1^{-/-}$, $Mst2^{-/-}$, $Vgll3^{-/-}$, and $Vgll4^{+/-}$ mice."

"This analysis revealed that Lats1$^{-/-}$ mice have 11% more slow type I fibers than age and sex-matched wild-type controls."

"The mRNA expression of slow Myh7 increased by 50%, and the concentration of type I myosin heavy chain is 80% higher in Lats1$^{-/-}$ mice than in age and sex-matched wild-type controls."

"We re-analyzed published datasets and found that Lats1 mRNA in muscle is 63% higher in muscular dystrophy, increases by 17–77% after cardiotoxin-induced muscle injury, by 41–71% in muscles during overload-induced hypertrophy, and by 19–21% after endurance exercise when compared to respective controls."

"Lats1 contributes to the regulation of muscle fiber type proportions, and its expression is regulated by physiological and pathological situations in skeletal muscle."

Introduction

"We use our ≈650 human skeletal muscles to move, stand, speak, generate heat, store glucose as glycogen, and amino acids as protein."

"Muscle size and function are regulated by many signaling pathways and depend on genetics."

"Two decades, the Hippo signal transduction network has been identified as a regulator of skeletal muscle development, regeneration, and size and is associated with diseases such as muscular dystrophy and rhabdomyosarcoma [48]."

"These mice have all been generally phenotyped but a specific, quantitative phenotyping of skeletal muscle is rarely performed and often muscles are not phenotyped at all."

Materials and Methods

"Following dissection, the soleus muscle samples were homogenized in 1 ml of QIAzol (QIAGEN, Germany) using a Precellys lysing kit (Bertin Instruments, USA) and Precellys homogenizer machine (Bertin Technologies, USA) with high speed for 40 s. C2C12 myotubes were first washed with phosphate-buffered saline (PBS) and then lysed in lysis T buffer (Peqlab Biotechnology GmbH, reference number 12-6634-01) and stored at −80°C."

"To extract total protein, we homogenized soleus muscles in lysis buffer (0.1% SDS,0,5 M sodium orthovanadate, 0.5% sodium deoxycholate, 50 mM NaF, 1 mM EDTA,150 mM NaCl, 1%Triton-X 100) supplemented with protease and phosphatase inhibitor cocktail (Peqlab Biotechnology GmbH, Germany) using a Precellys lysing kit (Bertin Instruments, USA) and a homogenizer machine (Bertin Technologies, USA) with high speed for 40 s. After centrifugation at 12,000 g for 10 min at 4 °C, the supernatant was transferred to a new tube and assayed to determine the protein concentration of each sample using the Bradford protein assay kit (BioRad Laboratories GmbH, Cat #5,000,111, Munich, Germany)."

Results

"Grip strength was 27% ($P = 2.92 \times 10^{-06}$) lower in Lats1$^{/-}$ male mice when compared to sex- and age-matched wild-type mice (Supplemental table."

"To study the effect of transgenesis on skeletal muscle phenotype, we phenotyped hindlimb muscles of Lats1$^{-/-}$, Mst2$^{-/-}$, Vgll3$^{-/-}$, and Vgll4$^{+/-}$ mice versus sex- and age-matched wild-type mice as controls."

"To better understand the regulation of Lats1 expression and phosphorylation in skeletal muscle we retrieved published datasets from Gene Omnibus or downloaded supplemental data from published papers."

"This revealed that Lats1 is more expressed in slow type I muscle fibers in mice than in fast type II fibers [49]."

Discussion

"This suggests that Lats1 regulates fiber type proportion in mice."

"The finding that Lats1 deletion increased the proportion of slow, type I fibers fits the earlier discovery that a loss of Vgll2 in mice reduced type I fibers when compared to wild-type mice [50]."

"The fact that Lats1 is an inhibitor of Yap/Taz-Tead1-4 and that Vgll2 is a Tead1-4 coregulator suggest that a change of Tead1-4 activity in muscle can shift the fiber type composition of skeletal muscle."

"This suggests that Lats1 expression responds to exercise-related stimuli that may affect muscle fiber type-specific gene expression."

"In DMD muscles, high miR-21 expression leads to an increase of Lats activity."

"Lats1$^{-/-}$ mice had 11% more type I fibers, and that Lats1$^{-/-}$, Mst2$^{-/-}$, Vgll3$^{-/-}$, Vgll4$^{+/-}$ transgenic mice have no pathological skeletal muscle phenotype."

Do Skeletal Muscle Composition and Gene Expression as Well as Acute Exercise-Induced Serum Adaptations in Older Adults Depend on Fitness Status? [51]

This is a machine-generated summary of:

Bizjak, Daniel A.; Zügel, Martina; Schumann, Uwe; Tully, Mark A.; Dallmeier, Dhayana; Denkinger, Michael; Steinacker, Jürgen M.: Do skeletal muscle composition and gene expression as well as acute exercise-induced serum adaptations in older adults depend on fitness status? [51]

Published in: BMC Geriatrics (2021)

Link to original: https://doi.org/10.1186/s12877-021-02666-0

If you want to cite the papers, please refer to the original.

For technical reasons we could not place the page where the original quote is coming from.

Abstract-Summary "We aimed to answer the question whether or not circulating and skeletal muscle biomarkers are differentially expressed depending on fitness status in a group of elderly individuals."

"A cardiopulmonary exercise test (CPX) and resting skeletal muscle biopsy were performed to determine individual physiological performance capacity."

"Participants were categorized into a high physical fitness group (HPF) and a low physical fitness group (LPF) depending on peak oxygen uptake (VO_2peak)."

"Skeletal muscle tissue was analyzed by silver staining to determine the myosin heavy chain (MyHC) composition and selected genes by qRT-PCR."

"HPF showed lower body weight and body fat, while skeletal muscle mass and oxygen uptake at the first ventilatory threshold (VO_2T1) did not differ between groups."

"qRT-PCR showed higher expression of BDNF and BRCA1 in LPF skeletal muscle while there were no differences in other examined genes regarding energy metabolism."

"Basal serum concentrations of Irisin were higher in HPF compared to LPF with a trend towards higher values in BDNF and HSP70 in HPF."

"Although no association between muscle composition/VO_2peak with fitness status in older people was detected, higher basal Irisin serum levels in HPF revealed slightly beneficial molecular serum and muscle adaptations."

Background

"Sarcopenia is defined as the gradual loss of skeletal muscle mass, quality and strength during ageing and/or immobility [52]."

"Harvey and others (2013) examined physical activity of older people in seven countries and found that approximately 67% of the age group >60 years showed sedentary behavior of 8.5 h per day [53]."

"Several studies showed the beneficial effect of physical exercise in slowing down the loss of skeletal muscle mass and to increase life expectancy throughout different age groups [54, 55]."

"Exercise referral schemes and self-management strategies were used to reduce sedentary behavior and improve health, quality of life and physical function, as well as psychosocial outcomes in the long term."

"In a sub-study, we aimed to test the hypothesis that fitter older individuals more likely have lower inflammation, improved cellular regulation processes involved in regeneration, metabolism and genomic health and a different muscle structure than unfit age-matched controls."

Materials and Methods

"For RNA isolation, muscle tissue was incubated for 24 h with RNAlater (QIAGEN GmbH, Hilden, Germany) at 4°C and then stored in cryotubes at − 80°C until further analysis."

"To determine the respective protein myosin heavy chain (MHC) isoform abundance, muscle samples were incubated with 250 µl Pierce™ RIPA buffer (Thermo Fisher Scientific, MA, USA), mixed with cOmplete™ Mini EDTA-free Protease Inhibitor Cocktail (Roche, Basel, Switzerland), homogenized and incubated for 10 min on ice in Pierce™ RIPA buffer to a final concentration of 5 µg/µl muscle protein."

"After centrifugation (4°C, 14000 rpm, 20 min), the supernatant was used for SDS-PAGE according to the manufacturer's instructions (Mini-Protean TGX Gels 4–20%, BIO-RAD, Berkeley, USA)."

"To determine associations between skeletal muscle mass (SMM), age, VO_2peak and VO_2T1, Spearman correlation analyses followed by two-tailored t-test were performed to determine statistical significance."

Results

"Gender analysis revealed no significant differences in the fat parameters or VO_2peak between female and male participants."

"As the results were based on the fitness level determined by VO_2peak, no gender effect is assumed to bias the data."

"A separate analysis showed no effect of sex on the tested parameters."

Discussion

"In this pilot study we found that although maximal oxygen uptake and body composition of fitter older adults were higher compared to unfit peers, most of the basal gene expression related to ageing, performance, and associated metabolism as well as fiber type composition did not significantly differ between HPF and LPF."

"Although a higher basal BDNF skeletal muscle mRNA expression was seen for LPF, serum protein concentrations showed non-significant elevated levels in HPF pre and post exercise compared to LPF."

"Skeletal muscle expression of BDNF in older people after exercise is incompletely understood and has been rarely examined: accumulating evidence on circulating protein levels seem to hint to transiently up-regulation of BDNF concentration after acute exercise, but with no long-term changes with regular training [56–58]."

"Il-6 serum basal levels of HPF and LPF were not significantly different and both groups showed increased values after the exercise test."

Conclusions

"Valenzuela and others stated that "The sooner the better, but never too late" in their seminal review regarding exercise adaptations and health effects in the older population [59]."

"Serum molecular adaptations in older individuals to an acute exercise test seem to depend on fitness level."

"Although we did not detect a significant association between basal muscle parameters like SSM or fiber type composition and VO_2peak/kg with fitness status, higher basal levels of Irisin in HPF revealed slightly beneficial long-term adaptations due to higher fitness level but most adaptations may be blunted during ageing."

"A significant lower inflammation after the acute test or a different muscle composition of fit- compared to unfit age-matched subjects could not be detected, suggesting that probably a still higher fitness status than our previously unactive participants has to be maintained to gain beneficial long-term adaptations in the older population."

Myogenic Regulatory Factor Response to Resistance Exercise Volume in Skeletal Muscle [60]

This is a machine-generated summary of:

Drummond, Micah J.; Conlee, Robert K.; Mack, Gary W.; Sudweeks, Sterling; Schaalje, G. Bruce; Parcell, Allen C.: Myogenic regulatory factor response to resistance exercise volume in skeletal muscle [60]

Published in: European Journal of Applied Physiology (2009)

Link to original: https://doi.org/10.1007/s00421-009-1279-z

If you want to cite the papers, please refer to the original.

For technical reasons we could not place the page where the original quote is coming from.

Abstract-Summary "This study examined the impact of resistance exercise volume on myoD and myogenin in rodent quadriceps muscle."

"Muscles were analyzed for myogenin and myoD mRNA and protein expression 6, 12, 24 and 48 h post-exercise."

"In red quadriceps (RQ), myogenin mRNA was significantly elevated at 6 h following LV and this response was greater than HV at 6 h, while myogenin protein was significantly increased at 6 and 12 h following LV but only at 12 h following HV ($P < 0.05$)."

"We conclude that acute resistance exercise can activate myogenin and myoD expression levels in RQ, but when exercise volume is doubled these myogenic responses are not proportional but delayed and blunted possibly because of excessive damage/injury."

Introduction

"The adaptive response of skeletal muscle to resistance exercise involves a series of molecular signaling pathways that lead to increased expression of contractile proteins (myosin and actin) and an eventual increase in muscle size and strength."

"Expression of myogenin and myoD is increased following a single bout of resistance exercise in humans and animals [61–65] supporting the hypothesis that MRF nuclear transcription factors play some role in skeletal muscle plasticity following resistance exercise."

"Although data is lacking on the specific molecular pathways involved in the increased expression of contractile proteins, we assume that myogenin and myoD expression should respond to resistance exercise."

"To test this hypothesis we used a rodent resistance exercise model [66] and evaluated the impact of exercise volume on myogenin and myoD mRNA and protein expression in the exercised skeletal muscle."

Methods

"Rats were randomly assigned to one of three treatment groups: non-exercised time-point control (CON; n = 24; body weight (BW) = 310 ± 3 g), low-volume resistance exercise (LV; n = 24; BW = 316 ± 3 g) and high-volume resistance exercise (HV; n = 24; BW = 320 ± 4 g)."

"Myogenin, myoD mRNA and their associated protein were measured at the above mentioned post-exercise time points in the red (RQ) and white (WQ) quadriceps muscle of the right hindlimb from each rat."

"The CON group rats were harnessed in the squat apparatus for a length of time equivalent to that of the HV group (~40 min) but did not receive electrical stimulation or perform resistance exercise."

"The blots were then rinsed in PBST solution three times for 10 min each, then incubated for 1 h in secondary antibody [goat anti-rabbit IgG (diluted 1:25,000); Santa Cruz Biotechnology, Santa Cruz, CA] at RT."

Results

"At 12 h post-exercise, myogenin mRNA expression returned to control levels in LV but remained elevated by 254 ± 67% ($P < 0.05$) in HV."

"From 24 to 48 h post-exercise, myogenin mRNA expression was similar to control levels for both LV and HV."

"By 12 h post-exercise myogenin protein expression was elevated in LV (268 ± 68%) and HV (282 ± 95%) similarly ($P < 0.05$)."

"At 12 h post-LV, myoD mRNA expression remained elevated by 288 ± 71% ($P < 0.05$), while at this same time point, HV increased myoD mRNA expression by 412 ± 105% ($P < 0.05$)."

"Myogenin and myoD mRNA and protein expression in WQ skeletal muscle was similar to the control throughout the 6–48 h post-exercise period for LV or HV (data not shown)."

Discussion

"The rodent model of resistance exercise used in this study increased myogenin and myoD mRNA and myogenin protein expression in active skeletal muscle within 6–12 h post-exercise."

"The important novel observations of this study were that the MRF response to resistance exercise, characterized by myogenin and myoD expression, was limited to the RQ muscle group and was not proportional to the volume of resistance exercise."

"HV resistance exercise appeared to blunt and delay myogenin mRNA and protein expression in the RQ muscle."

"A similar delay in myoD mRNA expression was seen with HV resistance exercise in RQ muscle."

"Exercise-induced muscle damage is a powerful stimulator of myoD and myogenin mRNA and protein expression in human [67–71] skeletal muscle; however, we propose that "excessive" muscle damage and the resultant inflammatory response may acutely impact muscle MRF gene and protein expression patterns."

Bibliography

1. Leal Junior ECP, Lopes-Martins RÁB, Vanin AA, Baroni BM, Grosselli D, De Marchi T, Iversen VV, Bjordal JM (2008) Effect of 830 nm low-level laser therapy in exercise-induced skeletal muscle fatigue in humans. Lasers Med Sci. https://doi.org/10.1007/s10103-008-0592-9
2. Torisu T, Wang K, Svensson P, De Laat A, Fujii H, Arendt-Nielsen L (2006) Effects of muscle fatigue induced by low-level clenching on experimental muscle pain and resting jaw muscle activity: gender differences. Exp Brain Res 174:566–574. https://doi.org/10.1007/s00221-006-0497-4
3. Chow RT, Heller GZ, Barnsley L (2006) The effect of 300 mW, 830 nm laser on chronic neck pain: a double-blind, randomized, placebo-controlled study. Pain 124:201–210. https://doi.org/10.1016/j.pain.2006.05.018
4. Gür A, Karakoc M, Nas K, Cevik R, Sarac J, Demir E (2002) Efficacy of low power laser therapy in fibromyalgia: a single-blind, placebo-controlled trial. Lasers Med Sci 17:57–61. https://doi.org/10.1007/s10103-002-8267-4
5. Avni D, Levkovitz S, Maltz L, Oron U (2005) Protection of skeletal muscles from ischemic injury: low-level laser therapy increases antioxidant activity. Photomed Laser Surg 23:273–277. https://doi.org/10.1089/pho.2005.23.273
6. Rizzi CF, Mauriz JL, Freitas Correa DS, Moreira AJ, Zettler CG, Filippin LI et al (2006) Effects of low-level laser therapy (LLLT) on the nuclear factor (NF)-kappaB signaling pathway in traumatized muscle. Lasers Surg Med 38:704–713
7. Lopes-Martins RA, Marcos RL, Leonardo PS, Prianti AC Jr, Muscara MN, Aimbire F et al (2006) Effect of low-level laser (Ga-Al-As 655 nm) on skeletal muscle fatigue induced by electrical stimulation in rats. J Appl Physiol 101:283–288. https://doi.org/10.1152/japplphysiol.01318.2005

8. Tullberg M, Alstergren PJ, Ernberg MM (2003) Effects of low-power laser exposure on masseter muscle pain and microcirculation. Pain 105:89–96. https://doi.org/10.1016/S0304-3959(03)00166-0
9. Andrade F, Trujillo X, Sánchez-Pastor E, Montoya-Pérez R, Saavedra-Molina A, Ortiz-Mesina M, Huerta M (2010) Glibenclamide increases post-fatigue tension in slow skeletal muscle fibers of the chicken. J Comparat Physiol B 181:403–412. https://doi.org/10.1007/s00360-010-0527-1
10. Inagaki N, Seino S (1998) ATP-sensitive potassium channels: structures, functions, and pathophysiology. Jpn J Physiol 48:397–412
11. Yokoshiki H, Katsube Y, Sunugawa M, Seki T, Sperelakis N (1997) Disruption of actin cytoskeleton attenuates sulfonylurea inhibition of cardiac ATP-sensitive K+ channels. Pflugers Arch 434:203–205
12. Weselcouch EO, Sargent C, Wilde MW, Smith MA (1993) ATP-sensitive potassium channels and skeletal muscle function in vitro. J Pharmacol Exp Ther 267:410–416
13. Light PE, French RJ (1994) Glibenclamide selectively blocks ATP-sensitive K+ channels reconstituted from skeletal muscle. Eur J Pharmacol 259:219–222
14. Wickenden AD, Prior H, Kelly E, Russell K, Poucher SM, Kumar P (1996) The effects of pharmacological modulation of KATP on the guinea pig isolated diaphragm. Eur J Pharmacol 302:79–88
15. Matar W, Nosek TM, Wong D, Renaud JM (2000) Pinacidil suppresses contractility and preserves energy but glibenclamide has no effect during muscle fatigue. Am J Physiol Cell Physiol 278:C404–C416
16. Matar W, Lunde JA, Jasmin BJ, Renaud JM (2001) Denervation enhances the physiological effects of the K(ATP) channel during fatigue in EDL and soleus muscle. Am J Physiol Regul Integr Comp Physiol 281:R56–R65
17. Pritschow BW, Lange T, Kasch J, Kunert-Keil C, Liedtke W, Brinkmeier H (2010) Functional TRPV4 channels are expressed in mouse skeletal muscle and can modulate resting Ca^{2+} influx and muscle fatigue. Pflügers Archiv Eur J Physiol 461:115–122. https://doi.org/10.1007/s00424-010-0883-4
18. Boittin FX, Petermann O, Hirn C, Mittaud P, Dorchies OM, Roulet E, Rüegg UT (2006) Ca2+-independent phospholipase A2 enhances store-operated Ca2+ entry in dystrophic skeletal muscle fibers. J Cell Sci 119:3733–3742
19. Ducret T, Vandebrouck C, Cao ML, Lebacq J, Gailly P (2006) Functional role of store-operated and stretch-activated channels in murine adult skeletal muscle fibres. J Physiol 575:913–924
20. Smyth JT, Dehaven WI, Jones BF, Mercer JC, Trebak M, Vazquez G, Putney JW Jr (2006) Emerging perspectives in store-operated Ca2+ entry: roles of Orai, Stim and TRP. Biochim Biophys Acta 1763:1147–1160
21. Stiber J, Hawkins A, Zhang ZS, Wang S, Burch J, Graham V, Ward CC, Seth M, Finch E, Malouf N, Williams RS, Eu JP, Rosenberg P (2008) STIM1 signalling controls store-operated calcium entry required for development and contractile function in skeletal muscle. Nat Cell Biol 10:688–697
22. Yeung EW, Whitehead NP, Suchyna TM, Gottlieb PA, Sachs F, Allen DG (2005) Effects of stretch-activated channel blockers on [Ca2+]i and muscle damage in the mdx mouse. J Physiol 562:367–380
23. Alderton JM, Steinhard RA (2000) How calcium influx through calcium leak channels is responsible for the elevated levels of calcium-dependent proteolysis in dystrophic myotubes. Trends Cardiovasc Med 10:268–272
24. Fraysse B, Liantonio A, Cetrone M, Burdi R, Pierno S, Frigeri A, Pisoni M, Camerino C, De Luca A (2004) The alteration of calcium homeostasis in adult dystrophic mdx muscle fibers is worsened by a chronic exercise in vivo. Neurobiol Dis 17:144–154
25. Iwata Y, Katanosaka Y, Arai Y, Komamura K, Miyatake K, Shigekawa M (2003) A novel mechanism of myocyte degeneration involving the Ca2+-permeable growth factor-regulated channel. J Cell Biol 161:957–967

26. Rolland JF, De Luca A, Burdi R, Andreetta F, Confalonieri P, Conte Camerino D (2006) Overactivity of exercise-sensitive cation channels and their impaired modulation by IGF-1 in mdx native muscle fibers: beneficial effect of pentoxifylline. Neurobiol Dis 24:466–474
27. Kunert-Keil C, Bisping F, Krüger J, Brinkmeier H (2006) Tissue-specific expression of TRP channel genes in the mouse and its variation in three different mouse strains. BMC Genomics 7:159
28. Krüger J, Kunert-Keil C, Bisping F, Brinkmeier H (2008) Transient receptor potential cation channels in normal and dystrophic mdx muscle. Neuromuscul Disord 18:501–513
29. Sánchez-Duarte E, Trujillo X, Cortés-Rojo C, Saavedra-Molina A, Camargo G, Hernández L, Huerta M, Montoya-Pérez R (2017) Nicorandil improves post-fatigue tension in slow skeletal muscle fibers by modulating glutathione redox state. Jo Bioenerg Biomembr 49:159–170. https://doi.org/10.1007/s10863-016-9692-6
30. Debska G, Kicinska A, Skalska J, Szewczyka A, May R, Elger C, Kunz W (2002) Opening of potassium channels modulates mitochondrial function in rat skeletal muscle. Biochim Biophys Acta 1556:97–105. https://doi.org/10.1016/S0005-2728(02)00340-7
31. Moses MA, Addison PD, Neligan PC, Ashrafpour H, Huang N, Zair M, Rassuli A, Forrest CR, Grover GJ, Pang CY (2005) Mitochondrial KATP channels in hindlimb remote ischemic preconditioning of skeletal muscle against infarction. Am J Physiol Heart Circ Physiol 288(2):559–567. https://doi.org/10.1152/ajpheart.00845.2004
32. Skalska J, Piwonska M, Wyroba E, Surmacz L, Wieczorek R, Koszela-Piotrowska I, Zielinska J, Bednarczyk P, Dolowy K, Wilczynski GM, Szewczyk A, Kunz WS (2008) A novel potassium channel in skeletal muscle mitochondria. Biochim Biophys 1777:651–659. https://doi.org/10.1016/j.bbabio.2008.05.007
33. Cahoon NJ, Naparus A, Ashrafpour H, Hofer SO, Huang N, Lipa JE, Forrest CR, Pang CY (2013) Pharmacologic prophylactic treatment for perioperative protection of skeletal muscle from ischemia reperfusion injury in reconstructive surgery. Plast Reconstr Surg 131(3):473–485. https://doi.org/10.1097/PRS.0b013e31827c6e0b
34. García MC, Hernández A, Sánchez JA (2009) Role of mitochondrial ATP-sensitive potassium channels on fatigue in mouse muscle fibers. Biochem Biophys Res Commun 385:28–32
35. César IC, Araujo DP, Oliveira FC, Menezes RR, Santos JR, Almeida MO, Dutra MM, Santos DA, Machado RR, Pianetti GA, Coelho MM, de Fátima A (2014) Synthesis, antinociceptive activity and pharmacokinetic profiles of nicorandil and its isomers. Bioorg Med Chem 22(9):2783–2790. https://doi.org/10.1016/j.bmc.2014.03.011
36. Chen Z, Chen X, Li S, Huo X, Fu X, Dong X (2015) Nicorandil improves myocardial function by regulating plasma nitric oxide and endothelin-1 in coronary slow flow. Coron Artery Dis 26(2):114–120. https://doi.org/10.1097/MCA.0000000000000179
37. Ratajczak-Wrona W, Jablonska E, Antonowicz B, Dziemianczyk D, Grabowska SZ (2013) Levels of biological markers of nitric oxide in serum of patients with squamous cell carcinoma of the oral cavity. Int J Oral Sci 5(3):141–145. https://doi.org/10.1038/ijos.2013.59
38. Gornall AG, Bardawill CJ, David MM (1949) Determination of serum proteins by means of the biuret reaction. J Biol Chem 177(2):751–766
39. Miranda KM, Espey MG, Wink DA (2001) A rapid, simple spectrophotometric method for simultaneous detection of nitrate and nitrite. Nitric Oxide 5(1):62–71. https://doi.org/10.1006/niox.2000.0319
40. Horvath DM, Murphy RM, Mollica JP, Hayes A, Goodman CA (2016) The effect of taurine and β-alanine supplementation on taurine transporter protein and fatigue resistance in skeletal muscle from mdx mice. Amino Acids 48:2635–2645. https://doi.org/10.1007/s00726-016-2292-2
41. McIntosh L, Granberg KE, Brière KM, Anderson JE (1998) Nuclear magnetic resonance spectroscopy study of muscle growth, mdx dystrophy and glucocorticoid treatments: correlation with repair. NMR Biomed 11:1–10
42. Terrill JR, Grounds MD, Arthur PG (2015) Taurine deficiency, synthesis and transport in the mdx mouse model for Duchenne Muscular Dystrophy. Int J Biochem Cell Biol 66:141–148

43. Vescovo G, Ravara B, Dalla Libera L (2007) Skeletal muscle myofibrillar protein oxidation and exercise capacity in heart failure. Basic Res Cardiol 103:285–290. https://doi.org/10.1007/s00395-007-0692-x
44. Coats AJS, Clark AL, Piepoli M, Volterrani M, Poole-Wilson PA (1994) Symptoms and quality of life in heart failure. The muscle hypothesis. Br Heart J 72:36–39
45. Dalla Libera L, Ravara B, Gobbo V, Danieli Betto D, Germinario E, Angelini A, Vescovo G (2005) Skeletal muscle myofibrillar protein oxidation in heart failure and the protective effect of Carvedilol. J Mol Cell Cardiol 38:803–807
46. Gielen S, Adams V, Mobius-Winkler S et al (2003) Anti-inflammatory effects of exercise training in the skeletal muscle of patients with chronic HF. J Am Coll Cardiol 42:861–868
47. Nezhad FY, Riermeier A, Schönfelder M, Becker L, de Angelis MH, Wackerhage H (2022) Skeletal muscle phenotyping of Hippo gene-mutated mice reveals that Lats1 deletion increases the percentage of type I muscle fibers. Transgenic Research 31:227–237. https://doi.org/10.1007/s11248-021-00293-4
48. Wackerhage H, Del Re DP, Judson RN, Sudol M, Sadoshima J (2014) The Hippo signal transduction network in skeletal and cardiac muscle. Sci Signal 7(337):re4. https://doi.org/10.1126/scisignal.2005096
49. Chemello F, Bean C, Cancellara P, Laveder P, Reggiani C, Lanfranchi G (2011) Microgenomic analysis in skeletal muscle: expression signatures of individual fast and slow myofibers. PLoS ONE 6(2):e16807. https://doi.org/10.1371/journal.pone.0016807
50. Honda M, Hidaka K, Fukada S, Sugawa R, Shirai M, Ikawa M, Morisaki T (2017) Vestigial-like 2 contributes to normal muscle fiber type distribution in mice. Sci Rep 7:7168
51. Bizjak DA, Zügel M, Schumann U, Tully MA, Dallmeier D, Denkinger M, Steinacker JM (2021) Do skeletal muscle composition and gene expression as well as acute exercise-induced serum adaptations in older adults depend on fitness status? BMC Geriatr 21:697. https://doi.org/10.1186/s12877-021-02666-0
52. Cruz-Jentoft AJ, Bahat G, Bauer J, Boirie Y, Bruyère O, Cederholm T et al (2019) Sarcopenia: revised European consensus on definition and diagnosis. Age Ageing. 48:16–31. https://doi.org/10.1093/ageing/afy169
53. Harvey JA, Chastin SFM, Skelton DA (2013) Prevalence of sedentary behavior in older adults: a systematic review. Int J Environ Res Public Health. 10:6645–6661. https://doi.org/10.3390/ijerph10126645
54. Tarpenning KM, Hamilton-Wessler M, Wiswell RA, Hawkins SA (2004) Endurance training delays age of decline in leg strength and muscle morphology. Med Sci Sports Exerc. 36:74–78. https://doi.org/10.1249/01.MSS.0000106179.73735.A6
55. Grimby G (1995) Muscle performance and structure in the elderly as studied cross-sectionally and longitudinally. J Gerontol A Biol Sci Med Sci 50.:Spec No:17–22. https://doi.org/10.1093/gerona/50a.special_issue.17
56. Knaepen K, Goekint M, Heyman EM, Meeusen R (2010) Neuroplasticity - exercise-induced response of peripheral brain-derived neurotrophic factor: a systematic review of experimental studies in human subjects. Sports Med. 40:765–801. https://doi.org/10.2165/11534530-000000000-00000
57. Tarassova O, Ekblom MM, Moberg M, Lövdén M, Nilsson J (2020) Peripheral BDNF response to physical and cognitive exercise and its association with cardiorespiratory fitness in healthy older adults. Front Physiol. 11:1080. https://doi.org/10.3389/fphys.2020.01080
58. Küster OC, Laptinskaya D, Fissler P, Schnack C, Zügel M, Nold V et al (2017) Novel blood-based biomarkers of cognition, stress, and physical or cognitive training in older adults at risk of dementia: preliminary evidence for a role of BDNF, Irisin, and the Kynurenine pathway. J Alzheimers Dis. 59:1097–1111. https://doi.org/10.3233/JAD-170447
59. Valenzuela PL, Castillo-García A, Morales JS, Izquierdo M, Serra-Rexach JA, Santos-Lozano A et al (2019) Physical exercise in the oldest old. Compr Physiol. 9:1281–1304. https://doi.org/10.1002/cphy.c190002

60. Drummond MJ, Conlee RK, Mack GW, Sudweeks S, Schaalje GB, Parcell AC (2009) Myogenic regulatory factor response to resistance exercise volume in skeletal muscle. Eur J Appl Physiol 108:771–778. https://doi.org/10.1007/s00421-009-1279-z
61. Kim JS, Kosek DJ, Petrella JK, Cross JM, Bamman MM (2005) Resting and load-induced levels of myogenic gene transcripts differ between older adults with demonstrable sarcopenia and young men and women. J Appl Physiol 99:2149–2158
62. Raue U, Slivka D, Jemiolo B, Hollon C, Trappe S (2006) Myogenic gene expression at rest and after a bout of resistance exercise in young (18–30 year) and old (80–89 year) women. J Appl Physiol 101:53–59
63. Tamaki T, Uchiyama S, Uchiyama Y, Akatsuka A, Yoshimura S, Roy RR, Edgerton VR (2000) Limited myogenic response to a single bout of weight-lifting exercise in old rats. Am J Physiol Cell Physiol 278:C1143–C1152
64. Willoughby DS, Nelson MJ (2002) Myosin heavy-chain mRNA expression after a single session of heavy-resistance exercise. Med Sci Sports Exerc 34:1262–1269
65. Yang Y, Creer A, Jemiolo B, Trappe S (2005) Time course of myogenic and metabolic gene expression in response to acute exercise in human skeletal muscle. J Appl Physiol 98:1745–1752
66. Krisan AD, Collins DE, Crain AM, Kwong CC, Singh MK, Bernard JR, Yaspelkis BB III (2004) Resistance training enhances components of the insulin signaling cascade in normal and high-fat-fed rodent skeletal muscle. J Appl Physiol 96:1691–1700
67. Bamman MM, Ragan RC, Kim JS, Cross JM, Hill VJ, Tuggle SC, Allman RM (2004) Myogenic protein expression before and after resistance loading in 26- and 64-year-old men and women. J Appl Physiol 97:1329–1337
68. McKay BR, O'Reilly CE, Phillips SM, Tarnopolsky MA, Parise G (2008) Co-expression of IGF-1 family members with myogenic regulatory factors following acute damaging muscle-lengthening contractions in humans. J Physiol 586:5549–5560
69. Hentzen ER, Lahey M, Peters D, Mathew L, Barash IA, Friden J, Lieber RL (2006) Stress-dependent and -independent expression of the myogenic regulatory factors and the MARP genes after eccentric contractions in rats. J Physiol 570:157–167
70. Okada A, Ono Y, Nagatomi R, Kishimoto KN, Itoi E (2008) Decreased muscle atrophy F-box (MAFbx) expression in regenerating muscle after muscle-damaging exercise. Muscle Nerve 38:1246–1253
71. Peters D, Barash IA, Burdi M, Yuan PS, Mathew L, Friden J, Lieber RL (2003) Asynchronous functional, cellular and transcriptional changes after a bout of eccentric exercise in the rat. J Physiol 553:947–957

The manufacturer's authorised representative in the EU is Springer Nature Customer Service Centre GmbH, Europaplatz 3, 69115 Heidelberg, Germany. If you have any concerns regarding our products, please contact ProductSafety@springernature.com

Printed and bound by CPI Group (UK) Ltd, Croydon, CR0 4YY
01/10/2026
02225897-0009